in Physics

W. Beiglböck
M. Goldhaber
E. H. Lieb
W. Thirring

Series Editors

Reiner Bass

Nuclear Reactions with Heavy Ions

With 176 Figures

Springer-Verlag
Berlin Heidelberg New York 1980

Reiner Bass

Institut für Kernphysik
J. W. Goethe-Universität Frankfurt
August-Euler-Straße 6
D-6000 Frankfurt a. M.
Federal Republic of Germany

Editors:

Wolf Beiglböck

Institut für Angewandte Mathematik
Universität Heidelberg
Im Neuenheimer Feld 5
D-6900 Heidelberg 1
Federal Republic of Germany

Maurice Goldhaber

Department of Physics
Brookhaven National Laboratory
Associated Universities, Inc.
Upton, NY 11973
USA

Elliott H. Lieb

Department of Physics
Joseph Henry Laboratories
Princeton University
P.O. Box 708
Princeton, NJ 08540
USA

Walter Thirring

Institut für Theoretische Physik
der Universität Wien
Boltzmanngasse 5
A-1090 Wien
Austria

ISBN 978-3-642-05718-2

Library of Congress Cataloging in Publication Data
Bass, Reiner. 1930 –
Nuclear reactions with heavy ions.
(Texts and monographs in physics)
Bibliography: p. Includes index. 1. Heavy ions.
2. Nuclear reactions. I. Title.
QC794.B37 539.7'5 79-21009

Preface

This book was conceived originally in 1972 as a written version of a series of lectures which I had given at Frankfurt University between 1967 and 1971. At that time preparations were under way in nearby Darmstadt and elsewhere for a new era of experimental research with heavy ions, involving projectiles of all masses up to uranium. There was a great deal of enthusiasm among students and senior researchers alike, and it was generally felt that a summary of the field should be useful to those entering active research. Consequently, when a former student who was then working for a well-known publishing company suggested that I convert my lecture notes into a textbook I didn't hesitate to agree – not knowing what was to come.

In the meantime, heavy-ion research saw an explosive development, and what appeared to be a well-defined task in 1972 has since turned into an almost hopeless race. Nevertheless, I believe that we have now again reached a point where a critical survey of what has been achieved should be helpful.

The enormous growth of the heavy-ion field in recent years necessitates concentration on a limited aspect. In this book I concentrate on the mechanisms of nuclear reactions induced by projectiles of fairly low energy – typically 5 to 10 MeV per nucleon. This choice is dictated by my own research interests and does not imply that I consider other areas less important. Thus topics like nuclear spectroscopy with heavy ions, heavy-ion collisions at relativistic energies, and the whole field of atomic phenomena are not discussed.

The book is written from an experimentalist's point of view. However, experimental details are included only occasionally, in order to illustrate differences from "conventional" nuclear physics research. The main emphasis is on the interpretation and understanding of experimental results in terms of simple, phenomenological models. Theoretical work is considered only to the extent that it connects with presently observable reality, or has stimulated experimental studies. This leaves out a large amount of more fundamental theoretical work which could not be covered adequately in the present context.

Even within the restricted range of topics discussed, it appeared neither possible nor desirable to give a complete survey of developments. Instead I have tried to strike a reasonable balance between a review and a critical selection, with the aim of presenting a coherent outline of the basic ideas and concepts. The result is undoubtedly biased by my own limited understanding and prejudice, and I should like to apologize sincerely to all colleagues whose work has not been discussed or quoted adequately.

The book is intended for advanced students with a basic knowledge in nuclear physics and quantum mechanics, and for nuclear physicists and chemists engaged in heavy-ion research. The material is organized in eight chapters which are arranged in order of increasing complexity of the topics considered. Each chapter has been made as self-contained as possible and is followed by its own bibliography. At the end there are several appendices explaining symbols and notation and giving information of a more general nature.

It gives me great pleasure to acknowledge help from a large number of people. Many colleagues from all over the world have given permission to use their published or unpublished results, and have sent preprints or figures for reproduction. Countless discussions with those working in or passing through the Frankfurt-Darmstadt area, and the stimulating atmosphere around the laboratory of the Gesellschaft für Schwerionenforschung and its accelerator UNILAC have provided important information and encouragement.

The advice of Dr. A. Reiter and Dr. H. Wiebking of Akademische Verlagsgesellschaft, Wiesbaden, and of Professor W. Beiglböck of Springer-Verlag, Heidelberg, has been most helpful. Mrs. A. M. Bussmann, Mrs. R. Diehl, and Mr. N. Kroker have given valuable assistance with tables, photographic reproductions, and figures. Special thanks are due to Mrs. U. von Graevenitz for performing an excellent job in typing the manuscript, including numerous additions and corrections.

I should like to express my sincere appreciation to Miss Elisabeth Steuer for her competent and untiring efforts in drawing figures, checking references, proof-reading, editing, and supervising the production of the manuscript. Her help has been essential.

Finally I want to thank my wife, Fatma, and son, Ashraf, for their patience and understanding during the last seven years. Without their support this book could not have been written.

Frankfurt/FRG, March 1980   Reiner Bass

Contents

1. Introduction

1.1 General Considerations

There is no well-defined borderline between light and heavy ions. The interaction of α particles with nuclei is known to exhibit features characteristic of much heavier complex projectiles, and therefore the α particle has sometimes been referred to as the "lightest heavy ion". On the other hand, helium beams of high quality and intensity have been widely available and have thus become a fairly conventional tool of nuclear physics. From a historical and technical standpoint, therefore, most nuclear physicists think of ions heavier than ^{4}He as heavy ions. In this book, we shall be interested mainly in projectiles considerably heavier than ^{4}He, but we shall include some reference to α-induced processes where necessary to illustrate a point.

In the introduction we present an elementary discussion of those kinematic and dynamic aspects which are typical of heavy-ion experiments, and do not depend on details of the nucleus–nucleus interaction. They are all related in some way to the larger mass, size, and electric charge of heavy ions compared to conventional light projectiles.

Let us consider first the Coulomb interaction between a heavy ion of nuclear charge Z_1 and a target nucleus of charge Z_2. It is associated with a repulsive potential proportional to the product $Z_1 Z_2$. In order to overcome this repulsion and to approach the target nucleus to within the range of nuclear forces, the kinetic energy of relative motion must exceed a minimum value called the Coulomb barrier. As a consequence, large ion energies are required for nuclear reaction studies with heavy ions, typically of the order of 5–10 MeV per nucleon (see Table 1.1). Both below and above the Coulomb barrier, the Coulomb interaction strongly distorts the ion trajectories at small internuclear separation; moreover, both target and projectile can be excited to several MeV by Coulomb excitation. In collisions of very heavy ions, the Coulomb repulsion may even be stronger than the nuclear attraction and thus prevent fusion of two nuclei in close contact.

Other very important aspects of heavy-ion collisions are the high linear and angular momenta involved. Since the projectile velocities required to overcome the Coulomb barrier for a given target nucleus do not vary strongly with projectile mass, the linear and angular momenta of relative motion are essentially proportional to the reduced mass of the colliding system. In inelastic collisions the orbital angular momentum—typically of the order of 100 $\hbar$—can be partly converted into internal rotation of the fragments, and thus provide a unique mechanism for the study of nuclear states with high spin. The possibility of high linear momentum transfer, on the other hand, is an essential ingredient of a number of spectroscopic techniques, which utilize such effects as recoil implantation into suitable environments or the Doppler shift of emitted radiation.

The compound systems produced in collisions between sufficiently heavy ions can have much larger mass, charge, and particle number than any of the known stable nuclei. As a consequence, the field of nuclear—and atomic— spectroscopy is vastly extended and predictions concerning the stability and structure of superheavy nuclei become accessible to experimental test. New modes of coexistence and excitation, characteristic of systems of 300 or 400 rather than 200 nucleons, may well be found. Modern experimental techniques should be capable of studying such systems if they live longer than 10^{-20}s.

We conclude this qualitative discussion of the basic features of heavy-ion collisions with some remarks on the kinematics of heavy ion motion. In this book we are interested in kinetic energies up to 10 MeV per nucleon, corresponding to velocities below about 15% of the speed of light. We shall therefore generally neglect relativistic effects, as is customary in low-energy nuclear physics. Moreover in many instances we shall use a classical rather than quantum-mechanical description of the orbital motion of heavy ions. The justification for this comes from the fact that the angular momenta involved are large compared with $\hbar$, as discussed above. Another way of describing the same physical situation is to state that the quantum mechanical wavelength $\lambdabar$ of the relative motion of two heavy ions is short compared with a characteristic geometrical dimension. The connection with angular momentum is obvious if we identify that dimension with the sum of the nuclear radii in nuclear scattering, or with the distance of closest approach in Coulomb scattering.

Finally we remark that the notion of a point projectile, as used in nucleon or α-particle scattering, is normally inadequate for a description of heavy-ion scattering. Effects of the finite size and internal structure of both the colliding nuclei have to be taken into account. Again basically classical concepts, like the liquid-drop model, are frequently used in this connection.

1.2   Kinematics of Ion–Ion Collisions

In this section we introduce some concepts which are needed for a quantitative discussion of heavy-ion collisions. The notation used here and throughout this book is summarized in Appendix A. Appendices B and C give a summary of the formulae relevant to the material discussed in this section.

We start by recalling briefly the relationship between the two natural frames of reference, the laboratory system and the center-of-mass system. In heavy-ion collisions, the two partners are of comparable mass; hence an appreciable fraction of the total kinetic energy is tied up in translational motion of the system as a whole, and the transformation from one system to the other has important consequences.

In the following we consider only collisions which result in the emission of two fragments, as in elastic or inelastic scattering, or simple transfer reactions.

a Before Collision

b After Collision

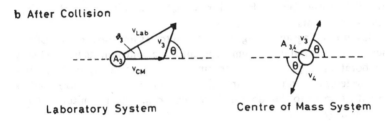

Laboratory System   Centre of Mass System

Fig. 1.1. Velocities and angles in the laboratory and centre-of-mass frames of reference

The basic features are illustrated in Fig. 1.1. The two upper diagrams represent the situation before the collision. In the laboratory system (left), the target (mass number A_2) is at rest, while the projectile (mass number A_1) approaches with an initial velocity v_∞. The kinetic energy and momentum are given by

$$E_{Lab} = \frac{m}{2} A_1 v_\infty^2, \quad P_{Lab} = m A_1 v_\infty, \tag{1.1}$$

where m is the nucleon mass and the significance of the other symbols is obvious. In the centre-of-mass system (right) the two ions move with initial velocities

$$v_1 = \frac{A_2}{A_1 + A_2} v_\infty, \quad v_2 = \frac{A_1}{A_1 + A_2} v_\infty, \tag{1.2}$$

and the kinetic energy and relative momentum are

$$E_{CM} = \frac{m}{2} (A_1 v_1^2 + A_2 v_2^2) = \frac{\mu}{2} v_\infty^2, \tag{1.3}$$

$$P_{CM} = m (A_1 v_1 + A_2 v_2) = \mu v_\infty. \tag{1.4}$$

The symbol μ denotes the reduced mass which is given by

$$\mu = m \frac{A_1 A_2}{A_1 + A_2} = A_{12} m, \tag{1.5}$$

where A_{12} may be termed the reduced mass number of the colliding system.

Another very useful quantity, which will frequently appear in this book, is the laboratory energy of the projectile divided by its mass number and normally measured in MeV per nucleon:

$$\varepsilon = \frac{E_{\text{Lab}}}{A_1} = \frac{E_{\text{CM}}}{A_{12}} = \frac{m}{2} v_\infty^2. \tag{1.6}$$

It should be noted that only the centre-of-mass values of energy and momentum are available for interactions between the two ions; they represent a fraction of the corresponding laboratory values equal to the ratio of reduced mass to projectile mass.

The lower left diagram in Fig. 1.1 refers to the asymptotic motion of any fragment with mass number A_3 emerging after the collision. Its final velocity in the laboratory, v_{Lab}, is the vector sum of its final velocity in the centre-of-mass system, v_3, and the translational velocity of the centre-of-mass system with respect to the laboratory system, v_{CM}. The magnitude of the latter is given by

$$v_{\text{CM}} = v_2 = \frac{A_1}{A_1 + A_2} v_\infty. \tag{1.7}$$

The emission angles of fragment 3 with respect to the incident beam direction, ϑ_3 (in the laboratory system) and Θ (in the centre-of-mass system), are related by

$$\tan \vartheta_3 = \frac{v_3 \sin \Theta}{v_3 \cos \Theta + v_{\text{CM}}} = \frac{\sin \Theta}{\cos \Theta + \gamma_3}, \tag{1.8}$$

where $\gamma_3 = v_{\text{CM}}/v_3$.

In the special case of elastic scattering (or, more generally, in collisions with negligible mass and energy transfer), the magnitude of the centre-of-mass velocity is not changed ($v_3 = v_1$), and hence we obtain from (1.2), (1.7), and (1.8)

$$\gamma_3 = \frac{v_{\text{CM}}}{v_3} = \frac{A_1}{A_2}. \tag{1.9}$$

If further the two colliding ions are of equal mass, we have $\gamma_3 = 1$, $\vartheta_3 = \Theta/2$. In the general case of a two-body reaction with product mass numbers A_3 and A_4, the quantity γ_3 is given by [Sc 55]

$$\gamma_3 = + \left(\frac{A_1 A_3}{A_2 A_4} \frac{E_{CM}}{E_{CM} + Q}\right)^{\frac{1}{2}} \tag{1.10}$$

where Q denotes the reaction Q value.

Next we consider scattering in the Coulomb field. Figure 1.2 shows the

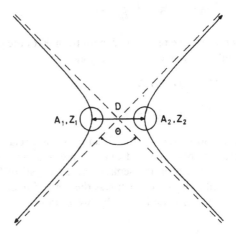

Fig. 1.2. Coulomb trajectories in the centre-of-mass system

classical trajectories of two colliding ions in the centre-of-mass system. The distance of closest approach between the two centres of gravity, D, and the scattering angle Θ are related by

$$D = a\left(1 + \csc \frac{\Theta}{2}\right), \tag{1.11}$$

where the parameter a is defined as one-half the distance of closest approach in a head-on collision ($\Theta = 180°$):

$$a = \frac{Z_1 Z_2 e^2}{\mu v_\infty^2} \tag{1.12}$$

Together with the asymptotic wavelength of relative motion at large separation, $\lambdabar = \hbar/\mu v_\infty$, we obtain the Sommerfeld parameter

$$n = \frac{a}{\lambdabar} = \frac{Z_1 Z_2 e^2}{\hbar v_\infty}. \tag{1.13}$$

According to our discussion in the preceding section, large values of n correspond to nearly classical motion.

For a given pair of colliding ions, the Coulomb trajectories are completely specified by the two kinematic parameters D and Θ. However, it is often more convenient to use instead the two constants of motion: the total energy

$E_{CM} = \frac{1}{2} \mu v_\infty^2$, and the (classical) angular momentum $L\hbar$ [here and in the following, we use L for classical and l for quantized angular momenta in units of $\hbar$; they are related by $L^2 = l(l+1) \approx (l + \frac{1}{2})^2$]. The requirement of energy conservation from infinity to the point of closest approach leads to the following relationship:

$$E_{CM} = \frac{Z_1 Z_2 e^2}{D} + \frac{L^2 \hbar^2}{2\mu D^2}, \tag{1.14}$$

which can be rearranged, with the help of the previously introduced Sommerfeld parameter n (1.13), to yield

$$L^2 = kD(kD - 2n), \tag{1.15a}$$

$$kD = n + (n^2 + L^2)^{1/2}. \tag{1.15b}$$

Equations (1.15) establish an important relationship between the distance of closest approach and angular momentum for Coulomb trajectories, which is often used in the semiclassical analysis of heavy ion scattering.

Likewise we can express the scattering angle Θ in terms of n and L; from (1.11), (1.13) and (1.15 b), we obtain

$$\sin \frac{\Theta}{2} = \frac{n}{kD - n} = \frac{n}{(n^2 + L^2)^{1/2}}. \tag{1.16}$$

At some internuclear distance $D \approx R_C$, the nuclear interaction will become effective. We define the Coulomb barrier E_C as the asymptotic kinetic energy in the centre-of-mass system at the (classical) threshold for nuclear reactions, and the Coulomb interaction distance $R_C = r_{0C}(A_1^{1/3} + A_2^{1/3})$ as the corresponding distance of closest approach in the absence of nuclear interactions. From (1.3) and (1.12) we obtain, with $2a = R_C$ and $E_{CM} = E_C$,

$$E_C = A_{12}\, \varepsilon_C = \frac{Z_1 Z_2 e^2}{R_C}, \tag{1.17}$$

which is, of course, just the Coulomb potential between two spherically symmetric, nonpenetrating charge distributions with total charges $Z_1 e$ and $Z_2 e$ at a distance R_C. Due to the diffuse nature of the nuclear surface and the finite range of the nuclear interaction, R_C will in general be significantly larger than the sum of the effective matter radii of the colliding nuclei; moreover, R_C will be larger than the actual distance of closest approach (R_{int}) at $\varepsilon = \varepsilon_C$, since we have neglected deflections by the attractive nuclear field in its definition. A quantitative discussion of the distances R_C and R_{int} will follow in Sect. 3.3.

At bombarding energies above the Coulomb barrier, we have to distinguish two different kinematic situations, depending on whether the distance of closest approach—assuming pure Coulomb scattering of point charges—is larger or smaller than R_C. For large impact parameters, corresponding to small scattering angles, we have Coulomb trajectories as shown in Fig 1.2. For small impact parameters, on the other hand, the collision is dominated by nuclear interactions and, in general, leads to inelastic processes. The limiting case of a "grazing collision" $(D = R_C)$ is of particular interest; from (1.16) we obtain for the corresponding Coulomb scattering angle Θ_{gr}

$$\sin \frac{\Theta_{gr}}{2} = \frac{n}{kR_C - n} = \frac{\varepsilon_C}{2\varepsilon - \varepsilon_C} . \tag{1.18}$$

In (1.18) we have made use of the identity

$$kR_C = 2n \frac{\varepsilon}{\varepsilon_C} = 2n_C \left(\frac{\varepsilon}{\varepsilon_C}\right)^{\frac{1}{2}} , \tag{1.19}$$

where n_C is the Sommerfeld parameter taken at the Coulomb barrier. It is often convenient to describe situations above the Coulomb barrier in terms of the parameters kR_C and $\varepsilon/\varepsilon_C$, rather than n, which is more appropriate for pure Coulomb scattering. A good example is (1.18), which shows that the grazing angle depends only on the ratio $\varepsilon/\varepsilon_C$.

The classical angular momentum L_{gr} corresponding to the grazing trajectory can be derived from (1.15a) and (1.19) as

$$L_{gr}^2 = (kR_C)^2 \left(1 - \frac{\varepsilon_C}{\varepsilon}\right) = 4n_C^2 \left(\frac{\varepsilon}{\varepsilon_C} - 1\right) \tag{1.20}$$

We conclude that for any given pair of ions the grazing angular momentum vanishes at the Coulomb barrier $(\varepsilon = \varepsilon_C)$, where grazing implies a head-on collision, and rises at higher energies proportional to $(\varepsilon - \varepsilon_C)^{1/2}$. For given values of ε_C and $\varepsilon > \varepsilon_C$, the angular momentum L_{gr} is proportional to the product $A_{12}R_C$.

From simple geometrical arguments we can derive the momentum transfer in an elastic grazing collision q_{gr} as

$$q_{gr} = 2k \sin \frac{\Theta_{gr}}{2} = 2k \frac{\varepsilon_C}{2\varepsilon - \varepsilon_C} . \tag{1.21}$$

In contrast to the angular momentum, this quantity decreases with increasing bombarding energy above the Coulomb barrier, as a consequence of the decrease in the scattering angle Θ_{gr}.

We conclude this discussion by noting that we have used the concept of a Coulomb interaction distance somewhat loosely, by calculating the Coulomb barrier (1.17) and the grazing trajectory (1.18) with the same quantity R_C. Since we are not interested in detailed predictions at this point, such a simplified description is considered adequate.

In order to give an impression of the magnitudes involved, we have collected in Table 1.1 numerical values of various quantities discussed in this section

Table 1.1.   Characteristic parameters of various collision pairs

Target	Proj.	R_c [fm]	E_c [MeV]	ε_c [MeV/u]	Q_{fu} [MeV]	$\varepsilon = \varepsilon_c$			$\varepsilon = 8$ MeV/u $(v/c = 0.13)$			
						λ [fm]	n	E_{ex} [MeV]	E_{ex} [MeV]	L_{gr}	Θ_{gr} [deg]	θ_{gr} [deg]
^{40}Ca	^{14}N	8.9	22.7	2.19	16.0	0.297	15	38.7	99	49	18.2	13.5
	^{40}Ar	10.1	51.3	2.56	2.0	0.142	36	53.3	162	103	22.0	11.0
^{120}Sn	^{14}N	10.6	47.4	3.78	−3.0	0.187	28	44.4	97	60	36.0	32.4
	^{40}Ar	11.9	109.0	3.63	−60.2	0.080	75	49.5	180	164	34.1	25.8
	^{136}Xe	13.9	279.0	4.38	−270.0	0.034	204	8.5	240	370	44.3	20.7
^{172}Yb	^{14}N	11.4	62.1	4.79	−17.2	0.161	35	44.9	86	58	50.6	47.2
	^{40}Ar	12.6	144.0	4.43	−94.0	0.067	95	49.5	166	170	45.0	37.0
	^{136}Xe	14.6	372.0	4.90	−388	0.027	270		220	429	52.4	29.5
^{238}U	^{14}N	12.1	76.7	5.80	−27.0	0.143	42	49.7	79	52	69.3	66.2
	^{40}Ar	13.3	179.0	5.22	−130.0	0.058	115	48.9	144	167	57.9	50.5
	^{136}Xe	15.4	466.0	5.38	−763	0.023	338			474	60.9	39.5
	^{238}U	16.6	735.0	6.18	−1214	0.015	538			585	78.0	39.0

for a number of colliding pairs. The distances R_C have been calculated with the equation (C.9) in Appendix C (see also Sect. 3.3). It can be seen that the projectile energy per nucleon at the Coulomb barrier, ε_C, is essentially determined by the heavier partner (usually the target nucleus) and is rather insensitive to the mass and charge of the lighter one. Typical barrier energies for medium to heavy target nuclei are in the range 4–6 MeV per nucleon.

Apart from kinematic parameters, we have included in Table 1.1 the Q values Q_{fu} for complete fusion of the various pairs, and excitation energies E_{ex} of the resulting compound nuclei. The Q values were deduced, whenever possible, from the mass compilation of Wapstra and Gove [Wa 71]. Masses of compound nuclei beyond the presently accessible region were either taken from calculations of Fiset and Nix [Fi 72] or—for the systems ^{238}U + ^{136}Xe and ^{238}U + ^{238}U—extrapolated from the liquid-drop model with parameters of Myers and Swiatecki [My 67]. It should be noted that the values obtained by the latter procedure are necessarily associated with large uncertainties of the order of 100 MeV. This has no consequence, however, for the qualitative trends to be discussed in the following.

The formation of comparatively light compound nuclei, with $A_{comp} \lesssim 100$, is usually an exothermic process, whereas the formation of heavier systems by fusion is endothermic. This fact is closely related to the well-known instability of nuclei with A > 90 with respect to the inverse process, namely spontaneous fission. Furthermore, as in fission, an energy barrier—the Coulomb barrier—must be overcome for fusion actually to occur.

With comparatively light projectiles, such as ^{14}N and ^{40}Ar, compound nucleus excitation energies of about 50 MeV are typically reached at the Coulomb barrier. With substantially heavier projectiles, on the other hand, compound nucleus formation involves lower excitation, or may be energetically forbidden at the Coulomb barrier. In the latter situation, which applies to compound systems with $A \gtrsim 300$, the interaction between the fragments at the point of contact is dominated by Coulomb repulsion, and the fusion barrier must be higher than the Coulomb barrier as defined in (1.17). Moreover, fusion may not occur for dynamical reasons, even if it is energetically allowed. We shall return to this interesting and important question in Chaps. 6 and 7.

1.3 Scattering and Reaction Cross Sections: Simple Formalism

In this section we turn from the kinematic to the dynamic aspects of heavy-ion collisions, and review briefly some elementary results of quantum-mechanical scattering theory [Mo 49, Bl 52]. We first neglect intrinsic spins and antisymmetrization but shall come back to these complications later. The wave function in the incident channel, which describes the relative motion of beam particle and target nucleus during the collision, can then be written simply as a superposition of an undisturbed (plane) incident wave and a scattered wave. Observable quantities, such as the scattering cross section, are determined by the asymptotic behaviour of this wave function for large values of the separation coordinate r, outside the range of interaction. This can be represented as

$$\psi(r) \xrightarrow{r \to \infty} N \left[\exp (ikz) + f(\Theta) \frac{1}{r} \exp (ikr)\right] \tag{1.22}$$

where N is a normalization factor and $f(\Theta)$ is called the scattering amplitude. The latter is related to the differential cross section for elastic scattering by

$$\frac{d\sigma}{d\Omega} = |f(\Theta)|^2, \tag{1.23}$$

as can be shown by calculating the probability flux corresponding to the incident and scattered waves in (1.22).

Expansion of the plane-wave term in (1.22) in terms of partial waves and comparison with an appropriate solution of the Schrödinger equation for the field-free region leads to the following expression for the scattering amplitude:

$$f(\Theta) = \frac{i}{2k} \sum_{l=0}^{\infty} (2l + 1) (1 - \bar{S}_l) \, P_l(\cos \Theta). \tag{1.24}$$

The Legendre polynomials $P_l(\cos \Theta)$ are essentially angular momentum eigenfunctions, and thus the scattering amplitude is decomposed into contributions corresponding to angular momentum l. It should be noted that this representation of $f(\Theta)$ is independent of the nature of the interaction, provided it is of finite range (see below); all information on the interaction is contained in the complex parameters $\bar{S}_l$, which are just asymptotic amplitude ratios of outgoing and ingoing partial waves with angular momentum l.

So far we have neglected the long-range nature of the Coulomb field, which falls off with the distance r as r^{-1}. Under these circumstances, the mathematical arguments leading to (1.24) have to be modified [see Mo 49], although the result remains unchanged. Experimentally, the Coulomb field of the target nucleus is cut off at large distances by the screening effect of atomic electrons, which can thus be invoked to justify the assumption of a finite range of interaction. Nevertheless, the long-range Coulomb interaction will usually affect a large number of partial waves in contrast to the short-range nuclear interaction. It is therefore customary and convenient to separate in (1.24) the Coulomb and nuclear effects by writing

$$\bar{S}_l = S_l \exp(2i\sigma_l) = A_l \exp[2i(\sigma_l + \delta_l)] \tag{1.25}$$

$$f(\Theta) = f_C(\Theta) + \frac{i}{2k} \sum_{l=0}^{\infty} (2l + 1) \exp(2i\sigma_l)(1 - S_l)P_l(\cos \Theta) \tag{1.26}$$

where $S_l, A_l, \delta_l, \sigma_l$ are now real, energy-dependent parameters. The quantities S_l, A_l, and δ_l describe purely nuclear scattering in the (hypothetical) absence of the Coulomb field. Absorption from the elastic channel into reaction channels leads to $A_l < 1$, whereas $A_l \equiv 1$ if elastic scattering is the only energetically allowed process. The symbols σ_l and δ_l denote the Coulomb and nuclear scattering phase shifts, respectively, and pure Coulomb scattering is described by

$$f_C(\Theta) = \frac{i}{2k} \sum_{l=0}^{\infty} (2l + 1)\,(1 - \exp 2i\sigma_l)\,P_l(\cos\Theta)$$

$$= - \frac{n}{2k}\left(\csc\frac{\Theta}{2}\right)^2 \exp\left[-2in\,\ln\left(\sin\frac{\Theta}{2}\right) + 2i\sigma_0\right], \qquad (1.27)$$

$$\sigma_l = \arg\{\Gamma(l + 1 + in)\}. \qquad (1.28)$$

The resulting Coulomb scattering cross section is given by the classical Rutherford formula

$$\left(\frac{d\sigma}{d\Omega}\right)_C = |f_C(\Theta)|^2 = \frac{n^2}{4k^2}\left(\csc\frac{\Theta}{2}\right)^4. \qquad (1.29)$$

Near and above the Coulomb barrier, the elastic scattering cross section arises from a coherent superposition of Coulomb and nuclear amplitudes (see (1.26)), but will always be dominated by Coulomb effects at sufficiently small scattering angles due to the divergence of the Coulomb amplitude. The integrated elastic scattering cross section is therefore not a useful concept from the nuclear point of view. The integrated total reaction cross section, on the other hand, reflects nuclear properties and can be calculated from the partial-wave representation of the asymptotic elastic wave function (1.22) by considering the ratio of absorbed to incident flux. The result is

$$\sigma_R = \frac{\pi}{k^2} \sum_{l=0}^{\infty} (2l + 1)\,(1 - |S_l|^2) = \pi \lambdabar^2 \sum_{l=0}^{\infty} (2l + 1)\,(1 - A_l^2). \qquad (1.30)$$

This relationship establishes an important link between scattering and reaction cross sections.

We now turn to a brief discussion of some consequences of the Pauli principle for heavy-ion scattering. The requirement of antisymmetry of the scattering wave function with respect to individual nucleon coordinates is irrelevant in the asymptotic region, but must be considered in treatments of the nucleus–nucleus interaction based on microscopic models of nuclear structure. In the special case of identical nuclei, however, the "macroscopic" wave function of relative motion (including spin coordinates) must be symmetric for even A (bosons) and antisymmetric for odd A (fermions) with respect to exchange of the two nuclei. This is a consequence of the basic inability of an observer to distinguish projectile and target nucleus. Similar considerations apply to the scattering of neighbouring nuclei, which can be described as consisting of identical cores plus few extra nucleons, if the latter can be exchanged during a collision (see Sect. 2.4).

Elastic scattering of identical, spinless nuclei has been studied extensively. In this case we are dealing with bosons, and the exchange symmetry of the relative wave function, which contains orbital motion only, must be even.

In the general case of identical nuclei with spin $I > 0$, the exchange symmetry of the spin part of the wave function has to be considered besides that of the orbital part. Consequently, the exchange symmetry of the orbital motion must be even, if the total spin S of the two nuclei is even, and odd if the total spin S is odd. This result holds, irrespective of whether we deal with bosons (even A) or fermions (odd A). Since the exchange of the two nuclei in orbital space is equivalent to the transformation $\Theta \rightarrow \pi - \Theta$, we can write for the scattering amplitude

$$f_S(\Theta) = f(\Theta) + (-1)^S f(\pi - \Theta), \tag{1.31}$$

where the amplitude $f(\Theta)$ describes the scattering of distinguishable nuclei with identical properties.

The scattering cross section is, in general, composed of contributions from different S states. A simplification arises if the scattering interaction can be assumed to be independent of spin, and spin polarization is not observed. In this case, the cross section is obtained by incoherent superposition of the different possible S contributions according to their statistical weights:

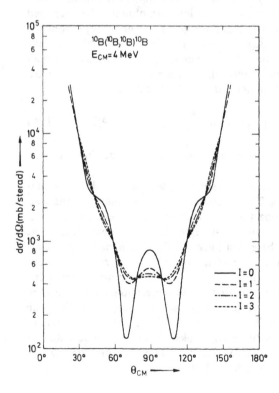

Fig. 1.3. Calculated angular distributions for $^{10}B + {}^{10}B$ Coulomb scattering, assuming different spin values I. (From Be 67)

$$\frac{d\sigma}{d\Omega} = \sum_{S=0}^{2I} \frac{2S+1}{(2I+1)^2} |f_S(\Theta)|^2$$

$$= |f(\Theta)|^2 + |f(\pi - \Theta)|^2 + \frac{(-1)^{2I}}{2I+1} \{f(\Theta)f^*(\pi - \Theta) \qquad (1.32)$$

$$+ f^*(\Theta)f(\pi - \Theta)\}$$

It is important to note that the interference term in (1.32) is of quantum-mechanical origin and has no classical analogue; it tends to zero for large values of the spin I.

Clearly the scattering cross section for identical nuclei is always symmetric with respect to $\Theta = 90°$ [see (1.31)] as it must be. A disadvantage of this situation is that the properties of the scattering amplitude $f(\Theta)$ at far back angles are usually obscured by effects of the strong forward Coulomb amplitude, and thus escape observation.

For spin-zero nuclei, only $S = 0$ is possible, and therefore, as mentioned above, the orbital part of the scattering wave function must have even exchange symmetry. It follows from the partial-wave decomposition of the scattering amplitude (1.24) and the relationship $P_l(\Theta) = (-1)^l P_l(\pi - \Theta)$, that in this case only even l values contribute to the scattering. This important result greatly simplifies the analysis of scattering data for identical even–even nuclei.

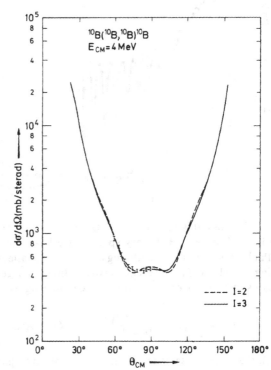

Fig. 1.4. Comparison of measured and calculated angular distribution for $^{10}\text{B} + {}^{10}\text{B}$ Coulomb scattering at $E_{CM} = 4$ MeV (I measured points; full line: calculation for I = 3, broken line: calculation for I = 2; from Be 67)

For pure Coulomb scattering of point charges (or homogeneously charged spheres below the Coulomb barrier), (1.32) assumes the form

$$\left(\frac{d\sigma}{d\Omega}\right)_I = \frac{n^2}{4k^2}\left\{\left(\csc\frac{\Theta}{2}\right)^4 + \left(\sec\frac{\Theta}{2}\right)^4\right.$$
$$\left. + \frac{(-)^{2I}}{2I+1} 2\cos\left(n\ln\tan^2\frac{\Theta}{2}\right)\left(\csc\frac{\Theta}{2}\right)^2\left(\sec\frac{\Theta}{2}\right)^2\right\}, \qquad (1.33)$$

which is also known as the Mott scattering formula for Coulomb scattering of identical particles. The sensitivity of this expression to the spin value is demonstrated in Figs. 1.3 and 1.4 for the case of ^{10}B nuclei at $E_{CM} = 4$ MeV [Be 67]. The precisely measured differential cross sections are in agreement with the known spin value $I = 3$ and exclude lower, but not higher spin values. With somewhat heavier nuclei of spin zero the interference term in (1.33) can lead to a very striking oscillatory behaviour of the cross section, as shown in Fig. 1.5 for ^{28}Si + ^{28}Si scattering at $E_{CM} = 20$ MeV [Fe 71].

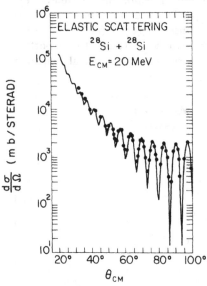

Fig. 1.5. Coulomb scattering of identical nuclei: ^{28}Si + ^{28}Si at $E_{CM} = 20$ MeV. The points are experimental and the solid line is the predicted angular distribution for pure Mott scattering. (From Fe 71)

To close this section we consider the total reaction cross section for identical nuclei of arbitrary spin I. Clearly the conditions imposed by the Pauli principle on the scattering amplitude will affect not only elastic scattering, but also the total reaction cross section. By arguments completely analogous to those given above for elastic scattering we obtain, instead of (1.30),

$$\sigma_R = \pi\lambdabar^2 \sum_{l=0}^{\infty}\left(1 + \frac{(-1)^{2I+l}}{2I+1}\right)(2l+1)(1-A_l^2). \qquad (1.34)$$

In the special case of zero spin, $I = 0$, only even partial waves contribute to

σ_R as expected; for finite values of I there is an odd–even modulation in l, which decreases with increasing spin I.

1.4 Strong Absorption Models of Heavy-Ion Scattering

In this section we discuss the consequences for scattering and reaction cross sections of two general properties of the nuclear interaction: its strength and its finite range. The former makes internal rearrangements of two colliding nuclei—and therefore absorption from the elastic channel—highly probable as soon as they penetrate to distances less than R_C. The latter means that there is practically no nuclear interaction at distances larger than R_C, corresponding to orbital angular momenta $L > L_{gr}$. The transition from Coulomb scattering to nuclear absorption within a narrow and well-defined region of distances near R_C leads to simple relationships for elastic scattering and reaction cross sections, which are of an essentially geometric nature and do not depend on details of nuclear structure. In this sense, nucleus–nucleus scattering is closely analogous to the diffraction of light by a totally absorbing disc or sphere, and models exploiting this analogy are often referred to as strong absorption models or diffraction models.

Let us consider first the classical limit, in which the nuclei move on classical trajectories, and are absorbed completely at distances $r < R_C$. The differential elastic scattering cross section is then

$$\left.\begin{aligned}\frac{d\sigma}{d\Omega} &= \frac{d\sigma_C}{d\Omega} \quad \text{for } \Theta \leq \Theta_{gr}\\ &= 0 \quad \text{for } \Theta > \Theta_{gr}\end{aligned}\right\}, \tag{1.35}$$

where the grazing angle Θ_{gr} is given by (1.18). Using equations (1.29) and (1.18), we obtain for the total reaction cross section

$$\left.\begin{aligned}\sigma_R &= 2\pi \int_{\Theta_{gr}}^{\pi} \frac{d\sigma_C}{d\Omega} \sin\Theta \, d\Theta\\ &= \begin{cases}\pi R_C^2 \left(1 - \dfrac{\varepsilon_C}{\varepsilon}\right) = \pi \lambdabar^2 L_{gr}^2 & \text{for } \varepsilon > \varepsilon_C\\ 0 & \text{for } \varepsilon \leq \varepsilon_C\end{cases}\end{aligned}\right\}. \tag{1.36}$$

In the quantal "sharp cut-off model" due to Blair [Bl 54], the quantum mechanical expressions (1.26) and (1.30) are used for the cross sections, and a sharp cut-off is introduced in angular momentum space:

$$\left.\begin{aligned}S_l &= A_l = 1; \quad \delta_l = 0 \quad \text{for } l > l_{max}\\ S_l &= A_l = 0 \quad\quad\quad\quad \text{for } l \leq l_{max}\end{aligned}\right\}. \tag{1.37}$$

Here l_{max} is an integer and related to the (classical) grazing angular momentum by

$$l_{max}(l_{max} + 1) \leq L_{gr}^2 < (l_{max} + 1)(l_{max} + 2). \tag{1.38}$$

The cross sections for $\varepsilon > \varepsilon_C$ are now given by

$$\frac{d\sigma}{d\Omega} = |f_C(\Theta) + \frac{i}{2k} \sum_{l=0}^{l_{max}} (2l + 1) \exp(2i\sigma_l) P_l(\cos\Theta)|^2, \tag{1.39}$$

$$\sigma_R = \pi \lambdabar^2 \sum_{l=0}^{l_{max}} (2l + 1) = \pi \lambdabar^2 (l_{max} + 1)^2. \tag{1.40}$$

In order to remove unphysical discontinuities resulting from the integer nature of l_{max}, it seems reasonable to replace in (1.40) the quantity $(l_{max} + 1)^2$ by its continuous average L_{gr}^2 [see (1.38)]. Thus the reaction cross section of the quantal sharp cut-off model becomes identical with the classical reaction cross section (1.36).

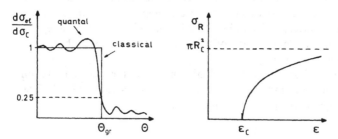

Fig. 1.6. Schematic representation of elastic scattering and total reaction cross sections in the classical and quantal sharp cut-off models

The elastic scattering and reaction cross section of the classical and quantal sharp cut-off models are shown schematically in Fig. 1.6. The most conspicuous difference between the two models is the oscillatory structure of the quantal scattering cross section as a function of angle, which is reminiscent of optical diffraction patterns. This feature, and especially the appearance of scattered flux in the classical shadow region, is a consequence of the wave-mechanical aspects of the model. It must be stressed, however, that the model is not equivalent to a rigorous quantum mechanical description of scattering by a totally absorbing sphere with sharp geometric radius R_C.

Early analyses of heavy-ion elastic scattering in terms of the Blair model have been reviewed by Zucker [Zu 60]. A typical set of results, taken from the work of Reynolds et al. [Re 60], is shown in Fig. 1.7. Here, as in other cases, the model yields qualitative overall agreement, but fails to reproduce the data in detail. The most conspicuous discrepancy occurs in the back-angle region ($\Theta > \Theta_{gr}$), where the experimental angular distributions in general exhibit a smooth exponential decrease with increasing angle, whereas the model predicts the persistence of a significant cross section with pronounced modulation. Exceptions to this trend occur for some light scattering systems, and will be discussed at the end of this section and in chapter 2.

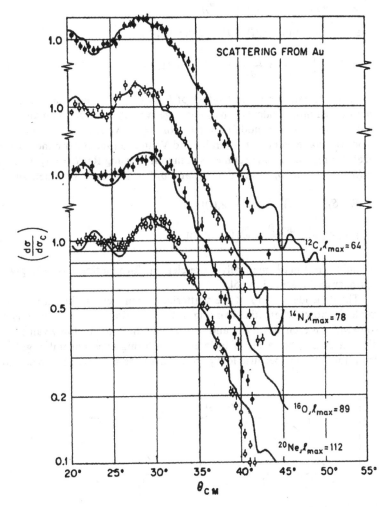

Fig. 1.7. Comparison of experimental angular distributions for heavy-ion elastic scattering with Blair model predictions. (From Zu 60)

The discrepancy mentioned above largely disappears if the sharp cut-off in l is replaced by a smooth l dependence of the parameters A_l, δ_l. This "smooth cut-off model" can be interpreted, on the basis of a semiclassical correspondence between angular momentum and distance, as reflecting the diffuse nature of the nuclear surface and the finite range of the nuclear interaction.

A number of different analytical expressions have been suggested for the smooth cut-off parametrization of the scattering phase shifts; one of the most widely applied is that due to McIntyre et al. (Mc 60, Mc 62):

$$A_l = g_A(l); \quad \delta_l = \delta(1 - g_\delta(l)), \tag{1.41}$$

where the functions $g_A(l)$, $g_\delta(l)$ are defined by

$$g_i(l) = \left\{ 1 + \exp \frac{l_i - l}{\varDelta_i} \right\}^{-1} \quad (i = A, \delta), \tag{1.42}$$

with, in general, nonintegral values of l_i and $\varDelta_i$. Conzett et al. have shown that optical model calculations yield negative values of δ_l for small l values, and have proposed a modified version of the above expression for δ_l, which produces that effect [Co 63]. Frahn and Venter, on the other hand, give physical and mathematical arguments in favour of using $\mathrm{Re}(S_l) = A_l \cos 2\delta_l$ and $\mathrm{Im}(S_l) = A_l \sin 2\delta_l$ rather than A_l and δ_l [Fr 63]; their parametrization is

$$\mathrm{Re}\,(S_l) = g_1(l) + \varepsilon_1\,[1 - g_1(l)] \tag{1.43}$$

$$\mathrm{Im}\,(S_l) = \mu \frac{dg_2(l)}{dl} + \varepsilon_2\,[1 - g_1(l)] \tag{1.44}$$

A similar, but mathematically more elegant, formulation has been given by Springer and Harvey [Sp 65].

The dependence on l of the scattering parameters in the smooth cut-off model is shown schematically in Fig. 1.8. The various parametrizations differ somewhat in their flexibility and in their predictions for low l values; these differences are, however, of little effect in fitting experimental angular distributions. This is because low partial waves are strongly absorbed and the

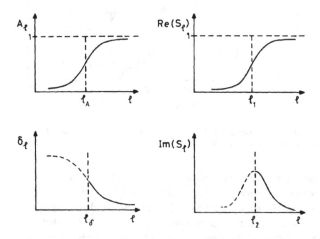

Fig. 1.8. Schematic representation of the scattering parameters A_l, δ_l; $\mathrm{Im}(S_l)$, $\mathrm{Re}(S_l)$ in the smooth cut-off model as functions of l

data are therefore insensitive to the corresponding phase shifts (as indicated by the broken lines in Fig. 1.8). As a consequence, three-parameter versions of the smooth cut-off model (e.g. $l_A = l_\delta$ and $\Delta_A = \Delta_\delta$ in the McIntyre parametrization, or $l_1 = l_2$, $\Delta_1 = \Delta_2$ and $\varepsilon_1 = \varepsilon_2 = 0$ in the Frahn–Venter parametrization) have in many cases—involving not-too-light target nuclei—been found adequate [Al 64, Ve 64]. The fitting power of the smooth cut-off model and its sensitivity to parameter variations are demonstrated in Fig. 1.9, which shows an analysis of Ta + ^{12}C scattering data with the McIntyre parametrization by Alster and Conzett [Al 64].

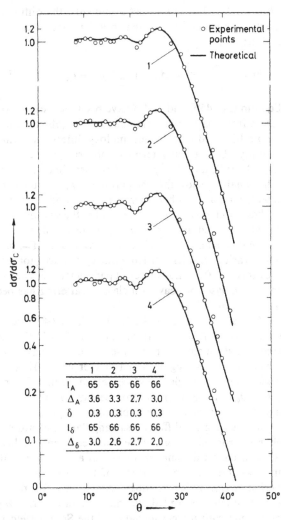

Fig. 1.9. Comparison of experimental angular distributions for Ta + ^{12}C elastic scattering with predictions of the smooth cut-off model. (From Al 64)

A somewhat unsatisfactory feature of the phenomenological strong-absorption models is that their parameters depend on energy in an unspecified manner. On physical grounds it is clear, however, that the angular momentum cut-off parameters l_t [see (1.42)] are closely related to the grazing angular momentum L_{gr} and thus to the Coulomb radius R_C of the interaction region. Semiclassically one expects [see (1.15)]

$$l_t(l_t + 1) = kR_C(kR_C - 2n), \tag{1.45a}$$

$$kR_C = n + [n^2 + l_t(l_t + 1)]^{1/2}. \tag{1.45b}$$

With similar, but less direct arguments, the diffuseness Δ_t in angular momentum space can be related to a "surface thickness" t of the interaction region:

$$kt = \Delta_t \frac{d(kR_C)}{dl_t} = \Delta_t\left(l_t + \frac{1}{2}\right)[n^2 + l_t(l_t + 1)]^{-1/2}. \tag{1.46}$$

Equations (1.45b) and (1.46) have been used extensively to extract R_C and t values from elastic scattering angular distributions [Zu 60, Al 64, Ve 64]. It should be noted that the terminology differs somewhat in the literature: our quantity R_C is usually termed "absorption radius" and denoted $R_a = r_a$ $(A_1^{1/3} + A_2^{1/3})$, and the symbol d is sometimes used instead of t. The published analyses show that the parameter r_a (r_{OC} in our notation) depends only weakly on mass number, energy, and method of analysis; values close to 1.44 fm have been consistently deduced for the scattering of projectiles with $A = 12$–20 from medium or heavy targets, using either the sharp cut-off model or different versions of the smooth cut-off model. The parameter t, on the other hand, varies considerably from one system to another, with typical values in the range 0.2–0.5 fm. A survey of parameters deduced from strong-absorption model analyses of heavy-ion elastic scattering has been given by Anni and Taffara [An 70].

In a series of papers [Fr 63, Ve 63, Ve 64], Frahn and Venter have discussed in detail elastic scattering angular distributions in the framework of the smooth cut-off strong-absorption model, and have given closed-form expressions for different regions in parameter space. They distinguish a "Coulomb region" at small angles ($\Theta < \Theta_{gr}$) and a "diffraction region" at larger angles ($\Theta > \Theta_{gr}$). In the latter, the cross section ratio $d\sigma/d\sigma_C$ decreases, on the average, with increasing angle approximately proportional to $\exp[-2\pi\Delta(\Theta - \Theta_{gr})]$. Thus a larger diffuseness in angular momentum space corresponds to a sharper drop-off in angle of the scattered intensity; this is a consequence of the wave-mechanical complementarity of angle and angular momentum. Superimposed on the average trend of the cross section is an oscillation with period π/L_{gr}, which is most pronounced for the lighter projectiles and targets at energies well above the Coulomb barrier ($n \lesssim 5$), and decreases in amplitude with increasing value of the Sommerfeld parameter n (Coulomb damping).

More recently, Frahn [Fr 71] has discussed analogies between nuclear and optical diffraction, and specified conditions for the occurrence of "Fresnel" or "Fraunhofer" diffraction. Superficially it might appear that nuclear scattering should always be of the Fraunhofer type, since source and detector are at practically infinite distance from the scattering centre. However, for a given classical impact parameter, the Coulomb field deflects the scattered particles in such a manner that they appear to originate from a virtual point source close to the interaction region. This is shown schematically in Fig. 1.10. As a consequence, angular distributions in elastic nucleus–nucleus scattering often exhibit a striking similarity to Fresnel-type diffraction patterns. Quantitatively one expects Fresnel diffraction for $L_{gr} \sin \Theta_{gr} \gtrsim 1$, $L_{gr} \gg 1$; both of these conditions are usually satisfied in heavy-ion scattering above the Coulomb barrier.

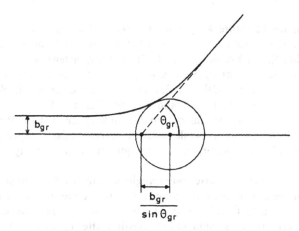

Fig. 1.10. Fresnel diffraction in heavy-ion elastic scattering: particles scattered through Θ_{gr} appear to come from a virtual point source at a distance $b_{gr}/\sin \Theta_{gr}$ from the centre of mass

In the limiting case of the sharp cut-off model (1.37), with pure Rutherford scattering for $l > l_{max}$, Fresnel diffraction is described by the following simple expression for the elastic scattering cross section [Fr 71, Fr 72]:

$$\frac{d\sigma}{d\sigma_C} = \frac{1}{2} \left\{ \left[\frac{1}{2} - S(y)\right]^2 + \left[\frac{1}{2} - C(y)\right]^2 \right\}. \tag{1.47}$$

Here $S(y)$ and $C(y)$ are the Fresnel integrals (see, for example, Ja 45) and the argument y is defined by

$$y = \left(\frac{L_{gr}}{\pi \sin \Theta_{gr}}\right)^{\frac{1}{2}} (\Theta - \Theta_{gr}) = \left(\frac{n}{2\pi}\right)^{\frac{1}{2}} \csc\left(\frac{\Theta_{gr}}{2}\right) (\Theta - \Theta_{gr}). \tag{1.48}$$

We note that for $\Theta = \Theta_{gr}$, $y = 0$, we have $S(y)$, $C(y) = 0$ and hence the right-hand side of (1.47) becomes equal to 0.25. The physical reason behind this result—wich applies also to Blair's sharp cut-off model (see Fig. 1.6)—is that diffraction by a large, totally absorbing object causes the amplitude at the edge of the geometrical shadow to drop to approximately one-half of the undisturbed amplitude, almost independently of wavelength and absorber size. Consequently the angle where the elastic scattering cross section is one-quarter of the Rutherford value ($\Theta_{1/4} = \Theta_{gr}$) can be used to obtain a simple and fast estimate of the effective strong absorption distance R_C and the grazing angular momentum L_{gr} [compare (1.11, 1.16)]:

$$R_C = a\left[1 + \csc\left(\tfrac{1}{2}\,\Theta_{1/4}\right)\right], \tag{1.49}$$

$$L_{gr} = n \cot\left(\tfrac{1}{2}\,\Theta_{1/4}\right). \tag{1.50}$$

In the literature $\Theta_{1/4}$ is usually referred to as "quarter-point angle" and the application of (1.49) and (1.50) is known as "Blair's quarter-point recipe" [Bl 54]. The Fresnel model (1.47) and the quarter-point recipe have been used extensively to analyse elastic scattering data, especially in situations involving heavy target nuclei. Practical examples are discussed in Sect. 3.3.

In a recent paper, Frahn has generalized (1.47) to include the effect of a smooth cut-off in angular momentum space, with finite width Δ [Fr 78]. This modification—while producing excellent fits to experimental data—does not affect, however, the simple quarter-point property of the elastic cross section at $\Theta = \Theta_{gr}$.

In relating quarter-point angles and interaction distances one makes use of the semiclassical nature of heavy-ion scattering in the Fresnel regime. In the same spirit it is possible to derive a simple relationship between the total reaction cross section and the elastic scattering cross section [Ho 65, Wo 76]:

$$\sigma_R = 2\pi \int_{\Theta_0}^{\pi} \left\{\frac{d\sigma_C}{d\Omega} - \frac{d\sigma}{d\Omega}\right\} \sin\Theta \, d\Theta. \tag{1.51}$$

Here Θ_0 is an angle such that for $\Theta < \Theta_0$ the elastic scattering cross section differs negligibly from the Rutherford cross section. Equation (1.51) essentially states that the flux going into reaction channels is missing from elastic (Rutherford) scattering. Its usefulness arises from the fact that its application is much simpler than either a direct experimental determination of σ_R or a rigorous calculation of σ_R via a phase shift analysis of the elastic angular distribution, based on (1.24) and (1.30). Since σ_R is related to the interaction distance R_C [compare (1.36)], (1.51) provides an alternative method of deducing the latter quantity from elastic scattering data.

Returning now to the smooth cut-off models in angular momentum space, we note that although many scattering systems are very well described by these models there are conspicuous exceptions. There the elastic angular distributions exhibit a striking enhancement and pronounced oscillations at larger angles, somewhat reminiscent of the predictions of the sharp cut-off model. This effect was first noticed by Gruhn and Wall [Gr 66] in the scattering of α particles from target nuclei with $A \approx 40$, and fitted empirically by adding a delta function anomaly ("Gruhn–Wall dip") at some l value to a smooth cut-off parametrization of A_l. A similar procedure—using a Gaussian instead of a delta function—was followed by Helb et al. [He 69] in an analysis of $^{12}C + {^{12}C}$ and $^{12}C + {^{14}N}$ scattering data, and later applied by the same group to $^{16}O + {^{16}O}$ scattering [Vo 71]. The effect of the dip is to remove partly an approximate cancellation which occurs in the scattering amplitude at back angles for partial-wave contributions with smoothly varying A_l and δ_l. Thereby significantly improved fits have been obtained.

The anomalous behaviour of the parameters S_l (also denoted as diagonal elements of the scattering matrix) for certain l values has been described by McVoy [Mc 71] in terms of singularities in the complex l plane, or "Regge poles". He suggested the following ansatz:

$$S_l = B(l) \left(1 + \frac{i\,D(l)}{l - L_0 - \frac{1}{2}\Gamma(l)}\right),\tag{1.52}$$

where $B(l)$ is a complex function of l, which contains the conventional smooth cut-off dependence, and the expression in brackets is a Breit-Wigner term in l space, which represents the influence of a "Regge pole" at $l = L_0 + i\Gamma(L_0)/2$. Note that L_0 and the weakly l-dependent functions $D(l)$ and $\Gamma(l)$, which describe the strength and width of the pole, in general depend on energy.

It should be emphasized that both the empirical approach of Gruhn and Wall [Gr 66] and of Helb et al. [He 69], and the (formally more appealing) Regge pole analysis of McVoy are convenient methods to identify and parametrize phase shift anomalies, but do not specify the underlying physical mechanism and the energy dependence of the parameters. Further insight can only be derived from models of the nuclear interaction, such as the optical model or more detailed microscopic models. Work along these lines has shown that two major mechanisms may be responsible for anomalies of the kind discussed above: one is a potential scattering phenomenon, sometimes referred to as "orbiting resonances", which occurs in the scattering of relatively light nuclei ($^{16}O + {^{16}O}$), if absorption into reaction channels is weak near the surface of the interaction region; the other mechanism occurs in the scattering of nuclei with unequal, but similar mass, and consists of an exchange of nucleons or nucleon clusters during the collision (elastic transfer). We shall return to both mechanims in detail in Chapter 2.

References Chapter 1

(For AR 71, AS 63 see Appendix D)

Al 64   Alster, J.; Conzett, H.E.: Phys. Rev. *136*, B 1023 (1964)
An 70   Anni, R.; Taffara, L.: Riv. Nuovo Cimento *2*, 1 (1970)
Be 67   Bethge, K. et al.; Z. Phys. *202*, 70 (1967)

Bl 52   Blatt, J.M.; Weisskopf, VF: *Theoretical Nuclear Physics* (Chapt. VIII). New York: Wiley 1952

Bl 54   Blair, J.S.: Phys. Rev. 95, 1218 (1954)

Co 63   Conzett, H.E.; Isoya, A.; Hadjimichael, E: AS 63 p. 26

Fe 71   Ferguson, A.I. et al.: AR 71 p. 187

Fi 72   Fiset, E.O.; Nix, J.R.: Nucl. Phys. *A193*, 647 (1972)

Fr 63   Frahn, W.E.; Venter, R.H.: Ann. Phys. (N.Y.) *24*, 243 (1963)

Fr 71   Frahn, W.E.: Phys. Rev. Lett. *26*, 568 (1971)

Fr 72   Frahn, W.E.: Ann. Phys. (N.Y.) *72*, 524 (1972)

Fr 78   Frahn, W.E.: Nucl. Phys. *A302*, 267 (1978)

Gr 66   Gruhn, C.R.; Wall, N.S.: Nucl. Phys. *81*, 161 (1966)

He 69   Helb, H.D. et al.: Phys. Rev. Lett. *23*, 176 (1969)

Ho 65   Holdeman, J.T.; Thaler, R.M.: Phys. Rev. *139B*, 1186 (1965)

Ja 45   Jahnke, E.; Emde, F.: *Tables of Functions*, 4th ed., p. 35. New York: Dover 1945

Mc 60   McIntyre, J. A.; Wang K. H.; Becker, L. C.: Phys. Rev. *117*, 1337 (1960)

Mc 62   McIntyre, J.A.; Baker, S.D.; Wang, K.H.: Phys. Rev. *125*, 584 (1962)

Mc 71   McVoy, K.W.: Phys. Rev. *C3*, 1104 (1971)

Mo 49   Mott, N.F.; Massey, H.S.W.: *The Theory of Atomic Collisions*, 2nd ed. Chap. II and III. Oxford: Clarendon Press 1949

My 67   Myers, W.D.; Swiatecki, W.J.: Ark. Fys. *36*, 343 (1967)

Re 60   Reynolds, H.L.; Goldberg, E.; Kerlee, D.D.: Phys. Rev. *119*, 2009 (1960)

Sc 55   Schiff, L. I.: *Quantum Mechanics*, 2nd ed., p. 99. New York, Toronto, London: McGraw Hill 1955

Sp 65   Springer, A.; Harvey, B.G.: Phys. Lett. *14*, 116 (1965)

Ve 63   Venter, R.H.: Ann. Phys. (N.Y.) *25*, 405 (1963)

Ve 64   Venter, R.H.; Frahn, W.E.: Ann. Phys. (N.Y.) *27*, 401 (1964)

Vo 71   Voit, H. et al.: AR 71 p. 303

Wa 71   Wapstra, A.H.; Gove, N.B.: Nucl. Data Tables *A9*, 265 (1971)

Wo 76   Wojciechowski, H. et al.: Phys. Lett. *63B*, 413 (1976)

Zu 60   Zucker, A.: Ann. Rev. Nucl. Sci *10*, 27 (1960)

2. Light Scattering Systems

2.1 Introduction and Historical Survey

In this chapter we concentrate on comparatively light projectiles and targets for several reasons: on the experimental side, a distinction arises from the fact that light ion beams of high quality and precisely controllable energy have been available from electrostatic accelerators since around 1960. This has lead to the accumulation of a large body of very detailed experimental results. An extension of these studies to medium and heavy target nuclei was prevented until much later by limitations in beam energy and the increase of the Coulomb barrier with atomic number of the target.

A more fundamental reason for our special interest in the interaction of light nuclei is the prominent role played by quantum effects and nuclear structure effects. Scattering situations involving at least one heavy partner are usually dominated by the Coulomb interaction at large impact parameters and by strong absorption at small impact parameters. The resulting scattering cross sections are in accord with semiclassical considerations and depend mainly on gross properties of the system, such as size and shape of the colliding nuclei. In contrast, the scattering of light nuclei is often clearly affected by details of internal structure, and this phenomenon has been the object of extensive experimental and theoretical investigation.

The history of modern heavy-ion scattering work starts in 1960 with the famous study by Bromley, Kuehner and Almqvist of the elastic scattering of ^{12}C from ^{12}C, ^{16}O from ^{12}C, and ^{16}O from ^{16}O, performed in Chalk River at the newly installed electrostatic tandem accelerator [Br 60, Br 61, Ku 63a]. The measurements comprised both angular distributions at selected energies and—for the first time—continuous excitation functions at selected angles with an energy resolution of about 100 keV. The results for a centre-of-mass angle $\Theta = 90°$ are shown in Fig. 2.1. At bombarding energies below the Coulomb barriers the data closely reproduce the predictions for pure Coulomb scattering of point charges, while at bombarding energies above the Coulomb barriers, the scattering cross sections fall below the Coulomb predictions due to absorption from the elastic channel by nuclear reactions. In addition, marked structure appears in the $^{12}C + {}^{12}C$ and $^{16}O + {}^{12}C$ excitation functions, but not in the $^{16}O + {}^{16}O$ excitation function. This striking and unexpected difference in the behaviour of the scattering systems was further investigated in a companion study by the same authors [Al 60], where they measured excitation functions for various nuclear reactions induced by the $^{12}C + {}^{12}C$ and $^{16}O + {}^{16}O$ interaction. The results, as shown in Fig. 2.2, follow the same qualitative pattern as the scattering data: smooth energy dependence in the $^{16}O + {}^{16}O$ system and structure in the $^{12}C + {}^{12}C$ system. A particularly noteworthy feature of Fig. 2.2 are three resonance-like structures, which appear in all reaction channels of the $^{12}C + {}^{12}C$ system at the same energies near the Coulomb barrier.

A more detailed study of these resonances with the help of angular correlation measurements in the reaction $^{12}C({}^{12}C, \alpha)^{20}Ne$ yielded spin assignments and reduced widths for decay into the α and carbon channels [Al 63]. The results indicate appreciable $^{12}C + {}^{12}C$ parentage and have been taken as evidence for a "quasi-molecular" interpretation of the three resonances.

The observation of structure in the $^{12}C + {}^{12}C$ and $^{16}O + {}^{12}C$ scattering excitation functions has prompted considerable experimental and theoretical activity. The possible existence of "nuclear molecules" has ever since been one of the most fascinating—if somewhat speculative—aspects of heavy-ion

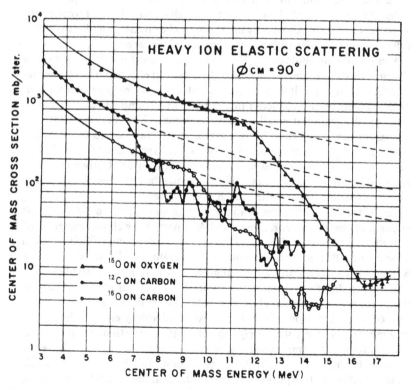

Fig. 2.1. Elastic scattering excitation functions at 90° for ^{12}C on ^{12}C, ^{16}O on ^{12}C, and ^{16}O on ^{16}O. (From Br 60)

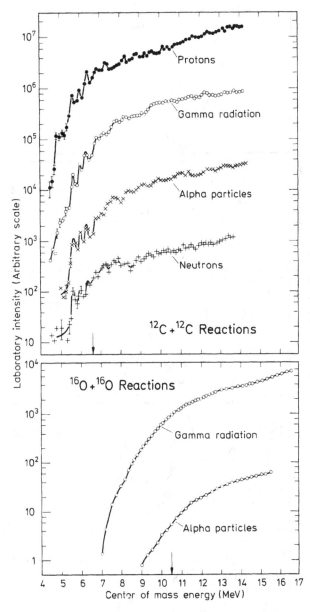

Fig. 2.2. Excitation functions for $^{12}C + ^{12}C$ and $^{16}O + ^{16}O$ reactions. (From Al 60)

research. We shall return to a more detailed discussion of this interesting problem in Sect. 2.3 and proceed here with our historical survey of experimental work.

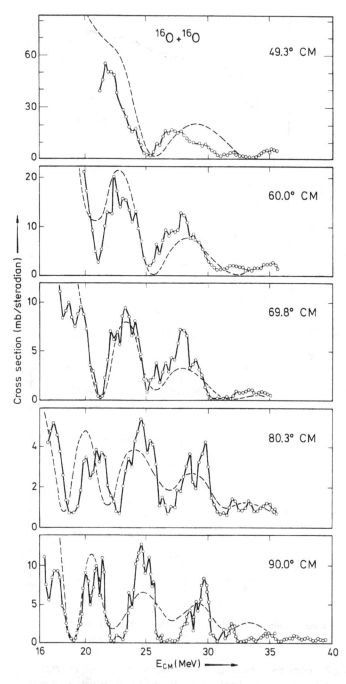

Fig. 2.3. Excitation functions for $^{16}O + {}^{16}O$ elastic scattering. The dashed curves are optical model predictions with parameters as given in Table 2.2. (From Ma 69)

Following a period of consolidation and steady development, the advent of the model FN and MP tandem accelerators with significantly increased beam energies in 1966/67 set the stage for a new era of heavy-ion-scattering work. Important advances in nuclear instrumentation and the beginning world-wide discussion about new accelerator projects contributed to an almost explosive growth of relevant information in the subsequent years.

The first and most famous contribution to the new period was the work of Siemssen, Maher, Weidinger, and Bromley on the elastic scattering of ^{16}O from ^{16}O [Si 67, Ma 69]. Using the MP tandem accelerator at Yale University these authors were able to extend the energy range up to 80 MeV in the laboratory system or 40 MeV in the centre-of-mass system—roughly three times the Coulomb barrier. The most remarkable feature of the results is a very pronounced, almost periodic gross structure in the excitation functions (Fig. 2.3), with peak widths of 2–3 MeV and a peak-to-valley ratio approaching 100. Superimposed on the gross structure is an intermediate structure with an apparent width of about 100–300 keV.

Using the optical model (see Sect. 2.2) to analyse their data, Siemssen et al. were able to explain both the gross structure of the excitation functions and the angular distributions as a consequence of potential scattering.

Following the work on ^{16}O + ^{16}O scattering, systematic investigations of other scattering systems in the same mass region were carried out at various laboratories, notably at Yale University [Br 69, Go 71] and Argonne National Laboratory [Si 71]. The results, although qualitatively similar, revealed characteristic differences from one scattering system to another with respect to the amplitude of the gross and intermediate structure in the excitation functions and the magnitude of the cross sections at high energies. This is demonstrated in Fig. 2.4 and 2.7, which compare 90° excitation functions for several pairs of identical nuclei. In the framework of the optical model (see Sect. 2.2) these differences are correlated with different magnitudes of the absorptive potential in the surface region [Sh 70, Si 74]; possible connections to nuclear properties of the individual systems are discussed in Sect. 2.2 and 2.3.

Another interesting effect, which can be observed in the scattering of non-identical light nuclei of not-too-different mass, is a significant rise of the differential cross section at back angles (compared to strong absorption predictions), often accompanied by pronounced structure as a function of angle. A striking illustration is provided by Fig. 2.5, taken from the work of von Oertzen et al. [Oe 68], in which the angular distributions for elastic scattering of ^{10}B, ^{16}O, and ^{19}F from ^{12}C are compared. The back angle effects are attributed to "elastic transfer", a process unique to heavy-ion scattering, in which target and projectile exchange their identity by transfer of the mass difference during the encounter. The very low back-angle cross section for the ^{19}F projectile is, in this picture, a natural consequence of the much smaller transfer probability for a ^{7}Li cluster compared to that for a deuteron or α-particle.

Elastic transfer has been studied by Gobbi et al. for the ^{13}C + ^{12}C system

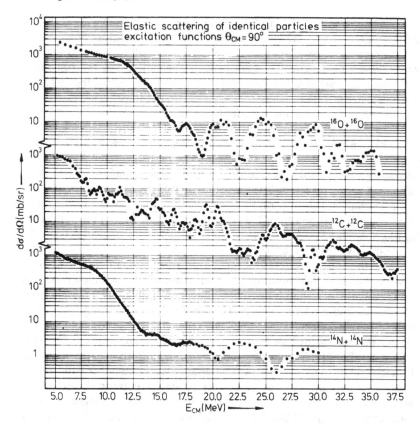

Fig. 2.4. Comparison of the 90° elastic scattering excitation functions for the $^{16}O + ^{16}O$, $^{12}C + ^{12}C$, and $^{14}N + ^{14}N$ systems. (From Br 69)

(Go 68, Go 71) and by the Heidelberg group of von Oertzen and collaborators in systematic investigations of many scattering systems [(Oe 71 b) and references given in Table 2.1]. Mechanisms other than elastic transfer, which may lead to back-angle enhancement and back-angle structure in elastic scattering, have been discussed by Robson [Ro 71, see also Si 71, Si 74]. We shall return to this question in Sect. 2.4.

In Table 2.1 we give a survey of experimental work on the elastic scattering of light nuclei with mass numbers $A < 20$. Included are all references published prior to July 1973, as far as they have come to our attention. In the column labelled "comments" we give information on the type of measurement performed (Ex: excitation functions at discrete angles; Ang: angular distributions at discrete energies; In: inelastic scattering data for resolved final states included) and on the type of analysis applied to the data (PP: parametrized phase-shift analysis; OM: optical model analysis; ET: superposition of scattering and elastic transfer). As the techniques of analysis have

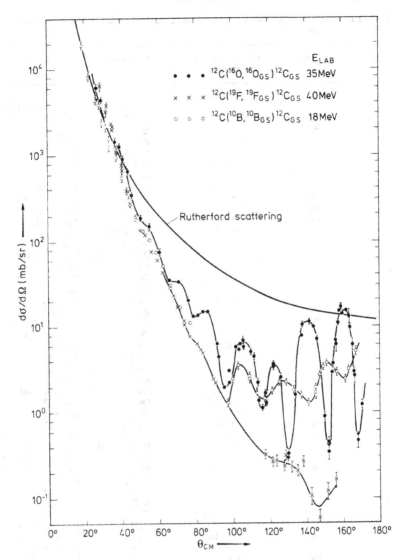

Fig. 2.5. Comparison of elastic scattering angular distributions for three systems, showing different degrees of back-angle enhancement. (From Oe 68)

been refined and standardized considerably since about 1968, the latter kind of information is usually omitted for earlier work.

In the following sections we discuss in turn the most significant phenomena observed in these studies: the gross structure and intermediate structure of excitation functions, and the back-angle enhancement of angular distributions. For further details concerning the topics treated in this and the fol-

Table 2.1. Experimental investigations of elastic scattering of light nuclei from light targets $(A < 20)$

Projectile	Target	Centre-of-mass energy range [MeV]	References	Comments[a]
^{6}Li	^{6}Li	1.6 – 3.5	Pi 66	Ang;
		5 – 17	Fo 71	Ex, Ang; OM, PP
		2 – 10	Gr 73	Ex, Ang; OM
	^{12}C	13.3	Be 68	Ang; OM
		4.5 – 13	Po 72	Ang; OM
^{7}Li	^{7}Li	3.6	Be 63	Ang;
		2.0 – 3.2	Pi 66	Ang;
	^{12}C	4.6	Be 63	Ang;
		4.5 – 13	Po 72	Ang; OM
^{10}B	^{12}C	9.8	Oe 68, Vo 69	Ang; OM, ET
^{11}B	^{11}B	2.7 – 15.5	Gü 67	Ex, Ang;
	^{12}C	14.6	Oe 68, Vo 69	Ang; OM,ET
^{12}C	^{12}C	3 – 14	Br 60, Br 61	Ex, Ang;
		63	Wa 62	Ang, In;
		13.5 – 37.5	Br 69, Re 69	Ex, Ang, In; OM
		5 – 19	Em 73	Ex, Ang, In; SM
	^{13}C	3 – 13	Go 68	Ex, Ang; PP, ET
		7.8 – 20.8	Go 71	Ang; ET
		7.8, 9.9	Bo 71	Ang; ET
	^{14}C	8.1	Bo 72	Ang; ET
	^{18}O	7 – 12.5	Go 68	Ex, Ang; PP
^{13}C	^{13}C	7 – 14	He 73	Ex, Ang; ET
	^{14}C	7.8, 9.8	Oe 71b	Ang; ET
^{14}N	^{9}Be	10.7	Ha 59	Ang;
		3 – 12	Ku 63 b, Ku 64	Ex, Ang; OM
	^{12}C	9.9 – 12.6	Ha 60	Ang;
		5 – 15	Ku 63b, Ku 64	Ex, Ang; OM
		9.2 – 12.6	Oe 71a, b	Ang; ET
	^{13}C	9.8	Oe 71a, b	Ang; ET
	^{14}C	10, 12.5	Oe 71b	Ang; ET
	^{14}N	7.5 – 10.8	Re 56	Ang;
		5 – 30	Br 69, Re 69	Ex, Ang; OM
		5 – 20	Ja 69	Ex, Ang; OM, PP
	^{16}O	8 – 18.5	Ja 69	Ex, Ang; OM, PP
		13 – 28	Si 70, Si 71	Ex, Ang; OM
		10.6, 13.3	Oe 71 b	Ang; ET
^{15}N	^{16}O	11 – 26	Si 70, Si 71	Ex, Ang; OM
	^{19}F	12.8 – 16.2	Ga 73	Ang; ET
^{16}O	^{9}Be	10.8	Kr 69	Ang; PP, OM
	^{10}B	5.8 – 10.6	Ok 68	Ex, Ang; OM
		10.0 – 12.5	Kr 69	Ang; PP, OM
	^{11}B	5.9 – 11.2	Ok 68	Ex, Ang; OM
		11.0 – 14.2	Vo 69	Ang; OM
	^{12}C	5.4 – 15.4	Br 60, Ku 63 a	Ex, Ang;
		72	Hi 64	Ang, In;
		8.6 – 18.0	Oe 68, Vo 69	Ang; OM, ET
		11.1 – 13.9	Kr 69	Ang; PP, OM
		9 – 26	Si 71, Ma 72	Ex, Ang; OM
	^{13}C	6 – 12.5	Go 68	Ex, Ang; PP
	^{14}N	13 – 28	Si 70	Ex, Ang; OM
	^{15}N	11 – 26	Si 70	Ex, Ang; OM
^{16}O	^{16}O	5 – 17.5	Br 60, Br 61	Ex, Ang;
		16 – 40	Si 67, Ma 69	Ex, Ang; OM
	^{17}O	12.4 – 16.5	Ge 73	Ang; ET
	^{18}O	7 – 19	Go 68	Ex, Ang; PP

Projectile	Target	Centre-of-mass energy range [MeV]	References	Comments[a]
		12 – 32	Si 70, Si 72	Ex, Ang; OM
		12.7 – 16.9	Ge 72	Ang; ET
^{18}O	^{18}O	7.5 – 27.5	Sh 70	Ex; OM
^{19}F	^{12}C	15.5, 23.2	Oe 68, Vo 69	Ang; OM, ET
	^{16}O	12.3 – 16.5	Ga 73	Ang; ET
	^{18}O	13.1 – 16.0	Ga 73	Ang; ET

[a] Ex: excitation function measurement;
Ang: angular distribution measurement;
OM: optical model analysis;
PP: parametrized phase-shift analysis;
ET: elastic transfer analysis.

lowing sections, the reader is referred to the proceedings of the conferences held in Heidelberg 1969 [HE 69] and in Argonne 1971 [AR 71], and to review articles by Siemssen [Si 74], Stokstad [St 73], von Oertzen and Bohlen [Oe 75], Bromley [Br 75], and Feshbach [Fe 76].

2.2 Gross Structure in Elastic Scattering and the Nucleus–Nucleus Interaction

Gross Structure with a characteristic width of 2–4 MeV has been observed in all elastic scattering excitation functions studied so far [Si 74, Go 73], most strikingly, however, in $^{16}O + {}^{16}O$ scattering [Si 67, Ma 69; see Fig. 2.3] The systematic variation in the width and position of the gross-structure peaks with nuclear size and scattering angle clearly indicates a diffraction phenomenon due to potential scattering.

Various methods have been applied to the analysis of the data. At first sight, it might seem most satisfactory to extract (complex) scattering phase shifts at each energy and then to search for an interaction which reproduces the "experimental" phase shifts. This approach is not feasible, however, because the large number of partial waves involved prevents an unambiguous determination of all phases from a single angular-distribution masurement. In addition, the existence of energy-dependent fluctuations throws doubt on the significance of results obtained at discrete energies. A meaningful phase-shift analysis therefore has to be based on a suitable parametrization of the scattering parameters (see Sect. 1.4) with respect to their dependence on angular momentum and energy. Thus a certain degree of model dependence cannot be avoided. Analyses of this type have been published by McVoy [Mc 71] and by Voit and Helb [Vo 71] for the $^{16}O + {}^{16}O$ system. In both cases the data [Ma 69] could be reproduced with a smooth cut-off parametrization of the phase shifts and a superimposed anomaly in l space, whose position was allowed to vary systematically with energy.

Voit and Helb have also analysed $^{16}O + {}^{16}O$ scattering between $E_{CM} =$

19 MeV and $E_{CM} = 22$ MeV without a priori assumptions about the l dependence of A_l and δ_l, treating these quantities for a limited number (four) of partial waves as free parameters [Vo 73a]. Again the existence of an anomaly near the grazing angular momentum was clearly established.

The appearance of resonance-like anomalies in phase-shift analyses for the $^{16}O + ^{16}O$ and other scattering systems has been interpreted as evidence for "orbiting resonances" [Mc 71] or "quasi-molecular rotational bands" [Ar 72]. As emphasized in Sect. 1.4, however, the phase-shift analysis yields only a parametrization of the asymptotic scattering wave function and does not imply any specific interaction mechanism.

The most popular and widely used approach to the analysis of heavy-ion elastic scattering has been the application of the optical model. This model, originally conceived for nucleon–nucleus scattering, reduces the interaction between two nuclei to a simple two-body potential, which must be complex in order to allow for both scattering and absorption of the incident flux. The application of the model to the interaction of composite systems has no rigorous theoretical foundation, but is nevertheless extremely successful in the systematic reproduction of experimental data.

The usefulness of the optical model has a number of reasons: firstly, and most significantly, it contains the basic physics of the scattering process—namely surface interaction in a well-defined region—in a quantum-mechanically correct manner. Consequently the gross features of the scattering cross section as a function of mass, energy, and angle are implicit in the model, and not imposed by externally prescribed parameter variation. Secondly, the elastic wave function is not only specified asymptotically, but is continued into the interaction region. Thus direct reaction cross sections can be calculated with DWBA or coupled channels methods. Last, but not least, the availability of suitable optical model computer codes has been essential for the power and flexibility of the method.

A word of caution is necessary, however. The actual values of the real and imaginary potential at small internuclear distances—corresponding to large overlap—and of the resulting scattering wave function well inside the interaction region have no physical significance. This follows from systematic investigations of ambiguities in the parameters of the real [Ku 64, Vo 69] and imaginary [Kr 69, Ma 69] potential (for a comprehensive discussion see Si 74). The results of these studies show clearly that large changes of the potential at small separation have no influence on the scattering cross sections, and that only the magnitude of the potential in the surface region is important. Thus absorption effectively screens the deep interior of the interaction region against observation, as far as elastic scattering is concerned. Just how far we can probe below the surface seems to vary from one system to another; this question will be taken up again below.

In practically all optical model analyses of heavy-ion scattering performed to date, the radial dependence of the potential has been assumed to be of Woods—Saxon form with volume absorption:

$$U(r) = V_C(r) - V\left(1 + \exp\frac{r - R_V}{a_V}\right)^{-1} - iW\left(1 + \exp\frac{r - R_W}{a_W}\right)^{-1} \quad (2.1)$$

where $R_V = r_{0V}(A_1^{1/3} + A_2^{1/3})$, $R_W = r_{0W}(A_1^{1/3} + A_2^{1/3})$, (2.2)

and the Coulomb potential $V_C(r)$ is usually approximated by the potential between a point charge $Z_1 e$ and a homogeneously charged sphere with charge $Z_2 e$ and radius $R_C = r_{0C}(A_1^{1/3} + A_2^{1/3})$:

$$V_C(r) = \begin{cases} \dfrac{Z_1 Z_2 e^2}{R_C} & \text{for } r > R_C, \\[2mm] \dfrac{Z_1 Z_2 e^2}{R_C}\left(\dfrac{3}{2} - \dfrac{1}{2}\dfrac{r^2}{R_C^2}\right) & \text{for } r < R_C. \end{cases} \quad (2.3)$$

Apart from the Coulomb radius parameter r_{0C}, which has little influence on the cross sections, there are six adjustable parameters in this potential: V, r_{0V}, a_V; W, r_{0W}, a_W.

As a consequence of the ambiguities mentioned above, six parameters are usually more than adequate; therefore, in many cases four-parameter fits have been performed, setting the geometrical parameters of the imaginary part of the potential equal to those of the real part. The potential depths, V and especially W, can in principle depend on conserved quantities like energy and angular momentum. This possibility has been exploited in a number of studies, some of which are discussed below. Table 2.2 gives a survey of potential parameters used in published analyses with the standard (angular-momentum-independent) optical model.

Table 2.2. Optical model potentials
For definition of symbols, see text and (2.1) and (2.2)

System	E_{CM}[MeV]	V [MeV]	r_{0V}[fm]	a_V[fm]	W [MeV]	r_{0W}[fm]	a_W[fm]	Refs
$^{16}O + {}^{16}O$	12 – 40	17	1.35	0.49	$0.4 + 0.1\ E$	1.35	0.49	Ma 69
$^{18}O + {}^{18}O$	10 – 27	17	1.35	0.57	$1.4 + 0.35\ E$	1.35	0.57	Sh 70
$^{12}C + {}^{12}C$		14	1.35	0.35	$0.4 + 0.1\ E$	1.40	0.35	Go 71
$^{14}N + {}^{14}N$		15	1.35	0.49	$0.4 + 0.125\ E$	1.35		
$^{16}O + {}^{14,15}N$	10 – 28	$7.5 + 0.5\ E$	1.35	0.49	$0.4 + 0.15\ E$	1.35	0.49	Si 71
$^{16}O + {}^{16}O$	12 – 40	$12 + 0.25\ E$	1.35	0.49	$0.4 + 0.1\ E$	1.35	0.49	
$^{16}O + {}^{18}O$	12 – 32	$12 + 0.25\ E$	1.35	0.49	$0.4 + 0.15\ E$	1.35	0.49	
$^{11}B + {}^{12}C$	15	50	1.29	0.48	15	1.18	0.30	Vo 69
		100	1.19	0.48	27	1.26	0.26	
$^{16}O + {}^{11}B$	11 – 13	50	1.21	0.63	6.0	1.46	0.52	
		100	1.21	0.55	$0.57\ E$	1.35	0.55	
$^{19}F + {}^{12}C$	23	50	1.26	0.48	27	1.26	0.26	
		100	1.19	0.48	27	1.26	0.26	

A remarkable feature of the optical model analysis of $^{16}O + {}^{16}O$ scattering [Si 67, Ma 69, Go 73] is that satisfactory fits over a large energy range could only be obtained with a real potential, which is unusually shallow ($V \approx 17$ MeV, see Table 2.2 and Fig. 2.6) by comparison with potentials based on extrapolation from light projectiles like hydrogen or helium ions. Similarly small values had to be chosen for the imaginary potential, leading to a significant degree of transparency of the interaction region for partial waves near the grazing angular momentum. It was concluded that the pronounced gross structure in the scattering excitation functions is enhanced by interference effects between waves reflected at the outer edge of the interaction region and from the centrifugal barrier in its interior [Br 69].

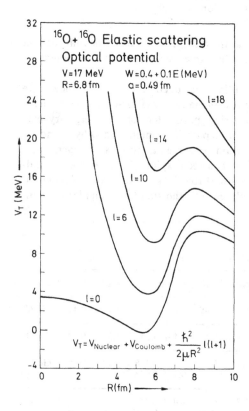

Fig. 2.6. Radial dependence of the $^{16}O + {}^{16}O$ optical potential for several partial waves. (From Ma 69)

The shallow potentials have yielded acceptable fits not only for $^{16}O + {}^{16}O$, but also for other systems in the same mass region [Go 71, Si 71; see Table 2.2]. However, the amplitude of the gross structure and the magnitude of the cross section at high energies vary appreciably from one system to another, and these differences are mainly reflected in different imaginary potentials required to fit the data. A particularly striking example is shown in Fig. 2.7, where the 90° excitation function for $^{16}O + {}^{16}O$ is compared with that for

^{18}O $+$ ^{18}O [Sh 70]. The latter falls much more steeply towards high energies, and exhibits much less structure; it is approximately reproduced by an optical model calculation, where the imaginary potential has been increased by a factor 3.5 compared with that used for ^{16}O $+$ ^{16}O [see Table 2.2].

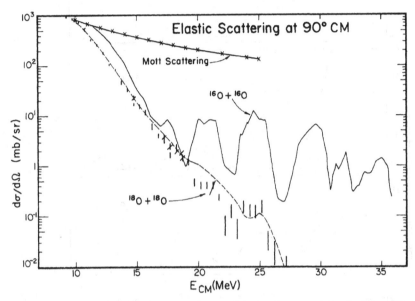

Fig. 2.7.   Comparison of elastic scattering excitation functions at 90° for the ^{16}O $+$ ^{16}O (solid curve) and ^{18}O $+$ ^{18}O (vertical bars) systems. The dashed line is an optical model calculation for ^{18}O $+$ ^{18}O with parameters given in Table 2.2. (From Sh 70)

Similar observations were made also in a later study of ^{12}C $+$ ^{20}Ne scattering [Va 74]. The latter system, which would produce the same compound nucleus as ^{16}O $+$ ^{16}O, exhibits much less structure and a smaller high-energy scattering cross section in accordance with a more strongly absorbing potential.

In spite of their success in reproducing experimental data, the calculations with the standard optical model are unsatisfactory in two respects: firstly, the small imaginary potentials found in a number of cases imply unphysically large mean free paths of the interpenetrating nuclei [Si 74]. Secondly, the model provides no insight into the question why some systems show much less absorption than others, and thus into the mechanism by which nuclear structure enters the scattering problem. A significant step forward in both directions was achieved by a modification of the standard optical model, first proposed by Bisson and Davis [Bi 69] for α scattering, and by Chatwin et al. [Ch 69] for heavy-ion scattering. It introduces an explicit dependence of the imaginary potential parameter W on orbital angular momentum according to the relationship

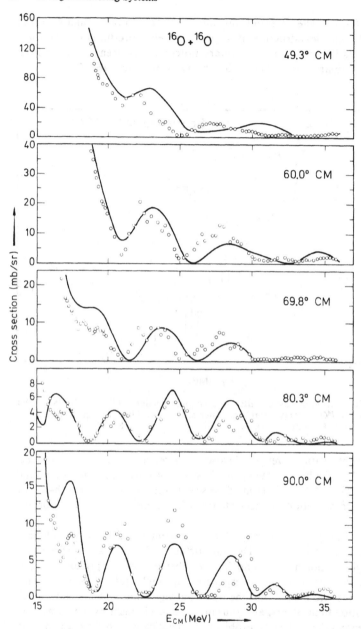

Fig. 2.8. Fit to $^{16}O + ^{16}O$ elastic scattering excitation functions with the angular-momentum-dependent optical model. (From Ch 70: compare Fig. 2.3)

$$W(l) = W\left(1 + \exp\frac{l - l_{\mathrm{w}}}{\Delta_{\mathrm{w}}}\right)^{-1}. \tag{2.4}$$

This version of the optical model was found superior to the standard model in fitting the very pronounced structure of the $^{16}O + {}^{16}O$ data, as shown in Fig. 2.8 [Ch 70]. Fits of a similar quality were obtained, on the other hand, by Gobbi et al. without l dependence by simply reducing the radius of the imaginary potential [Go 71, Go 73]. The apparent equivalence of the two approaches can be understood semiclassically by noting that the near-grazing partial waves are localized at the outer edge of the interaction region. The question arises, however, whether the effective surface transparence of these potentials is intrinsically connected with angular momentum [as implied by (2.4)] or is a geometrical property of the entrance channel. A crucial point of the angular-momentum-dependent optical model is therefore the significance of the parameter l_w, i.e. its relationship to known nuclear properties of the individual scattering system.

Chatwin et al. [Ch 69] and Eberhard [Eb 70] have related l_w to the availability of compound nucleus decay channels, and were thus able to produce estimates consistent with experimental observation on α and ^{16}O scattering from different targets Shaw et al. [Sh 70], on the other hand, have shown that this interpretation of l_w cannot be correct, as it disregards the specific relevance of the optical model to the entrance channel and the independence of compound nucleus formation and decay. The situation is illustrated schematically in Fig. 2.9, which demonstrates that the processes responsible for abosrption in the optical model are compound nucleus formation and direct reactions. The decay of the compound nucleus is only indirectly related to the entrance channel via the excitation energy and angular momentum of the compound system, which determine both the number of states available for its formation and the number of open decay channels.

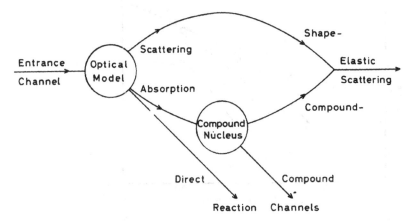

Fig. 2.9. Schematic representation of nuclear scattering, showing the relationship of the optical model with various reaction mechanisms

According to Shaw et al. and Vandenbosch [Sh 70, Va 71] the difference in absorption between the $^{16}O + {}^{16}O$ and $^{18}O + {}^{18}O$ systems can be understood by considering the maximum angular momentum, which can be carried away by direct reaction channels like α transfer, nucleon transfer, and inelastic scattering. As shown in Fig. 2.10, this is lower than the grazing angular momentum for $^{16}O + {}^{16}O$, and higher than the grazing angular momentum for $^{18}O + {}^{18}O$ in the energy region of interest. Thus the angular momentum cut-off l_W should be effective in the former system, but not in the latter, in agreement with observation. The underlying physical reason for this differ-

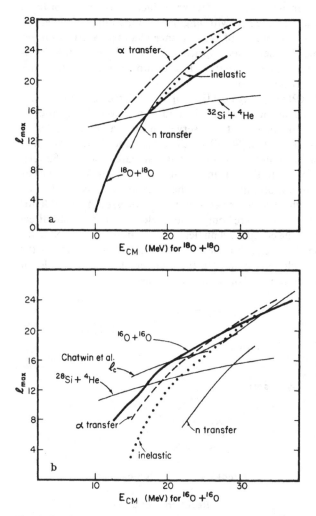

Fig. 2.10. Comparison between the grazing angular momentum in the elastic channel and maximum angular momenta in various reaction channels for the $^{16}O + {}^{16}O$ and $^{18}O + {}^{18}O$ systems. (From Sh 70)

ence is the more strongly bound structure of doubly magic ^{16}O, which results in unfavourable Q values for the transfer and inelastic channels. A detailed analysis of $^{16}O + {}^{16}O$ scattering with the l-dependent model [Ch 70] shows that l_W becomes larger than the grazing value, and thus ineffective, both above and below the energy range of pronounced gross structure (approximately 20–33 MeV). The arguments developed above indicate that the absorption is dominated by compound nucleus formation at lower energies and by direct processes at higher energies.

We can summarize this discussion of angular-momentum-dependent absorption by stating that the cut-off arises primarily from a lack of coupling between the entrance channel and surface reaction channels due to kinematic mismatch. This conclusion is supported by a considerable body of experimental evidence which shows that the interaction of strongly bound fragments, like ^{12}C and ^{16}O, is described by effectively surface-transparent potentials. While compound nucleus formation necessarily involves hard contact and hence penetration to small distances, the absorption at intermediate distances is determined by surface reactions which differ in importance from system to system. In this sense it seems irrelevant whether weak surface absorption is ascribed to a cut-off of the interaction in angular momentum space or in radial space. It is important to note, however, that the problem is not entirely a geometrical one, but is closely related to the internal structure of the fragments.

More recently a number of careful studies of fusion excitation functions have revealed that characteristic gross structure can exist not only in elastic scattering or reaction cross sections, but also in fusion cross sections. Again the systems involving either ^{12}C or ^{16}O fragments are those which show the effect. As an example, we present in Fig. 2.11 results obtained by Sperr et al. at Argonne for the system $^{12}C + {}^{16}O$ [Sp 76a]. Superimposed on a smooth energy dependence, the fusion excitation function clearly exhibits a modulation with a relative amplitude of the order of 10% and a period of about 3–4 MeV. Qualitatively similar characteristics were observed also for $^{12}C + {}^{12}C$ [Sp 76b; see Fig. 7.14] and $^{16}O + {}^{16}O$ [Fe 78, Ts 78], but not for other systems. It seems plausible to associate this phenomenon with the same mechanism which is also responsible for the pronounced gross structure in elastic scattering, namely weak coupling to surface reaction channels, although the connection is not immediately obvious. We shall return to this interesting point in the context of intermediate structure in Sect. 2.3.

The success of the phenomenological analysis of heavy-ion scattering with the optical model has encouraged numerous attempts to deduce the nucleus–nucleus interaction from microscopic theories of nuclear structure. Clearly the complex problem of two interacting many-body systems requires extreme simplification and therefore—although interesting qualitative insights have been obtained—a direct quantitative comparison of the theoretical results with experimental data does not seem meaningful in the majority of cases. We shall not try here to review relevant theoretical work, but briefly discuss some illustrative examples.

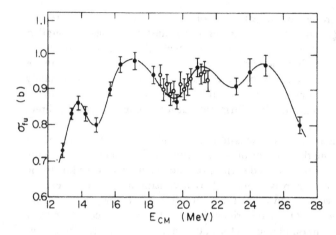

Fig. 2.11. Total fusion cross section as a function of centre-of-mass bombarding energy for the system $^{16}O + {}^{12}C$. The solid curve has been drawn to connect the experimental points. (From Sp 76a)

A first step towards a microscopic description of the nucleus–nucleus interaction is provided by the so-called "folding potentials". "Single folding potentials" are generated by folding the optical model potential for scattering of a nucleon from nucleus 2 with the nucleon density distribution of nucleus 1, or vice versa. Alternatively, the density distributions of both interacting nuclei may be folded together with an effective nucleon–nucleon interaction to produce a "double folding potential". These approaches have been very popular in the literature; their successes and problems were discussed by, among others, Satchler [Sa 74, Sa 76], Brink [Br 76], and Mosel and collaborators [Zi 75, Fl 77].

The influence of nuclear shell structure on the nucleus–nucleus interaction may be studied by means of the "macroscopic–microscopic method", or "shell correction method", which has been very successful in the field of nuclear fission [St 67, Br 72, Ni 72]. In this approach one calculates separately the smooth, structure-indepedent part of the interaction and a shell correction which takes microscopic properties of the system into account. The smooth part is usually obtained by some kind of folding prescription. Scheid and Greiner, for example, have proposed an "extended liquid-drop model" which allows the total potential energy of the nucleus–nucleus system E to be expressed as a functional of the total density ρ [Sc 68, Sc 69]. For two identical nuclei with mass number A and atomic number Z, they set

$$E(\rho) = W_0 A + \frac{C}{2\rho} \int (\rho - \rho_0)^2 d\tau + \frac{V}{8\pi} \int \rho(\boldsymbol{r}_1) \frac{\exp\left(-\dfrac{|\boldsymbol{r}_1 - \boldsymbol{r}_2|}{\mu}\right)}{|\boldsymbol{r}_1 - \boldsymbol{r}_2|} [\rho(\boldsymbol{r}_2)$$
$$- \rho(\boldsymbol{r}_1)] d\tau_1 d\tau_2 + \frac{1}{2}\left(\frac{eZ}{A}\right)^2 \int \rho(\boldsymbol{r}_1) \frac{1}{|\boldsymbol{r}_1 - \boldsymbol{r}_2|} \rho(\boldsymbol{r}_2) d\tau_1 d\tau_2$$

$$+ \frac{G}{2\rho_0}\left(2\frac{Z}{A} - 1\right)^2 \int \rho^2 d\tau. \tag{2.5}$$

In (2.5) the surface, Coulomb, and symmetry terms are obtained by explicit integration over the combined nuclear density distribution $\rho(\mathbf{r})$; in addition a compression term is included to account for repulsion effects caused by high densities in the overlap region. The nuclear attraction between the fragments arises in this model from a reduction of the surface term with decreasing fragment separation. The parameters W_0, V, and G are determined by experimental binding energies, whereas C and μ are adjustable; the density distribution of an isolated fragment is obtained by minimizing the energy $E(\rho)$.

The calculation of the interaction potential, which is the difference between $E(\rho)$ at a given fragment separation, and the same expression for infinite separation, requires a prescription for the superposition of fragment densities during the scattering process. Two extreme cases have been studied: the adiabatic potential, considered typical of slow collisions, and the sudden potential, considered typical of very fast collisions. In the former case the total volume is conserved and no compression occurs, whereas in the latter the individual fragment densities are added, resulting in highly compressed configurations and strong repulsion at small separations, due to the compression term in (2.5). Examples of sudden and adiabatic potentials for the $^{16}O + ^{16}O$ system are shown in Fig. 2.12.

The repulsive core implied by the sudden potential can be regarded as a consequence of the Pauli exclusion principle, which forces part of the nucleons into higher momentum states when compressed to high density. There has been much speculation about the possible existence of such a repulsive core

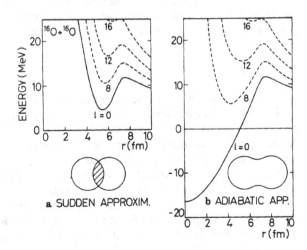

Fig. 2.12. Comparison of effective $^{16}O + ^{16}O$ potentials in the sudden and adiabatic limits for several partial waves. (From Gr 71)

and its effect on scattering. The results of optical model analyses discussed previously show, however, that elastic scattering is completely insensitive to assumptions concerning the existence and magnitude of a potential at small internuclear separation.

In actual scattering situations a density distribution intermediate between the adiabatic and sudden limits is expected, depending on bombarding energy. This question has been discussed by Fliessbach [Fl 71] using microscopic (antisymmetrization) arguments, and by Scheid et al. [Sc 72] using arguments based on classical hydrodynamics.

The effects of nuclear structure are incorporated into the real potential by adding a "shell correction term" to the liquid-drop potential:

$$V(r) = V_{LD}(r) + \delta U\ (r). \tag{2.6}$$

The shell correction $\delta V(r)$ is obtained by a renormalization procedure due to Strutinsky [St 67, St 68]. This is based on the idea that a straight summation of single particle energies will not yield correct absolute values of the total energy of a nucleus of given shape, but may be adequate to calculate fluctuations around some average behaviour. Thus one sets

$$\delta U(r) = U(r) - \bar{U}(r), \tag{2.7}$$

where $U(r)$ is obtained by summation over single particle energies calculated with an appropriate shell model, and $\bar{U}(r)$ differs from $U(r)$ by the use of an energy-averaged distribution of single-particle levels. The resulting correction depends sensitively on the density of single-particle levels at the Fermi energy of the deformed nuclear system under consideration.

The single particle levels of the two-nucleus system may be calculated in the "two-centre shell model". In the simplest case, using isotropic harmonic oscillator potentials and neglecting angular momentum dependent terms, the model Hamiltonian for identical nuclei can be written [Ho 69]

$$h = \frac{p^2}{2m} + \frac{m\omega^2}{2}(x^2 + y^2 + (|z| - z_0)^2) \tag{2.8}$$

where the distance between the potential centres is $r = z_0$.

In shell model calculations, the oscillator frequency ω is inversely proportional to the nuclear radius; hence the dependence of fragment radius on centre separation, as implied by the assumption of an adiabatic (slow) or sudden (fast) collision, can be accommodated by an appropriate variation of ω with z_0. During the collision, the nucleons are assumed to fill always the lowest available states of the two-centre potential.

Using these methods, C + C and O + O potentials were computed by Prüss and Greiner [Pr 70] and by Mosel, Thomas, and Riesenfeldt [Mo 70]. The former authors calculated both slow and fast potentials, starting from the extended liquid-drop model of Scheid and Greiner, whereas the latter authors used a standard liquid-drop model [My 67] and derived adiabatic (slow) potentials only. The results of Mosel et al. are shown in Figs. 2.13 and 2.14: Fig. 2.13 gives the single particle levels as functions of centre separation, and Fig. 2.14 gives potentials for the $^{12}C + ^{12}C$, $^{16}O + ^{16}O$, and $^{18}O + ^{18}O$ systems with and without shell correction. It is interesting to note the comparatively shallow potential calculated for the $^{16}O + ^{16}O$ case, which arises from a large negative shell correction for the isolated, doubly magic ^{16}O clusters.

A comprehensive review of the semimicroscopic treatment of nucleus–nucleus potentials has been given by Greiner and Scheid [Gr 71]. Experimentally, the mutual interaction of fragment shells in heavy-ion scattering has not yet been clearly established (apart from the special case of exchange scattering, see Sect. 2.4), due to the difficulties in deducing the nucleus–nucleus potential at strong penetration from scattering data.

In the theoretical discussion so far we have implied reversibility and have neglected coupling to inelastic channels. This is not consistent with experimental observation, however, which clearly shows the strong influence of absorptive processes. In the spirit of the optical model, we can identify the semimicroscopic potentials with the real part of the potential, and introduce absorption via an imaginary part. Scheid and Greiner [Sc 69] tried to estimate the imaginary potential from hydrodynamical arguments, considering the outflow time of nuclear matter from the compressed region. This approach cannot explain, however, the strong nuclear structure dependence of the absorption.

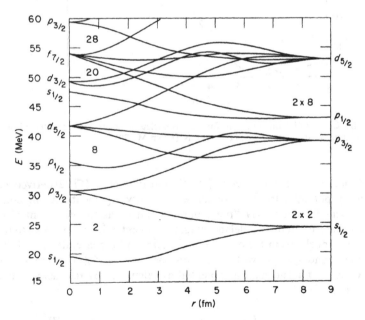

Fig. 2.13. Single particle levels in the two-centre shell model for adiabatic $^{16}O + ^{16}O$ collisions. (From Mo 70)

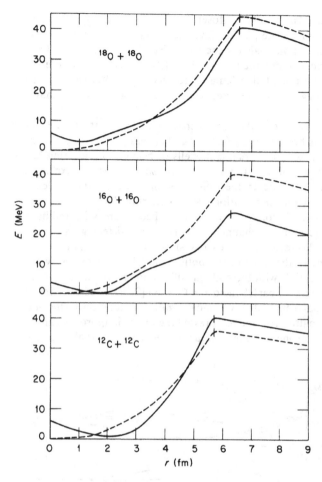

Fig. 2.14. Adiabatic potentials for various nucleus–nucleus systems with (full curves) and without (dashed curves) shell corrections. (From Mo 70)

More recently, Helling et al. [He 71] and Fink et al. [Fi 72] derived an imaginary potential from statistical level density arguments, taking compound nucleus formation via "precompound states" as the dominant absorptive mechanism. In this model the imaginary potential is proportional to the number of nucleons in the overlap region $N(r)$ and the level density of the "precompound system" with excitation energy $E - V_{ad}(r)$, where $V_{ad}(r)$ is the adiabatic nucleus-nucleus potential as calculated by the methods discussed above:

$$W(r,E,l) = \alpha N(r) \frac{2l + 1}{\sigma^3} \exp \left\{ 2\sqrt{a[E - V_{ad}(r)]} - \frac{(l + \frac{1}{2})^2}{2\sigma^2} \right\} \qquad (2.9)$$

The spin cut-off parameter σ is related to the moment of inertia $\mathscr{I}(r)$ and the excitation energy by

$$\sigma^2(r) = \frac{\mathscr{I}(r)}{\hbar^2} \sqrt{\frac{E - V_{ad}(r)}{\beta}}, \tag{2.10}$$

and the parameters α and β are adjusted to fit experimental data. The explicit dependence of this imaginary potential on energy and angular momentum is reminiscent of the E- and l-dependent phenomenological potentials, and may be regarded as a justification of the latter from somewhat more fundamental arguments. A defect of the model in its present form is its neglect of the coupling to direct reaction channels as a source of absorption. Nevertheless, good fits have been achieved to $^{16}O + ^{16}O$ [He 71] and $^{12}C + ^{12}C$ [Fi 72] scattering. Work along similar lines was published by Low and Tamura [Lo 72] and by Gobbi et al. [Go 73]. The latter authors proposed including absorption due to direct reactions by means of an additional surface term in the imaginary potential:

$$W(r, E, l) = C \, \rho(E, l) V_{real}(r) + W_{surf}(E, r), \tag{2.11}$$

where $\rho(E, l)$ is the level density in the compound nucleus and W_{surf} was taken to be of derivative Woods–Saxon form and proportional to the square of the excess energy above the Coulomb barrier. With this ansatz for the imaginary potential, significantly improved fits were obtained to forward angle ($\leq 70°$) excitation functions in $^{16}O + ^{16}O$ scattering, although the simultaneous fit to the $90°$ data was less satisfactory than in previous analyses [Go 73].

Significant progress towards a microscopic theory of the imaginary optical potential was achieved by Fink and Toepffer [Fi 73]. These authors introduced an absorption coefficient κ (equal to the inverse of the mean free path), which is related semiclassically to the imaginary potential by

$$W(r, E) = \frac{\hbar^2}{2\mu} \kappa \left\{ \frac{\kappa^2}{4} + \frac{2\mu}{\hbar^2} [E - V(r)] \right\}^{\frac{1}{2}}. \tag{2.12}$$

The nucleons in the two colliding nuclei were assumed to be distributed spatially according to density distributions ρ_1 and ρ_2, and to occupy appropriate Fermi spheres in momentum space. The absorption coefficient was then calculated from the relation

$$\kappa(r, k) = \int d^3r' \, \rho_1(r')\rho_2(r - r')\sigma(r, r', k), \tag{2.13}$$

where $\sigma(r, r', k)$ is an average cross section for nucleon–nucleon collisions and r, k are the relative distance and momentum of the fragments. The cross section σ was obtained by integration over the nucleonic momentum distributions, using experimental nucleon–nucleon scattering cross sections and taking the Pauli principle into account. Thus essentially parameter-free imaginary potentials were derived, which resulted in good fits to experimental scattering data for the $^{12}C + {}^{12}C$ and $^{16}O + {}^{16}O$ systems. This approach appears promising, especially for heavier nuclei; it seems less suitable, however, to explain the striking dependence of absorption on nuclear structure observed in light scattering systems.

The latter problem has been studied by introducing dynamic features into the two-centre shell model. Glas and Mosel [Gl 74] calculated transition probabilities between nucleonic levels at level crossings in a "correlation diagram" of the $^{16}O + {}^{16}O$ system ("Landau–Zener effect"; see also Fig. 2.13). Their results indicate that the coupling between relative and intrinsic motion should set in sharply at some critical distance which depends sensitively on shell structure. A number of authors have analysed nucleus–nucleus collisions in the framework of dynamic particle-core models, where the interacting fragments are considered to be composed of a heavy core and one or a few loosely bound valence nucleons. This approach has been applied to elastic and inelastic scattering [Pa 72, Te 78], and especially to situations involving nucleon transfer or exchange [see, for example, Oe 75].

Important progress has been made recently—since about 1975—in applying the time-dependent Hartree–Fock approximation (TDHF) to nucleus–nucleus collisions [see, for example, Bo 76]. In this theory the macroscopic collective motion of the fragments and the microscopic degrees of freedom are not treated independently, but are coupled by the time-dependent mean field in a self-consistent manner. Applications to date have dealt mainly with simple systems, like $^{16}O + {}^{16}O$ or $^{40}Ca + {}^{40}Ca$, and with certain average properties, like time-dependent density distributions and various characteristics of the classical trajectories [see Fl 78, Bo 78 and references quoted there]. Although many questions are still open concerning the interpretation and accuracy of the results, the method seems a potentially powerful tool for probing the connections between microscopic nuclear structure and nucleus–nucleus scattering.

We close this section by summarizing the most important results. The gross structure in the elastic scattering of light nuclei as a function of energy and angle is well understood as a diffraction phenomenon, arising from an interaction region of finite size and describable in terms of a complex potential model. The scattering data yield information on the potential near the surface of the interaction region, but leave the interaction in the interior largely undetermined due to strong absorption of low partial waves. The transparency of the surface region for the higher partial waves varies strikingly from one scattering system to another. This can be explained semi-quantitatively by considerations of energy balance and angular momentum matching between the elastic channel on one side, and direct reaction channels or compound nucleus formation on the other. Arguments of this nature throw light on the influence of fragment structure and stability on the scattering cross sections, which is further explored in various theories of the nucleus–nucleus interaction.

2.3   Intermediate Structure and the Nuclear Molecule

In this section we consider in more detail the energy dependence of light-nucleus scattering and reaction cross sections, and review proposed interpretations from a phenomenological point of view. The relevant observations can be classified in two groups:

i) "Resonance" structure in reaction cross sections at energies near and below the Coulomb barrier with characteristic widths between 50 and 200 keV. Following the work of Almqvist, Bromley, and Kuehner on the $^{12}C + ^{12}C$ system [Al 60, Al 63; see Sect. 2.1] similar structure has been identified by many other groups in $^{12}C + ^{12}C$ and $^{12}C + ^{16}O$ reactions. A summary of "resonances" observed for the former system is presented in Table 2.5. A distinct feature of these structures is that they are strongly correlated in different decay channels of the compound nucleus, and at different angles of observation. No such effect could be detected with other target–projectile combinations in spite of an extensive search [Vo 72, Ha 74].

ii) "Intermediate" or "fine" structure in elastic scattering and reaction cross sections at energies well above the Coulomb barrier with characteristic widths in the range 100–500 keV. Again this type of structure is most striking in the $^{12}C + ^{12}C$ system and completely absent in some other systems like $^{14}N + ^{14}N$ or $^{18}O + ^{18}O$. It appears as a modulation of the gross structure in elastic scattering, as shown in Fig. 2.15 for $^{16}O + ^{16}O$. In contrast to the structure of class i), there is often no apparent cross-correlation between different reaction channels or significantly different angles of observation.

Prominent structure of both types is indicative of intermediate states with comparatively long lifetimes and appreciable decay widths into the appropriate heavy ion channels. Difficulties in reconciling this phenomenon with traditional concepts of strong absorption and statistical compound nucleus decay have led to the hypothesis of a "quasi-molecular" interaction between complex nuclei [Br 60, Al 60, Vo 60]. A basic feature of this interaction was thought to be a secondary potential minimum near the surface of the interaction region, giving rise to resonant molecule-like configurations of two nuclei in close contact.

Much of the early discussion about the physical nature of the quasi-molecular interaction concentrated on the different behaviour of the $^{12}C + ^{12}C$ and $^{16}O + ^{16}O$ systems near the Coulomb barrier. Different but not necessarily unrelated arguments were given by different authors. Vogt and McManus [Vo 60] stressed the influence of nuclear deformability in producing a long-range attraction between the fragments. The same authors, and especially Wildermuth and Carovillano [Wi 61], have drawn attention to the Pauli exclusion principle, which tends to preserve the identity of the clusters near the point of contact and to prevent strong interpenetration. Virtual nucleon transfer [Br 60] and the correlated motion of nucleon pairs or clusters in the common potential of the two fragments have also been considered as a source of nuclear attraction, in analogy to the homeopolar bond in molecular physics. All of these arguments indicate a stronger surface interaction between

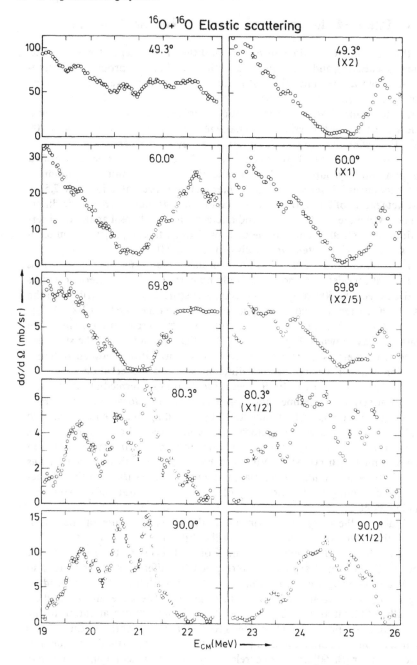

Fig. 2.15. $^{16}O + ^{16}O$ elastic scattering excitation functions, showing intermediate structure. (From Ma 69)

two ^{12}C nuclei, with their unfilled p shells, compared to two ^{16}O nuclei. Davis, on the other hand, has pointed out that a secondary potential minimum arises naturally from the superposition of conventional Coulomb, nuclear, and centrifugal forces, without explicitly invoking new mechanisms [Da 60]. A review of early work on quasi-molecular interactions has been given by Bromley [Br 69]. Regardless of its relevance to intermediate structure phenomena, this work has led the way to more quantitative theories of the nucleus–nucleus interaction like those mentioned in Sect. 2.2.

It was soon recognized, however, that at least part of the observed "intermediate" structure—especially that of class ii)—might find a less spectacular explanation than the existence of nuclear molecules. This explanation is based on the existence of compound-nucleus contributions to elastic scattering and reaction cross sections, which are expected to exhibit statistical fluctuations in situations of strong overlap of the compound-nucleus levels [Er 63]. The effect should depend critically on two features: the angular momentum distribution in the entrance channel and the excitation energy of the compound nucleus, which in turn determines the availability of compound levels and open exit channels for each angular momentum. In Table 2.3 we have collected relevant parameters for a number of systems.

Table 2.3. Reaction data for different target–projectile combinations (all energies in the CM system)

Target/ Projectile	Coulomb barrier [MeV][a]	Excitation energy [MeV][b]	Maximum energy of reaction products [MeV]		
			n	p	α
^{12}C + ^{12}C	6.3	20.2	3.7	8.5	10.9
^{12}C + ^{16}O	8.0	24.7	7.6	13.2	14.8
^{14}N + ^{14}N	8.1	35.4	18.2	23.8	25.4
^{16}O + ^{16}O	10.2	26.8	11.7	17.9	19.8
^{18}O + ^{16}O	10.0	34.4	23.0	23.5	26.5
^{18}O + ^{18}O	9.8	38.9	29.0		29.9

[a] Calculated from $E_C = \dfrac{Z_1 Z_2 e^2}{r_0(A_1^{1/3} + A_2^{1/3})}$ with $r_0 = 1.80$ fm.

[b] Excitation energy of the compound nucleus at the Coulomb barrier.

At energies near the Coulomb barrier the incident angular momenta are small and the absorption goes largely into compound nucleus formation. The compound contribution to each exit channel, including elastic scattering, drops off exponentially with increasing excitation energy or—more accurately—with increasing energy available for the dominant decay modes, in inverse proportion to the number of open channels. The figures given in Table 2.3 thus predict the largest compound contribution for the ^{12}C + ^{12}C system, followed by ^{12}C + ^{16}O and other systems in excellent qualitative correlation with the magnitude of the observed structure.

Detailed statistical analyses of compound nucleus cross sections and their fluctuation properties have been performed for a number of heavy-ion systems, notably those involving ^{12}C and ^{16}O. Table 2.4 gives a survey of relevant work. In most cases the mean level width or "coherence width" of the compound

Table 2.4.   Published work on the statistical analysis of fluctuating cross sections

Reaction	E_{CM} [MeV]	E_{ex} [MeV][a]	Γ[keV][b]	References
$^{12}C(^{12}C, ^{12}C)^{12}C$	8 – 16	22 – 30	~ 100	Bo 73
	7 – 18	21 – 32	111 ± 11	Em 73
$^{12}C(^{12}C, ^{12}C)^{12}C$	16 – 21	30 – 35	277 ± 68	Sh 74
	13.5 – 37.5	27.5 – 51.5	377 ± 60	
$^{12}C(^{12}C, ^{12}C)^{12}C*$	10 – 18	24 – 32	102 ± 13	Em 73
$^{12}C(^{12}C, ^{12}C)*^{12}C*$	15 – 18	29 – 32	105 ± 21	Em 73
$^{12}C(^{12}C, \alpha)^{20}Ne$	10 – 13	24 – 27	80 – 120	Al 64, Vo 64
$^{12}C(^{12}C, \alpha)^{20}Ne$	16 – 21	30 – 35	263 ± 42	Sh 74
$^{12}C(^{16}O, ^{16}O)^{12}C$	12 – 25	28.5 – 41.5	110	Ma 72
$^{12}C(^{16}O, \alpha)^{24}Mg$	8.5 – 18.5	25 – 35	118 ± 17	Ha 67
	19 – 25	35.5 – 41.5	90 – 150	Gr 72
$^{16}O(^{16}O, ^{16}O)^{16}O$	17.5 – 19.5	34 – 36	76 ± 17	Sh 69
	12 – 16	28.5 – 32.5	~ 100	Bo 73
$^{16}O(^{16}O, \alpha)^{28}Si$	17.5 – 19.5	34 – 36	76 ± 17	Sh 69
	12.5 – 14.5	29 – 31	73 ± 7	Le 72

[a] Excitation energy of compound nucleus
[b] Average level width of compound nucleus

nucleus has been extracted by calculating the autocorrelation function [Er 63] of the excitation functions. Coherence widths Γ of the order of 100 keV have been obtained for the systems in question. In addition, the fraction of each reaction proceeding via either a direct or a compound nucleus mechanism has been estimated from the mean-square fluctuation amplitudes or by statistical model calculations of the compound-nucleus cross sections. A calculation of the latter type by Shaw et al. [Sh 69] of the compound elastic contribution to $^{16}O + ^{16}O$ scattering is shown in Fig. 2.16. The results of these analyses support a statistical interpretation of the intermediate and fine structure for the investigated systems. They indicate a predominant compound nucleus mechanism for the (HI, α) reactions, although direct contributions up to 50% and more cannot be excluded in some cases. The average compound elastic cross sections are of the order of a few per cent of the total elastic cross sections, but produce significant structure by interference with the shape elastic part.

It might appear from the preceding discussion that the intermediate structure in light heavy-ion systems above the Coulomb barrier is well understood and of statistical origin. Such a statement would be misleading, however. All that can be said at this point is that existing statistical models predict structure of the observed magnitude. Other and more interesting mechanisms

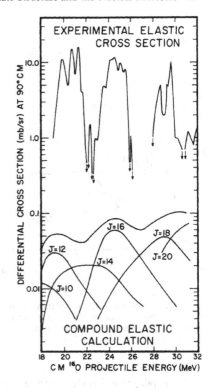

Fig. 2.16. Differential cross section for $^{16}O + ^{16}O$ elastic scattering at 90°: Comparison between experimental cross section and calculated compound elastic part. Contributions to the latter associated with various angular momenta are shown separately. (From Sh 69)

—specifically related to nuclear structure—may well be involved, but are difficult to identify in the presence of statistical "noise". Moreover, the very application of statistical methods to compound systems with $A = 24 - 32$ and angular momenta of 10–20 $\hbar$ may be open to question.

There are, in fact, strong indications for nonstatistical effects both in the experimental results and on theoretical grounds. "Anomalous" structures have been observed in various reactions, most prominently, however, in the $^{12}C + ^{12}C$ system. An important characteristic of these structures is that they are correlated in different exit channels, in contrast to expectation for statistical fluctuations. The experimental identification of nonstatistical structure has usually been based on two pieces of evidence: the persistence of the structure, when data for many final states or different exit channels are added, and the appearance of a well-defined angular distribution pattern indicating a unique angular momentum of the compound nucleus. The former criterion is illustrated in Fig. 2.17 [Ga 77], and the latter in Fig. 2.18 [Er 76]. It should be noted that the analysis of angular distributions becomes especially simple in cases where the entrance and exit channel are composed of spinless particles, as in ^{12}C (^{12}C, $\alpha)^{20}Ne$. An intermediate state of (pure) angular momentum J then gives rise to an angular distribution which is simply proportional to $|P_J(\cos \Theta)|^2$, where P_J denotes the Legendre polynomial of order J.

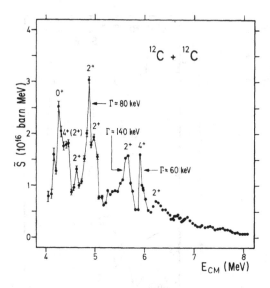

Fig. 2.17. Total cross section for transitions to the six lowest states in ^{20}Ne in the reaction ^{12}C(^{12}C, α)^{20}Ne, plotted as "nuclear structure factor" $\tilde{S}$ [for definition see (7.15)] versus centre-of-mass bombarding energy. Spins, parities, and widths of the most prominent "resonances" are shown (from Ga 77)

The experimental evidence on resonance-like structures in light heavy-ion systems has been reviewed at different occasions by Stokstad [St 73], Bromley [Br 75] and Feshbach [Fe 76]. Table 2.5 gives a summary for the most thoroughly studied system, ^{12}C + ^{12}C. Qualitatively similar, but less complete results are available for the system ^{12}C + ^{16}O, where in particular an "anomaly" near $E_{CM} = 19.7$ MeV ($J^\pi = 14^+$) has been studied by many groups [Ma 72, Co 72, St 72]. No evidence has been found in other systems, however, for a systematic occurence of nonstatistical structures.

The resonances listed in Table 2.5 span an energy range from well below the Coulomb barrier to about four times the barrier. It seems likely that many of these "resonances" do not correspond to well-defined states of the compound nucleus, but rather to groups of close-lying states with the same spin and parity. This is true, in particular, for the higher-energy structures which form part of what was referred to earlier as structures of class two.

In discussing possible interpretations of the observed resonance structure it is interesting to recall the strong correlation which exists between gross structure in elastic scattering and fusion cross sections on one hand, and intermediate structure (both statistical and nonstatistical) on the other hand. The gross structure has been described in Sect. 2.2 as an entrance-channel effect, caused by the unusual stability of ^{12}C and ^{16}O and the resulting lack of coupling to direct reaction channels. For the same reason, the excitation energy brought into the compound system, and hence the number of states

Fig. 2.18. Legendre-polynomial fits to ground-state angular distributions in the reaction $^{12}C(^{12}C, \alpha)^{20}Ne$ at different bombarding energies. The cross section scale is linear and saturates at 25 mb/sr. (From Er 76)

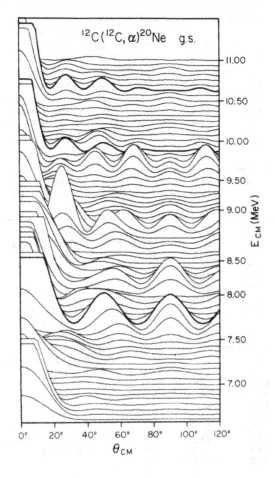

$^{12}C(^{12}C, \alpha)^{20}Ne$ g.s.

available at a given angular momentum, is relatively small compared to other systems under similar bombarding conditions (see Table 2.3). This feature would tend to enhance the possibility of intermediate structure irrespective of the detailed mechanism involved. At the same time the number of open decay channels is reduced; this means that the compound nucleus part of the cross section for a given channel is increased, and so is the amplitude of statistical fluctuations. We conclude that the systems composed of ^{12}C and (or) ^{16}O nuclei are preferred candidates for all types of structure, for simple reasons based on nuclear stability and the balance of energy and angular momentum. This does not rule out the possibility, however, that the existence of α substructures or other microscopic phenomena are essential for producing particular intermediate structures.

The clustering of resonances with a given spin and parity, as shown for the $^{12}C + ^{12}C$ system in Table 2.5, indicates that the interaction in the entrance

Table 2.5. Resonances in $^{12}C + ^{12}C$ reactions

E_{CM} ($^{12}C + ^{12}C$)	E_{ex} (^{24}Mg)	$J^{\pi a}$	Observed decay (References)
3.2	17.1		$^{20}Ne + \alpha$ (Ke 77)
4.25	18.2	0^+	$^{20}Ne + \alpha$, $^{23}Na + p$ (Pa 69, Sp 72,
4.46	18.4	$4^+(2^+)$	Ma 73, Sp 74, Ga 77, Ke 77)
4.62	18.55	2^+	
4.88	18.8	2^+	
5.00	18.9	2^+	
5.64	19.6	2^+	$^{20}Ne + \alpha$, $^{23}Na + p$ (Br 61, Al 63, Pa 69,
5.92	19.85	4^+	Sp 72, Sp 74, Ga 77, Ke 77)
6.25	20.2	2^+	
6.87	80.8	4^+	$^{20}Ne + \alpha$ (Vo 77)
7.71	21.6	4^+	$^{20}Ne + \alpha$ (Vo 77, Er 76)
7.90	21.8	4^+	$^{20}Ne + \alpha$ (Vo 77)
8.26	22.2	4^+	
8.46	22.4	4^+	
8.86	22.8	6^+	
9.06	23.0	6^+	
9.84	23.8	8^+	$^{20}Ne + \alpha$ (Vo 77, Er 76)
10.63	24.6	8^+	$^{20}Ne + \alpha$ (Vo 77)
10.90	24.8	8^+	
11.4	25.3	8^+	$^{16}O + ^8Be$(Fl 76), $^{20}Ne + \alpha$(Vo 77), $^{23}Na + p$ (Co 75)
(11.8)	25.7	(8^+)	$^{16}O + ^8Be$ (Fl 76)
12.0	25.9	8^+	
12.35	26.3	8^+	
12.85	26.8	8^+	
13.35	27.3	10^+	
13.85	27.8	10^+	
14.3	28.2	10^+	$^{16}O + ^8Be$ (Fl 76), $^{23}Na + p$ (Co 75)
15.3	29.2	10^+	$^{16}O + ^8$ Be (Fl 76)
16.2	30.1	10^+	
17.15	31.1	10^+	
17.75	31.7	12^+	
18.4 – 18.5	32.4	12^+	$^{16}O + ^8Be$ (Eb 75, Fl 76)
18.8	32.7	$(10,12)^+$	$^{16}O + ^8Be$ (Fl 76)
19.3 – 19.5	33.3	12^+	$^{16}O + ^8Be$ (Fl 76), $^{23}Na + p$ (Co 75)
25.5	39.4	14^+	$^{16}O + ^8Be$ (Eb 75)

[a] J^π assignments for $E_{ex} = 4.2 - 11.4$ MeV from $^{20}Ne + \alpha$ (Ga 77, Vo 77, Al 63) and for $E_{ex} = 11.4$–25.5 MeV from $^{16}O + ^8Be$ (Fl 76, Eb 75).

channel is dominated at each bombarding energy by the grazing partial wave. It is tempting to interpret this observation as due to a transmission resonance, occurring at an energy close to the top of the effective barrier for each partial wave. The existence of such resonances is expected for weakly absorbing potentials on the basis of simple barrier transmission arguments (see, for example, Sc 55) and is supported by optical model and phase-shift analyses for the system $^{16}O + ^{16}O$ (see Sect. 2.2). It seems likely that the steep rise of the fusion cross section produced by these resonances near the effective barrier of a new partial wave is responsible for the gross structure observed in the fusion excitation functions of the systems $^{12}C + {}^{12}C$ [Sp 76b], $^{12}C + {}^{16}O$ [Sp 76a], and $^{16}O + {}^{16}O$ [Fe 78, Ts 78].

In the limit of very weak absorption, the transmission resonances mentioned above correspond to standing waves in the "potential pocket" between the outer and inner barriers. They may then be interpreted as forming a "quasi-molecular rotational band", as discussed by Arima, Scharff-Goldhaber, and McVoy [Ar 72]. The location of this band in ^{24}Mg, formed in the $^{12}C + {}^{12}C$ interaction, is illustrated in Fig. 2.19.

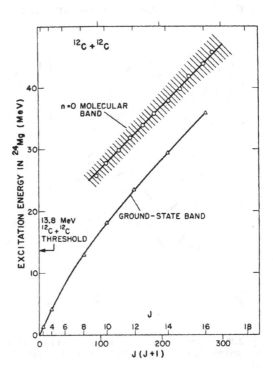

Fig. 2.19. Excitation energy versus angular momentum for the ground-state rotational band in ^{24}Mg and the $^{12}C + {}^{12}C$ quasi-molecular band. (From Ar 72)

The intermediate structure may now be interpreted as a "doorway state" phenomenon, which arises from the coupling of the quasi-molecular entrance channel configuration to somewhat more complicated states of the composite system, which in turn are coupled to the statistical states of the compound nucleus. This point of view has been stressed, for example, by Greiner and Scheid [Gr 71], and by Feshbach [Fe 76]. Imanishi [Im 68, Im 69] and Scheid, Greiner, and collaborators [Sc 70, Fi 72] have proposed a specific mechanism which gives rise to intermediate structure via inelastic excitation of one or both fragments.

In the Imanishi–Scheid–Greiner model the fragments are assumed to enter the interaction region in the elastic channel, where they have sufficient relative kinetic energy to pass over the Coulomb barrier, and to be subsequently trapped in their mutual potential well by inelastic excitation of one or both partners into low-lying collective states (see Fig. 2.20). The latter are described by surface coordinates, which explicitly enter the interaction potential and thus give rise to a system of coupled differential equations for the scattering process. Resonance effects are obtained at bombarding energies correspond-ing to "quasi-bound states" in the inelastic well. As emphasized by Scheid et al. [Sc 70], matching of energies and angular momenta between orbiting resonances in the elastic channel ("virtual molecules") and quasi-bound levels in the inelastic channel results in a double-resonance effect, which greatly enhances the structure in the scattering cross section.

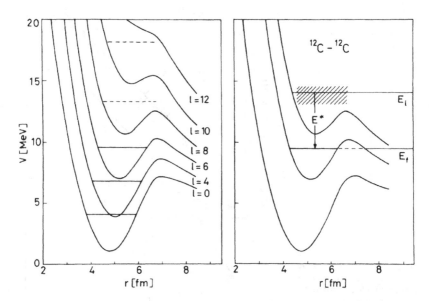

Fig. 2.20. Potentials illustrating the excitation mechanism of quasi-molecular resonances by coupling of elastic and inelastic channels in the case of $^{12}C + ^{12}C$ scattering. (From Fi 72)

This model was applied by Imanishi to explain the three pronounced re-sonances observed in $^{12}C + ^{12}C$ reactions just below the Coulomb barrier (see Fig. 2.2 and Table 2.5). Coupling the lowest 2^+ state in ^{12}C, which was interpreted as a rotational state, to the elastic channel, he was able to reproduce positions and widths of the resonances. Based on the same ap-proach, but different in detail, Scheid et al. [Sc 70] and Fink et al. [Fi 72] analysed intermediate structures in $^{16}O + ^{16}O$ scattering and $^{12}C + ^{12}C$ scattering, respectively, at bombarding energies above the Coulomb barrier. In the work of these authors, the inelastic channels were treated as surface

vibrations and the coupling strengths derived from experimental electromagnetic transition probabilities. Figure 2.21 gives an example of their results.
Obviously detailed fits should not be expected, and we conclude that the
model is capable of explaining structure with the observed characteristics.

It is interesting to note that the success of the Imanishi–Scheid–Greiner
model depends in large measure on two assumptions: the existence of a repulsive core in the potential, and the absence of absorption in the inelastic

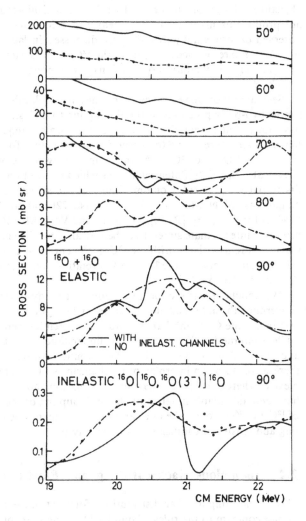

Fig. 2.21. Comparison of experimental (dashed) and calculated (solid) excitation functions for elastic $^{16}O + ^{16}O$ scattering at different angles. The calculation takes inelastic excitation of one ^{16}O nucleus to states at 6.13 MeV (3^-) or 6.92 MeV (2^+) into account. The
dash-dotted curve shown for a scattering angle of 90° only represents a calculation without
channel coupling (From Sc 70)

channel. The former assumption indicates the possibility of learning more about the interaction in the interior, if more specific tests of the model become possible in the future. Arguments based on the angular momentum dependence of the absorptive potential have been presented in support of the latter assumption [Fi 72, Lo 72]. A limitation of the model (as discussed so far) seems to be that it will only produce intermediate structure over a certain range of bombarding energies, extending from just below the Coulomb barrier upwards by about 10–15 MeV in the centre-of-mass system. Earlier objections—based on the original work of Imanishi—that the model could not produce as many resonances near and below the Coulomb barrier as observed experimentally [Mi 72] were dismissed in later studies, where the sub-Coulomb resonance structure of the $^{12}C + {}^{12}C$ system was fitted remarkably well with only minor modifications of the model parameters [Ca 75, Ko 75, Ca 76].

Several authors have also analysed the sub-Coulomb structure in terms of simple two-body potentials without coupling to inelastic channels. Michaud has shown that potentials with a repulsive core are required in order to fit the gross structure of the total reaction cross section for the systems $^{12}C + {}^{12}C$, $^{12}C + {}^{16}O$, and $^{16}O + {}^{16}O$ near and below the Coulomb barrier (Mi 73; see also Sec. 7.2). Potential models which in addition produce qualitative fits to parts of the sub-Coulomb intermediate structure were suggested by Nagorcka and Newton for $^{12}C + {}^{16}O$ [Na 72] and by Park, Scheid, and Greiner for $^{12}C + {}^{12}C$ [Pa 74]. Michaud and Vogt [Mi 72], on the other hand, proposed intermediate states described by an α-particle model to account for the structure associated with the $^{12}C + {}^{12}C$ entrance channel. Experimental evidence in support of the latter interpretation was presented by Voit et al. [Vo 72, Vo 73b, Vo 74].

We conclude this section by noting that many questions are still open concerning the intermediate structure in heavy-ion reactions, both above and below the Coulomb barrier. Much has been learned about the nucleus–nucleus interaction from attempts to understand the origin of the structure, and yet no definite answers have been found. Above the Coulomb barrier there seems to be an interplay of mechanisms with different degrees of complexity, whereas below the barrier the situation should be simpler and perhaps understandable in terms of channel coupling or the α-particle structure of ^{12}C and ^{16}O. So far, the nuclear molecule remains one of the most fascinating and elusive phenomena of nuclear physics.

2.4   Anomalous Backscattering and Elastic Transfer

Elastic scattering angular distributions for light nuclear systems frequently do not conform to the rules of standard "strong absorption models," which (at moderate energies) call for a rather smooth, essentially exponential drop-off in scattered intensity at angles larger than the grazing angle. The phenomenon of anomalous backscattering has been discussed qualitatively in Sec.

2.1, and from the point of view of phase shift analyses, in Sect. 1.4 and 2.2. In this section we describe in somewhat more detail a microscopic model—that of "elastic transfer"—which accounts quantitatively for the observed back-angle effects in many systems of almost identical nuclei.

Elastic transfer can be described by the following formula:

$$C_1 + (C_2 + x) \Bigg\langle {\begin{array}{l} C_1 + (C_2 + x) \\ (C_1 + x) + C_2 \end{array}}$$

where C_1 and C_2 are identical cores and x is a nucleon or nucleon cluster, which can be transferred during the collision. The transfer can be viewed as an exchange of identity between the two fragments; it leads back to the initial fragmentation and is therefore experimentally indistinguishable from elastic scattering. Two qualitative features are immediately obvious: firstly, elastic transfer should be most important in situations where the transferred particle x is loosely bound to the cores C_1, C_2, but has a tightly bound internal struc-ture; secondly, the transfer mechanism should be most noticeable at backward angles, whereas the forward cross section should be dominated by direct scattering. The connection between scattering angle and reaction mechanism is illustrated diagrammatically in Fig. 2.22; assuming zero angular momentum transfer, we can formally write for the scattering amplitude

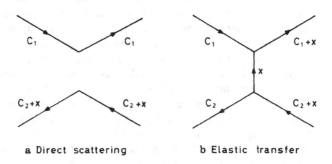

Fig. 2.22. Reaction mechanisms in the elastic scattering of almost identical nuclei: **a** Direct scattering; **b** Elastic transfer

$$f_S(\Theta) = f_{Di}(\Theta) + (-1)^S f_{Tr}(\pi - \Theta) \tag{2.14}$$

in close analogy to the case of elastic scattering of identical nuclei [see (1.31)]. In (2.14), f_{Di} and f_{Tr} denote the direct and transfer amplitudes for distinguish-able cores C_1 and C_2. The quantity S is the resultant spin quantum number of the core–core system, and the factor $(-1)^S$ ensures the proper symmetry of the scattering amplitude with respect to an exchange of the two cores. For spin--independent interaction and observation, the scattering cross section is ob-tained by incoherent superposition of contributions from the different possible S states according to their statistical weights. The result is, denoting by I the spin quantum number of the individual core,

$$\frac{d\sigma}{d\Omega}(\Theta) = |f_{Di}(\Theta)|^2 + |f_{Tr}(\pi - \Theta)|^2$$

$$+ \frac{(-1)^{2I}}{2I + 1} [f_{Di}(\Theta)f_{Tr}^*(\pi - \Theta) + f_{Di}^*(\Theta)f_{Tr}(\pi - \Theta)]. \qquad (2.15)$$

An important consequence of (2.14) is the so-called "odd–even staggering" of the scattering parameters as functions of angular momentum l. This effect can be demonstrated by writing the scattering amplitudes $f_{Di}(\Theta)$ and $f_{Tr}(\Theta)$ in terms of Legendre polynominals and making use of the relationship $P_l(\pi - \Theta) = (-1)^l P_l(\Theta)$:

$$f_{Di}(\Theta) = f_C(\Theta) + \sum_{l=0}^{\infty} a_{Di}(l) P_l(\cos\Theta), \qquad (2.16)$$

$$f_{Tr}(\Theta) = \sum_{l=0}^{\infty} a_{Tr}(l) P_l(\cos\Theta), \qquad (2.17)$$

$$f_s(\Theta) = f_C(\Theta) + \sum [a_{Di}(l) + (-1)^{s+l} a_{Tr}(l)] P_l(\cos\Theta). \qquad (2.18)$$

Equation (2.18) shows that the interference between direct and transfer amplitudes occurs with different sign for odd and even partial waves. Assuming a smooth variation of a_{Di} and a_{Tr} with l we conclude that the dependence of the scattering parameters A_l and δ_l on l is described by two distinct functions for odd and even l values, as shown schematically in Fig. 2.23. This odd–even staggering is reminiscent of the scattering of identical nuclei (Sect. 1.3), where for á given resultant spin S either only even or only odd l values contribute. The close relationship between the two cases is formally evident from (2.14), (2.15), and (2.18), which are transformed into the corresponding equations for identical nuclei by the substitution $f_{Tr} \rightarrow f_{Di}$. Physically it is a consequence of the indistinguishable nature of the cores. Indeed it is not surprising that certain features of the core–core scattering amplitude are retained after addition of a loosely bound valence nucleon or cluster.

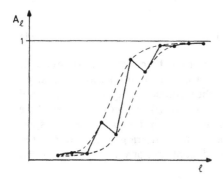

Fig. 2.23. Odd–even staggering of scattering parameters due to interference between direct scattering and elastic transfer

In the quantitative analysis of scattering data involving elastic transfer, two different methods can be distinguished. The more conventional approach consists of separately calculating the direct and transfer amplitudes by means of the optical model for the former and the distorted-wave Born approximation (DWBA; see also Chap. 5) or a related procedure for the latter. The two amplitudes are then added to obtain the total elastic scattering amplitude. A more elegant approach is that of "linear combination of nuclear orbitals" (LCNO), which is well known in molecular physics and has been applied to heavy-ion scattering by von Oertzen and collaborators [Oe 70, Bo 71, Bo 72, Oe 73, Oe 75]. Here scattering and transfer are treated on an equal footing and their combined effect is obtained—in simple cases—by solving a two-body scattering problem.

The LCNO model has been applied successfully to scattering data for many systems involving light nuclei, especially at bombarding energies near the Coulomb barrier. Most of this work has been done at Heidelberg (for surveys see Oe 71b, Oe 75). A classical example—the scattering of ^{12}C from ^{13}C—is shown in Fig. 2.24. In this case the experiment was done by Gobbi et al. [Go 68] at Zürich, and the analysis with the LCNO model by von Oertzen [Oe 70]; a fit of similar quality was obtained by Gobbi et al. by superposition of conventional direct and transfer amplitudes. The angular distribution exhibits a very pronounced oscillating structure, which can be explained quantitatively as due to interference between direct and exchange scattering. The

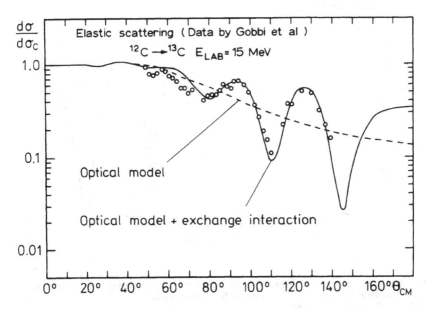

Fig. 2.24. Elastic scattering of ^{12}C on ^{13}C at $E_{CM} = 7.8$ MeV: Experimental data from Go 68, analysis with the LCNO model. (From Oe 70)

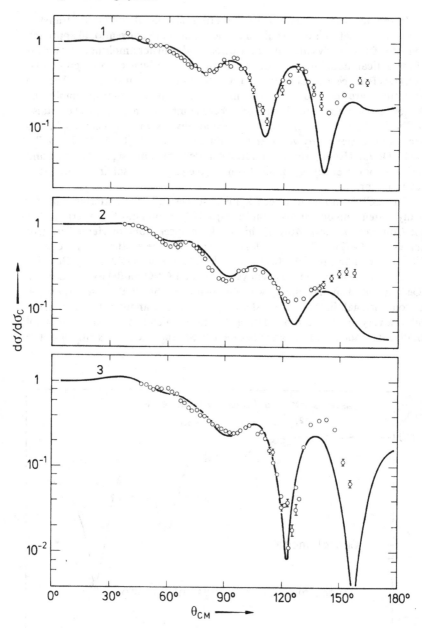

Fig. 2.25. Comparison of angular distributions for elastic scattering among different carbon isotopes together with LCNO model fits. (From Bo 72)
1: $^{12}C + {}^{13}C$, $E_{CM} = 7.8$ MeV
2: $^{12}C + {}^{14}C$, $E_{CM} = 8.1$ MeV
3: $^{13}C + {}^{14}C$, $E_{CM} = 7.8$ MeV

influence of symmetry properties of the scattering system on the sign of the interference term is strikingly demonstrated in Fig. 2.25, which compares angular distributions for elastic scattering among different carbon isotopes [Bo 72]. The sign of the interference term depends on the core spin I and on the parity of the bound-state function of the valence particle. The former influence is different for the ^{12}C $(I = 0)$ and ^{13}C $(I = \frac{1}{2})$ cores, and the latter for transfer of one and two $(p\frac{1}{2})$ neutrons. Consequently the phases of the interference patterns for ^{12}C + ^{14}C and ^{13}C + ^{14}C scattering should be equal, and opposite to that for ^{12}C + ^{13}C scattering, in agreement with observation.

Most analyses performed so far with the LCNO model refer to cases where the bound state of the valence particle has angular momentum 0 or $\frac{1}{2}$, and the angular momentum transfer is zero. For higher angular momenta, mixing of different intrinsic states due to Coriolis interaction occurs, and the usual "two-state approximation" is no longer applicable. In principle, this situation could be treated by a suitably generalized two-centre shell model calculation (see Sect. 2.2), which, however, should not be performed with oscillator wave functions because of their incorrect asymptotic behaviour. Such calculations are too involved to be feasible as a general tool for scattering analyses.

The DWBA approach to the calculation of the transfer amplitude, on the other hand, has been shown to give useful results without restriction of the angular momenta involved. Baur and Gelbke [Ba 73] compared DWBA and LCNO calculations in cases where both methods are applicable (single nucleon transfer with $j = \frac{1}{2}$, $L = 0$) and obtained essentially equivalent results. The scattering of ^{16}O on ^{17}O was studied by Gelbke et al. [Ge 73]; here a $d\frac{5}{2}$ neutron can be transferred, with possible values of the angular momentum transfer $L = 0, 2$, and 4 (neglecting recoil effects). The differential cross section now contains a coherent superposition of the amplitudes for direct scattering and $L = 0$ transfer, and incoherent contributions from higher angular momentum transfers:

$$\left(\frac{d\sigma}{d\Omega}\right)_{el} = |f_{Di}(\Theta) + f_{Tr}^{L=0}(\pi - \Theta)|^2 + \sum_{L>0} |f_{Tr}^{L}(\pi - \Theta)|^2. \qquad (2.19)$$

The latter contributions peak at angles near $\pi - \Theta_{gr}$, where Θ_{gr} is the grazing angle. They tend to wash out the structure due to interference between the direct and the $L = 0$ transfer amplitudes, especially at higher bombarding energies, where higher angular momentum transfers are kinematically favoured [Ge 73].

The question whether repeated transfers can occur during elastic nucleus-nucleus collisions has been the object of some speculation in connection with analogies to molecular phenomena (see Oe 70). Several authors have investigated the problem quantitatively by comparing results of (first-order) DWBA calculations with those of other models (like the LCNO), which were considered to account in some measure for the possibility of repeated transfer [Ba 73, Mc 73]. In the cases studied so far, no conclusive evidence for repeated transfers has been obtained.

We conclude this section by returning to the general question of back-angle enhancement in elastic scattering data. As pointed out previously, exchange scattering is not the only conceivable—although perhaps the best--understood—mechanism, which can produce such enhancement. In particular, scattering by a potential with weak or angular-momentum-dependent absorption can produce effects very similar to those due to exchange scattering with an otherwise strongly absorbing potential (see Ro 71 and Sect. 2.2). Normally it is therefore not possible to decide which of the two phenomena is effective or dominating by analysing isolated angular distributions. Rather conclusive evidence has been obtained, however, for elastic transfer from systematic comparative studies of related systems (such as those shown in Fig. 2.25) or of angular distributions at different energies. An example of the latter kind is provided by a careful study of $^{16}O + ^{18}O$ scattering due to Gelbke, Bock, and Richter [Ge 74]. The data were analysed both with an angular-momentum-dependent optical model and by adding a transfer amplitude to the direct amplitude, using energy-independent parameters except for the angular momentum cut-off l_W. The phase of the interference pattern was observed to be rather stable with energy; this feature is well reproduced in the transfer calculation, but not in the optical model analysis. Improved fits were obtained with the latter only, when the potential-well depth was allowed to vary with energy.

Studies like those discussed above point rather convincingly to the importance of exchange scattering for a number of light nucleus–nucleus systems, where the fragments differ by one or two loosely bound nucleons. Systems like $^{16}O + ^{16}O$ and $^{12}C + ^{12}C$, on the other hand, seem to be adequately described by potential scattering and weak (or angular-momentum-dependent) absorption. In these cases, transfers should be inhibited by the strongly bound nature of the fragments, which also is ultimately responsible for the reduced absorption of higher partial waves. Other systems, like $^{12}C + ^{16}O$, may be affected by both mechanisms.

References Chapter 2

(For AR 71, AR 73, AR 76, AS 63, CO 75, HE 69, NA 74 see Appendix D)

Al   60   Almqvist, E.; Bromley, D.A.; Kuehner, J.A.: Phys. Rev. Lett. *4*, 515 (1960)
Al   63   Almqvist, E. et al.: Phys. Rev. *130*, 1140 (1963)
Al   64   Almqvist, E. et al.: Phys. Rev. *136*, B 84 (1964)
Ar   72   Arima, A.; Scharff-Goldhaber, G.; McVoy, K.W.: Phys. Lett. *40B*, 7 (1972)
Ba   73   Baur, G.; Gelbke, C.K.: Nucl. Phys. *A204*, 138 (1973)
Be   63   Bennet J.R.J.: AS 63 p. 50
Be   68   Bethge, K.; Meyer-Ewert, K.; Pfeiffer K.O.: Z. Phys. *208*, 486 (1968)
Bi   69   Bisson, A.E.; Davis R.H.: Phys. Rev. Lett. *22*, 542 (1969)

Bo 71   Bohlen, H.G.; von Oertzen W.: Phys. Lett. *37B*, 451 (1971)
Bo 72   Bohlen, H.G. et al.: Phys. Lett. *41B*, 425 (1972)
Bo 73   Bondorf, J.P.: Nucl. Phys. *A202*, 30 (1973)
Bo 76   Bonche, P.: J. Phys. (Paris) *37*, C5–213 (1976)
Bo 78   Bonche, P.; Grammaticos, B.; Koonin, S.: Phys. Rev. *C17*, 1700 (1978)
Br 60   Bromley, D.A.; Kuehner, J.A.; Almqvist, E.: Phys. Rev. Lett. *4*, 365 (1960)
Br 61   Bromley, D.A.; Kuehner, J.A.; Almqvist, E.: Phys. Rev. *123*, 878 (1961)
Br 69   Bromley, D.A.: HE 69 p. 27
Br 72   Brack, M. et al.: Rev. Mod. Phys. *44*, 320 (1972)
Br 75   Bromley, D.A.: CO 75 p. 465
Br 76   Brink, D.M.: J. Phys. (Paris) *37*, C5-47 (1976)
Ca 75   Castro, J.; Federman, P.: CO 75 p. 526
Ca 76   Castro, J.; Federman, P.: Phys. Rev. *C14*, 332 (1976)
Ch 69   Chatwin, R.A. et al.: Phys. Rev. *180*, 1049 (1969)
Ch 70   Chatwin, R.A. et al.: Phys. Rev. *C1*, 795 (1970)
Co 72   Cosman, E.R. et al.: Phys. Rev. Lett. *29*, 1341 (1972)
Co 75   Cosman, E.R. et al.: Phys. Rev. Lett. *35*, 265 (1975)
Da 60   Davis, R.H.: Phys. Rev. Lett. *4*, 521 (1960)
Eb 70   Eberhard, K.A.: Phys. Lett. *33B*, 343 (1970)
Eb 75   Eberhard, K.A. et al.: Phys. Lett. *56B*, 445 (1975)
Em 73   Emling, H. et al.: Nucl. Phys. *A211*, 600 (1973)
Er 63   Ericson, T.: Ann. Phys. (N.Y.) *23*, 600 (1963)
Er 76   Erb, K.A. et al.: Phys. Rev. Lett. *37*, 670 (1976)
Fe 76   Feshbach, H.: J. Phys. (Paris) *37*, C5-177 (1976)
Fe 78   Fernandez, B. et al.: Nucl. Phys. *A306*, 259 (1978)
Fi 72   Fink, H.J.; Scheid, W.; Greiner, W.: Nucl. Phys. *A188*, 259 (1972)
Fi 73   Fink, B.; Toepffer, C.: Phys. Lett. *45B*, 411 (1973)
Fl 71   Fliessbach, T.: Z. Phys. *242*, 287 (1971)
Fl 76   Fletcher, N.R. et al.: Phys. Rev. *C13*, 1173 (1976)
Fl 77   Fleckner, J.; Mosel, U.: Nucl. Phys. *A277*, 170 (1977)
Fl 78   Flocard, H.; Koonin, S.E.; Weiss, M.S.: Phys. Rev. *C17*, 1682 (1978)
Fo 71   Fortune, H.T.; Morrison, G.C.; Siemssen, R.H.: Phys. Rev. *C3*, 2133 (1971)
Ga 73   Gamp, A. et al.: Z. Phys. *261*, 283 (1973)
Ga 77   Galster, W. et al.: Phys. Rev. *C15*, 950 (1977)
Ge 72   Gelbke, C.K. et al.: Phys. Rev. Lett. *29*, 1683 (1972)
Ge 73   Gelbke, C.K. et al.: Phys. Lett. *43B*, 284 (1973)
Ge 74   Gelbke, C.K.; Bock, R.; Richter, A.: Phys. Rev. *C9*, 852 (1974)
Gl 74   Glas, D.; Mosel, U.: Phys. Lett. *49B*, 301 (1974)
Go 68   Gobbi, A. et al.: Nucl. Phys. *A112*, 537 (1968)
Go 71   Gobbi, A.: AR 71 p. 63
Go 73   Gobbi, A. et al.: Phys. Rev. *C7*, 30 (1973)
Gr 71   Greiner, W.; Scheid, W.: J. Phys. (Paris) *32*, C6–91 (1971)
Gr 72   Greenwood, L.R. et al.: Phys. Rev. *C6*, 2112 (1972)
Gr 73   Gruber, G. et al.: Z. Phys. *265*, 411 (1973)
Gu 67   Günther, G.; Bethge, K.: Nucl. Phys. *A101*, 288 (1967)
Ha 59   Halbert, M.L.; Zucker, A.: Phys. Rev. *115*, 1635 (1959)
Ha 60   Halbert, M.L.; Hanting, C.E.; Zucker, A.: Phys. Rev. *117*, 1545 (1960)
Ha 67   Halbert, M.L. et al.: Phys. Rev. *162*, 899, 919 (1967)
Ha 74   Hanson, D.L. et al.: Phys. Rev. *C9*, 1760 (1974)
He 71   Helling, G.; Scheid, W.; Greiner, W.: Phys. Lett. *36B*, 64 (1971)
He 73   Helb, H.D. et al.: Nucl. Phys. *A206*, 385 (1973)
Hi 64   Hiebert, J.C.; Garvey, G.T.: Phys. Rev. *135*, B 346 (1964)
Ho 69   Holzer, P.; Mosel, U.; Greiner, W.: Nucl. Phys. *A138*, 241 (1969)
Im 68   Imanishi, B.: Phys. Lett. *27B*, 267 (1968)
Im 69   Imanishi, B.: Nucl. Phys. *A125*, 33 (1969)
Ja 69   Jacobson, L.A.: Phys. Rev. *188*, 1509 (1969)
Ke 77   Kettner, K.U. et al.: Phys. Rev. Lett. *38*, 337 (1977)
Ko 75   Kondo, Y.; Matsuse, T.; Abe, Y.: CO 75 p. 532
Kr 69   Krubasik, E.H. et al.: Z. Phys. *219*, 185 (1969)
Ku 63a Kuehner, J.A.; Almqvsit, E.; Bromley, D.A.: Phys. Rev. *131*, 1254 (1963)
Ku 63b Kuehner, J.A.; Almqvist, E.: AS 63 p. 11

Ku 64 Kuehner, J.A.; Almqvist, E.: Phys. Rev. *134*, B 12 (1964)
Le 72 Leachman, R.B.; Fessenden, P.; Gibbs, W.R.: Phys. Rev. *C6*, 1240 (1972)
Lo 72 Low, K.S.; Tamura, T.: Phys. Lett. *40B*, 32 (1972)
Ma 69 Maher, J.V. et al.: Phys. Rev. *188*, 1665 (1969)
Ma 72 Malmin, R.E. et al.: Phys. Rev. Lett. *28*, 1590 (1972)
Ma 73 Mazarakis, M.G.: Stephens, W.E.: Phys. Rev. *C7*, 1280 (1973)
Mc 71 McVoy, K.W.: Phys. Rev. *C3*, 1104 (1971)
Mc 73 McMahan, C.A.; Tobocman, W.: Nucl. Phys. *A202*, 561 (1973)
Mi 72 Michaud, G.J.; Vogt, E.W.: Phys. Rev. *C5*, 350 (1972)
Mi 73 Michaud, G.: Phys. Rev. *C8*, 525 (1973)
Mo 70 Mosel, U.; Thomas, T.D.; Riesenfeldt, P.: Phys. Lett. *33B*, 565 (1970)
My 67 Myers, W.D.; Swiatecki, W.J.: Ark. Fys. *36*, 343 (1967)
Na 72 Nagorcka, B.N.; Newton, J.O.: Phys. Lett. *41B*, 34 (1972)
Ni 72 Nix, J.R.: Ann. Rev. Nucl. Sci. *22*, 65 (1972)
Oe 68 von Oertzen, W. et al.: Phys. Lett. *26B*, 291 (1968)
Oe 70 von Oertzen, W.: Nucl. Phys. *A148*, 529 (1970)
Oe 71a von Oertzen, W. et al.: Phys. Lett. *34B*, 51 (1971)
Oe 71b von Oertzen, W.: AR 71 p. 121
Oe 73 von Oertzen, W.; Nörenberg, W.: Nucl. Phys. *A207*, 113 (1973)
Oe 75 von Oertzen, W.; Bohlen, H.G.: Phys. Rep. *19*, 1 (1975)
Ok 68 Okuma, Y.: J. Phys. Soc. Jpn. *24*, 677 (1968)
Pa 69 Patterson, J.R.; Winkler, H.; Zaidins, C.S.: Astrophys. J. *157*, 367 (1969)
Pa 72 Park, J.Y.; Scheid, W.; Greiner, W.: Phys. Rev. *C6*, 1565 (1972)
Pa 74 Park, J.Y.; Scheid, W.; Greiner, W.: Phys. Rev. *C10*, 967 (1974)
Pi 66 Pinsonneault, L.L.; Blair, J.M.: Phys. Rev. *141*, 961 (1966)
Po 72 Poling, J.E.; Norbeck, E.; Carlson, R.R.: Phys. Rev. *C5*, 1819 (1972)
Pr 70 Pruess, K.; Greiner, W.: Phys. Lett. *33B*, 197 (1970)
Re 56 Reynolds, H.L.; Zucker, A.: Phys. Rev. *102*, 1378 (1956)
Re 69 Reilly, W. et al.: HE 69 p. 93
Ro 71 Robson, D.: AR 71 p. 239
Sa 74 Satchler, G.R.: NA 74 Vol.2 p. 171
Sa 76 Satchler, G.R.: AR 76 Vol. 1 p. 33
Sc 55 Schiff, L.I.: *Quantum Mechanics*, 2nd ed., p. 92–95. New York, Toronto, London: McGraw Hill 1955
Sc 68 Scheid, W.; Ligensa, R.; Greiner, W.: Phys. Rev. Lett. *21*, 1479 (1968)
Sc 69 Scheid, W.; Greiner, W.: Z. Phys. *226*, 364 (1969)
Sc 70 Scheid, W.; Greiner, W.; Lemmer, R.: Phys. Rev. Lett. *25*, 176 (1970)
Sc 72 Scheid, W.; Fink, H.S.; Müller, H.: In *Proceedings of the Europhysics Study Conference on Intermediate Processes in Nuclear Reactions*, ed. by N. Cindro, P. Culišić, Th. Mayer-Kuckuk. Berlin, Heidelberg, New York: Springer, 1972
Sh 69 Shaw, R.W. et al.: Phys. Rev. *184*, 1040 (1969)
Sh 70 Shaw, R.W.; Vandenbosch, R.; Mehta, M.K.: Phys. Rev. Lett. *25*, 457 (1970)
Sh 74 Shapira, D.; Stokstad, R.G.; Bromley, D.A.: Phys. Rev. *C10*, 1063 (1974)
Si 67 Siemssen, R.H. et al.: Phys. Rev. Lett. *19*, 369 (1967)
Si 70 Siemssen, R.H. et al.: Phys. Rev. Lett.; *25*, 536 (1970)
Si 71 Siemssen, R.H.: AR 71 p. 145
Si 72 Siemssen, R.H. et al.: Phys. Rev. *C5*, 1839 (1972)
Si 74 Siemssen, R.H.: In *Nuclear Spectroscopy and Reactions*, Part B, ed. by J. Cerny p. 234. New York, London: Academic Press 1974
Sp 72 Spinka, H.; Winkler, H.: Astrophys. J. *174*, 455 (1972)
Sp 74 Spinka, H.; Winkler, H.: Nucl. Phys. *A233*, 456 (1974)
Sp 76a Sperr, P. et al.: Phys. Rev. Lett. *36*, 405 (1976)
Sp 76b Sperr, P. et al.: Phys. Rev. Lett. *37*, 321 (1976)
St 67 Strutinsky, V.M.: Nucl. Phys. *A95*, 420 (1967)
St 68 Strutinsky, V.M.: Nucl. Phys. *A122*, 1 (1968)
St 72 Stokstad, R. et al.: Phys. Rev. Lett. *28*, 1523 (1972)
St 73 Stokstad, R.G.: AR 73 Vol. I p. 325
Te 78 Terlecki, G. et al.: Phys. Rev. *C18*, 265 (1978)
Ts 78 Tserruya, I. et al.: Phys. Rev. *C18*, 1688 (1978)
Va 71 Vandenbosch, R.: AR 71 p. 103
Va 74 Vandenbosch, R.; Webb, M.P.; Zisman, M.S.: Phys. Rev. Lett. *33*, 842 (1974)

Vo 60   Vogt, E.W.; McManus, H.: Phys. Rev. Lett. *4*, 518 (1960)
Vo 64   Vogt, E.W. et al.: Phys. Rev. *136B*, B 99 (1964)
Vo 69   Voos, U.C.; von Oertzen W.; Bock, R.: Nucl. Phys. *A135*, 207 (1969)
Vo 71   Voit, H. et al.: AR 71 p. 303
Vo 72   Voit, H. et al.: Nucl. Phys. *A179*, 23 (1972)
Vo 73a   Voit, H.; Helb, H.-D.: Nucl. Phys. *A204*, 196 (1973)
Vo 73b   Voit, H.; Ischenko, G.; Siller, F.: Phys. Rev. Lett. *30*, 564 (1973)
Vo 74   Voit, H. et al.: Phys. Rev. *C10*, 1331 (1974)
Vo 77   Voit, H. et al.: Phys. Lett. *67B*, 399 (1977)
Wa 62   Wang, K.H.; Baker, S.D.; McIntyre, J.A.: Phys. Rev. *127*, 187 (1962)
Wi 61   Wildermuth, K.; Carovillano, R.L.: Nucl. Phys. *28*, 636 (1961)
Zi 75   Zint, P.G.; Mosel, U.: Phys. Lett. *56B*, 424 (1975)

3. Quasi-Elastic Scattering from Heavier Target Nuclei

3.1 Coulomb Excitation

Throughout this chapter we discuss nuclear collisions which differ from those considered in Chap. 2 by the presence of at least one heavier partner ($A > 20$). As pointed out previously, and shown in detail in the following, the scattering cross sections for such systems depend mainly on gross properties of the fragments, like size, shape, and charge, but give usually little evidence of their microscopic structure. This means that the elastic scattering is well described by the diffraction models discussed in Sect. 1.4 and the inelastic spectra are dominated by collective excitations of rotational or vibrational character. The Coulomb interaction is much more important for these heavier systems, and can compete on an equal footing with the nuclear interaction in typical grazing situations. It can thus no longer be considered a small perturbation, but enters in an essential way in the description of scattering processes even at energies well above the Coulomb barrier. Last, but not least, semiclassical or classical theoretical approaches are more appropriate and successful due to the larger dimensions and smaller wavelengths involved.

The material of this chapter has been subdivided mainly on the basis of the internuclear separations, at which the relevant interactions take place. We start in this section with (pure) Coulomb excitation, a process which occurs at large distances outside the range of nuclear forces. In the following Sect. 3.2 we move close to and slightly inside the borderline of nuclear interactions, in order to discuss macroscopic effects of Coulomb excitation—like distortion and break-up of the fragments—and their possible influence on the Coulomb barrier. Subsequent sections will then be devoted to situations where both Coulomb and nuclear interactions are important.

A number of excellent review articles [Al 56, Al 60, St 63, Mc 74] and books [Bi 65, Al 66, Al 75] have been published on Coulomb excitation, and we shall therefore not attempt here to cover the field in a systematic way.

Spectroscopic applications—expecially the very detailed information available on collective nuclear excitations from Coulomb excitation studies—will be largely disregarded, although this has undoubtedly been the most important aspect in the past. With increasing mass and charge of the projectile, multiple excitations become more probable and more nuclear states become involved; as a consequence, an unambiguous and model-independent extraction of electromagnetic matrix elements between definite nuclear states is no longer possible. In this situation it seemed best to focus attention on general properties of the excitation mechanism, with particular emphasis on its possibilities and limitations as a tool for the transfer of energy and angular momentum from orbital to intrinsic motion.

The theory of Coulomb excitation is usually based on the semiclassical approximation, in which the source of the electromagnetic field is treated classically as a point charge moving along a Rutherford trajectory, whereas the nuclear excitation is calculated by quantum-theoretical methods. The semiclassical treatment is applicable whenever the Sommerfeld parameter defined in (1.13),

$$n = \frac{a}{\lambda} = \frac{Z_1 Z_2 e^2}{\hbar v_\infty} = 0.15 \frac{Z_1 Z_2}{\varepsilon^{1/2}} \qquad (\varepsilon \text{ in MeV/u}) \tag{3.1}$$

is much larger than unity, as is usually the case in heavy-ion collisions (see Table 1.1 and Fig. 3.3). In (3.1) the quantity a is one-half the distance of closest approach in a head-on collision, or [see (1.12)]

$$a = \frac{Z_1 Z_2 e^2}{A_{12} m v_\infty^2} = 0.72 \frac{Z_1 Z_2}{A_{12} \varepsilon} \qquad (a \text{ in fm}, \varepsilon \text{ in MeV/u}). \tag{3.2}$$

Since Coulomb excitation converts relative energy into intrinsic energy, the asymptotic velocity v_∞ will be smaller after the collision than before, and the Rutherford trajectory to be used in the semiclassical method is not uniquely defined. By comparison with exact quantum calculations it has been shown, however, that the results of the semiclassical treatment can be optimized by using suitably defined average values of v_∞, n, and a. For heavy projectiles, the relative change in velocity is usually small ($\leq 1 \%$), and the geometric mean of the velocities is a sufficiently good approximation to more elaborate symmetrization prescriptions [see Bi 65]. Denoting the asymptotic velocities before and after the collison by v_i and v_f, respectively, we introduce the symmetrized parameters [Al 56]

$$\varepsilon_{if} = \frac{m}{2} v_i v_f , \tag{3.3a}$$

$$a_{if} = \frac{Z_1 Z_2 e^2}{A_{12} m v_i v_f} = \frac{Z_1 Z_2 e^2}{2 A_{12} \varepsilon_{if}} , \tag{3.3b}$$

$$n_{if} = \frac{Z_1 Z_2 e^2}{\hbar (v_i v_f)^{1/2}} = \frac{Z_1 Z_2 e^2}{(2 \hbar^2 \varepsilon_{if}/m)^{1/2}} , \tag{3.3c}$$

which define an effective Rutherford trajectory. The probability of exciting a state at excitation energy ΔE_{if} depends further on the "adiabaticity parameter"

$$\xi_{if} = n_f - n_i = \frac{Z_1 Z_2 e^2}{\hbar} \frac{v_i - v_f}{v_i v_f} \tag{3.4a}$$

$$\approx \frac{n_{if} \Delta E_{if}}{2 A_{12} \varepsilon_{if}} = 0.079 \frac{Z_1 Z_2 \Delta E_{if}}{A_{12} \varepsilon_{if}^{3/2}} \qquad (\Delta E_{if} \text{ in MeV}, \ \varepsilon_{if} \text{ in MeV/u}), \tag{3.4b}$$

where n_f and n_i are related to v_f and v_i, respectively. The physical significance of ξ_{if} may be more clearly exhibited by writing

$$\xi_{if} = \frac{2 a_{if}/(v_i + v_f)}{\hbar/\Delta E_{if}} = \frac{\text{collision time}}{\text{nuclear period}} \tag{3.4c}$$

The excitation probability is largest for $\xi_{if} \to 0$ and decreases exponentially with increasing ξ_{if}—approximately proportional to $\exp(-2\pi\xi_{if})$—for $\xi_{if} > 1$. As we shall show more quantitatively below, this exponential decrease limits severely the amount of energy that can be transferred in Coulomb excitation.

We now proceed to quote some formulae for the excitation cross section, which have been derived in first-order perturbation theory. It should be emphasized that this approach yields quantitatively reliable results only when the first-order excitation probabilities are small compared to unity, e.g. under conditions which are frequently not satisfied in heavy-ion collisions. Nevertheless, due to their simplicity, the first-order results are useful for a discussion of trends associated with various parameters, and of the conditions under which single or multiple excitation should be dominant. For an electric transition of multipole order $E\lambda$ with a reduced transition probability $B(E\lambda)$ (magnetic transitions are unimportant in practice and will not be discussed in this book), the first-order differential and integrated excitation cross sections are given by [Al 56, Bi 65]

$$\frac{d\sigma_\lambda}{d\Omega} = \frac{n_i^2}{a_{if}^{(2\lambda-2)}} \frac{B(E\lambda)}{(Z_2 e)^2} \frac{df_\lambda(\xi_{if}, \Theta)}{d\Omega}, \tag{3.5a}$$

$$\sigma_\lambda = \frac{n_i^2}{a_{if}^{(2\lambda-2)}} \frac{B(E\lambda)}{(Z_2 e)^2} f_\lambda(\xi_{if}), \tag{3.5b}$$

where the functions $df_\lambda/d\Omega$ and f_λ are tabulated in [Al 56]. For the most important case of an $E2$ transition, (3.5 b) may be rewritten as [see (3.1) and (3.2)]

$$\sigma_2 = 4.8 \frac{\varepsilon_f A_{12}^2}{Z_2^2} B(E2) f_2(\xi_{if}) \tag{3.5c}$$

(ε_{if} in MeV/u, $B(E2)$ in $e^2 \times 10^{-48}$ cm^4, σ_2 in b).

As an example, Fig. 3.1 shows excitation functions calculated with (3.5c) for the first excited 2^+ state in ^{114}Cd using ^{4}He, ^{16}O, and ^{40}Ar projectiles. Two features of this figure are particularly noteworthy: the steep increase of the cross section with increasing bombarding energy, and its dramatic change

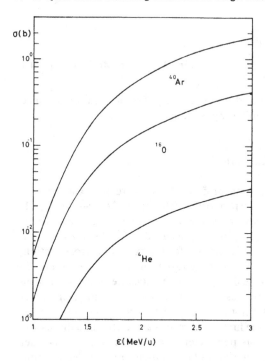

Fig. 3.1. First-order E2 Coulomb excitation cross section of the lowest 2^+ state in ^{114}Cd ($E_{ex} = 0.558$ MeV, $B(E2) = 0.51$ $e^2 \times 10^{-48}$cm^4) for various projectiles versus bombarding energy

with projectile mass. The latter influence can be separated quantitatively from effects of the orbital motion by comparing cross sections for different projectiles at a constant value of the parameter ξ_{if}; from (3.1), (3.2), and (3.5) we obtain for a given transition of arbitrary multipole order λ

$$\sigma_\lambda \sim Z_1^2 \left(\frac{Z_1 Z_2}{A_{12}}\right)^{-2\lambda/3} \tag{3.6a}$$

$$\text{for } \varepsilon_{if} \sim \left(\frac{Z_1 Z_2}{A_{12}}\right)^{2/3} \quad (\xi_{if} = \text{const}) \tag{3.6b}$$

As the term in brackets does not depend strongly on the projectile mass, we see that the cross section is approximately proportional to Z_1^2 and that a constant adiabaticity parameter ξ_{if} implies roughly constant incident energy per nucleon for a given target nucleus. It should be noted that (3.6a) is based on first-order perturbation theory; for heavy projectiles, deviations may occur, which are then indicative of the presence of significant higher-order effects.

   In order to assess the range of applicability of the first-order treatment, it is convenient to reformulate the differential excitation cross section of (3.5a) as a product of a Coulomb scattering cross section and a first-order excitation probability $P^{(1)}$

$$\frac{d\sigma_\lambda}{d\Omega} = P^{(1)} \frac{d\sigma_C}{d\Omega} = |\chi_\lambda|^2 F_\lambda(\Theta, \xi) \frac{d\sigma_C}{d\Omega} \tag{3.7}$$

where

$$\frac{d\sigma_C}{d\Omega} = \frac{v_f}{v_i} \frac{a_{if}^2}{4} \csc^4 \frac{\Theta}{2} \tag{3.8}$$

is the symmetrized Rutherford cross section, and the function $F_\lambda(\Theta, \xi)$ attains its maximum value of unity for $\Theta = \pi$, $\xi \to 0$. The dimensionless parameter χ_λ is given by [Al 60]

$$\chi_\lambda = (16\pi)^{1/2} \frac{(\lambda - 1)!}{(2\lambda + 1)!!} \frac{n_{if}}{a_{if}^\lambda} \frac{[B(E\lambda)]^{1/2}}{Z_2 e}, \tag{3.9}$$

and is a measure of the strength of the transition, or excitation probability. For values of $|\chi_\lambda|^2$ small compared to unity, first-order perturbation theory is adequate; if, however, $|\chi_\lambda|^2$ is comparable to unity, then higher-order effects (multiple excitation) must be taken into account. For $E2$ transitions, we obtain the following numerical relationship:

$$\chi_2 = 14.3 \frac{A_{12}^2 \varepsilon_{if}^{3/2}}{Z_1 Z_2^2} [B(E2)]^{1/2} \tag{3.10}$$

(ε_{if} in MeV/u, $B(E2)$ in $e^2 \times 10^{-48}\text{cm}^4$).

The parameters χ_λ and ξ_{if} depend on energy in such a way that the cross sections increase strongly with increasing bombarding energy. In experimental work—especially if high precision is required—one is usually looking for the highest possible yield, and thus for the highest possible bombarding energy at which the excitation under study can still be assumed to take place by purely electromagnetic interaction. It is therefore important to obtain an estimate of the "maximum safe bombarding energy" ε_s, or—in the spirit of the semiclassical approach—of the minimum internuclear distance (R_s), at which nuclear excitation is still insignificant. These quantities will depend somewhat on specific experimental conditions and can thus not be precisely defined; nevertheless useful estimates can be made semiempirically. Interference effects between Coulomb and nuclear excitation in α and heavy--ion bombardments have been studied experimentally by a number of authors [Pr 69, Cl 69, Ch 73, Sa 74a] and will be discussed in Sect. 3.4. The results lead to the conclusion that nuclear effects of more than about 1 % can be excluded, as long as the internuclear distance at the cassical turning point is larger than

$$R_s = R_1 + R_2 + 7.0 \text{ fm}. \tag{3.11}$$

Here and in the following we define the nuclear radii R_1, R_2 of the fragments by the point where the nuclear charge density has dropped to one-half of its central value. Equation (3.11) is based on the following approximate rela-

tionship for R_i, which has been deduced from electron scattering results
[El 65]

$$R_i = 1.12\, A_i^{1/3} - 0.94\, A_i^{-1/3} \quad [\text{fm}]. \tag{3.12}$$

We prefer this definition of the nuclear radius to others found in the literature, as it is relatively unambiguous and provides a convenient basis for a
unified discussion of scattering and fusion phenomena [see Sect. 3.3 and Chap.
7]. It should be realized, however, that it corresponds to appreciably smaller
radii than most other definitions. To illustrate this we note that the distance
R_S of (3.11) is only about 4 fm larger than the nuclear strong absorption
distance deduced from elastic scattering and reaction cross-section measurements [see Sect. 3.3]. The corresponding "safe bombarding energy" for
scattering through a centre of mass angle Θ is given by [see (1.11)]

$$\varepsilon_S(\Theta) = \frac{Z_1 Z_2 e^2}{A_{12} R_S} \frac{2}{\left(1 + \csc \dfrac{\Theta}{2}\right)} = \varepsilon_S(\pi) \frac{2}{\left(1 + \csc \dfrac{\Theta}{2}\right)} \tag{3.13a}$$

$$\varepsilon_S(\pi) = \frac{Z_1 Z_2}{A_{12}} \frac{1.44}{(R_1 + R_2 + 7.0)} \quad (\varepsilon_S \text{ in MeV/u}, R_1, R_2 \text{ in fm}) \tag{3.13b}$$

In Fig. 3.2 we show $\varepsilon_S(\pi)$, the safe bombarding energy for head-on collisions

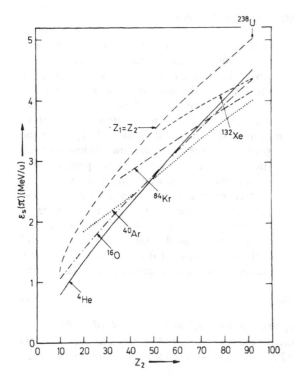

Fig. 3.2. Safe bombard-
ing energies $\varepsilon_S(\pi)$ for various projectiles according
to (3.13b) versus target
atomic number Z_2

as calculated from (3.13b), for various target–projectile combinations. For the same combinations, the Sommerfeld parameter n and the E2-interaction strength χ_2 [assuming an average set of values for B(E2) and ΔE_{if}] have also been calculated at the bombarding energy $\varepsilon_S(\pi)$; the results are shown in Fig. 3.3 and 3.4, respectively, and support our earlier statements that heavy-ion

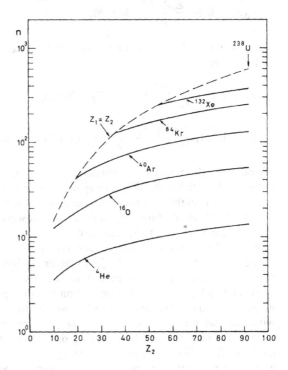

Fig. 3.3. Sommerfeld parameter n for various projectiles and the bombarding energies $\varepsilon_S(\pi)$ shown in Fig. 3.1 versus target atomic number Z_2

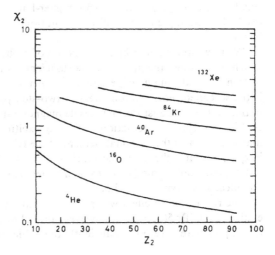

Fig. 3.4. The E2-interaction strength χ_2 for various projectiles [$\varepsilon = \varepsilon_S(\pi)$, B(E2) = 1.0×10^{-48} e^2cm^4] versus target atomic number Z_2

Coulomb excitation should be amenable to semiclassical methods of analysis ($n \gg 1$), and that multiple E2-excitation may be important ($\chi_2 \gtrsim 1$).

A convenient estimate of the parameters ξ_{if} and χ_2 may be obtained making use of the following approximate relationship

$$\varepsilon_s(\pi) \approx 0.24 \left(\frac{Z_1 Z_2}{A_{12}}\right)^{2/3} \quad \text{MeV/u,} \tag{3.14}$$

which generally holds within about $\pm 10\%$. Inserting (3.14) in (3.4b) and (3.10) we have

$$\xi_{if} \approx 0.67 \left(\frac{\varepsilon_s(\pi)}{\varepsilon_{if}}\right)^{3/2} \Delta E_{if} \quad (\Delta E_{if} \text{ in MeV}), \tag{3.15}$$

$$\chi_2 \approx 1.7 \, A_{12} \left(\frac{\varepsilon_{if}}{\varepsilon_s(\pi)}\right)^{3/2} \frac{[B(E2)]^{1/2}}{Z_2} \quad (B(E2) \text{ in } e^2 \times 10^{-48} \text{cm}^4). \tag{3.16}$$

We note that the numerical factors in (3.14), (3.15), and (3.16) result from our choice of the safe bombarding energy $\varepsilon_s(\pi)$, as given by (3.13b), and may have to be adjusted slightly for other choices of that quantity. Such an adjustment would not affect the following conclusions, however.

According to (3.15) the quantity ξ_{if} always exceeds unity for energy transfers above about 1.5 MeV; such transfers are therefore strongly suppressed by the factor $\exp(-2\pi\,\xi_{if})$ in the excitation probability. A practical upper limit for ξ_{if} might be expected near $\xi_{if} = 2$, corresponding to $\Delta E_{if} \approx 3$ MeV, at which point the E2 cross section has dropped by about 4 orders of magnitude and the E3 cross sections by about 3 orders of magnitude compared with their value at $\xi_{if} = 0$. Experimentally, states up to about 2.5 MeV excitation have been reached by direct E2 or E3 excitation of almost spherical nuclei with various projectiles. In strongly deformed nuclei, on the other hand, collective transitions with small energy transfers are favoured by small ξ_{if} and large χ_2 values, and higher-lying states may be more easily reached by multiple than by direct excitation. We shall return to this point below.

We now proceed to a brief discussion of multiple excitation phenomena, which occur when the interaction is sufficiently strong to produce a cascade of transitions during a single encounter. An important process of this type is the so-called reorientation effect, a two-step process, where the final (or initial) state of a transition changes its magnetic substate under the influence of the quadrupole interaction with the exciting fragment. The relevant matrix element is the diagonal element of the E2 operator, taken for the state in question, and therefore the reorientation effect can be used to measure static quadrupole moments of excited states. A schematic representation of the effect is shown in Fig. 3.5.

The reorientation effect was predicted and studied theoretically by Breit et al. in 1956 [Br 56]; it took, however, until about 1965 before experimental techniques were developed sufficiently to enable significant measurements to

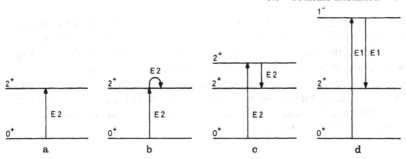

Fig. 3.5. Interfering amplitudes in the Coulomb excitation of the first excited 2⁺ state in an even–even nucleus

be made. De Boer and Eichler [Bo 68] and Smilanski [Sm 69] have reviewed in detail the theory and the experimental problems, and have given surveys of early experimental results. More recently, a large number of reorientation measurements have been performed and static quadrupole moments have been deduced for first excited 2⁺ states of nuclei in practically all regions of the periodic table [Ch 72, Hä 74]. A detailed account of this work would be beyond the scope of this book; in the following, we merely summarize the basic arguments involved.

The theroetical treatment of the reorientation effect requires, of course, at least second-order perturbation theory. In most cases of practical interest, however, contributions of higher than second order are significant. It is then necessary to compute the time-dependent occupation amplitudes of the various states involved by numerical solution of a set of coupled differential equations. Winther and De Boer have developed a computer program for multiple E2 excitation [Wi 65], by which this task can be performed and which has been used in the analysis of the majority of available experimental data.

The second-order perturbation treatment, although usually not adequate for quantitative analyses, is nevertheless useful for obtaining an estimate of the magnitude of the reorientation effect. In this approximation the only significant terms are the first-order excitation probability $P^{(1)}$ and the interference term $P^{(1,2)}$ between first and second order; for a transition to the first excited 2⁺ state in an even–even nucleus, and neglecting transitions through other intermediate levels [see Fig. 3.5], the differential cross section may be written [Bo 68, Sm 69]

$$\frac{d\sigma_{0-2}}{d\Omega} = (P_{if}^{(1)} + P_{if}^{(1,2)} + \ \ldots \ ) \frac{d\sigma_C}{d\Omega}$$

$$= P_{if}^{(1)} \left[1 + 1.32 \frac{A_{12} \Delta E_{if} Q(2^+)}{Z_2} K(\Theta, \xi_{if}) + \ \ldots \right] \frac{d\sigma_C}{d\Omega} \qquad (3.17)$$

(ΔE_{if} in MeV, $Q(2^+)$ in 10^{-24}cm^2).

Here $Q(2^+)$ is the static quadrupole moment of the excited 2⁺ state, and the function $K(\Theta, \xi_{if})$ is shown in Fig. 3.6. Several important qualitative conclusions are immediately obvious from (3.17) and Fig. 3.6: firstly the relative magnitude of the effect increases strongly with increasing mass of the exciting nucleus, and with increasing scattering angle, but depends only weakly on bombarding energy (ξ_{if}); secondly, the effect depends linearly on $Q(2^+)$ and can thus be employed to determine not only the magnitude, but also the sign of the quadrupole moment. This means that prolate and oblate charge distributions can be distinguished.

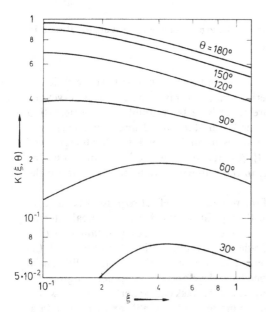

Fig. 3.6.
The function K (ξ,Θ) in (3.17), which gives an estimate of the relative strength of the reorientation effect. (From Bo 68)

An experimental problem arises from the fact that the reorientation term typically amounts to about 10–20% of the first-order term in the differential cross section. In order to obtain 10% accuracy in the quadrupole moment, the cross section must therefore be measured with an accuracy of about 1%. An absolute cross-section measurement with this precision is practically impossible; instead one observes the change in cross section with either scattering angle or projectile mass, keeping all other relevant parameters constant. Such a relative measurement can give sufficient accuracy, provided it covers situations in which the reorientation contribution is both weak (small scattering angle or light projectile) and strong (large scattering angle or heavy projectile). Even so, great care must be exercised to avoid or correct for a number of small effects, which would otherwise cause large errors in the deduced quadrupole moments. Such effects may be associated with background contributions in particle or gamma-ray spectra, with charge-state distributions of scattered projectiles detected by magnetic analysis, or with disturbed angular correlations due to hyperfine interactions, if gamma rays are observed. A most important requirement is further to avoid nuclear excitation by working at a "safe" bombarding energy (see discussion above).

While these experimental problems can be solved in principle, there exist uncertainties of a more basic nature in the analysis of reorientation data. These arise from the presence of interference terms due to indirect transitions through other (generally unobserved) levels. Just as the reorientation effect, these interference terms depend on the magnitudes and signs of the relevant electromagnetic matrix elements. In favourable cases the magnitudes can be obtained by measuring transition probabilities; however, the signs are not independently measurable. In even–even nuclei the most important effect of this kind usually arises from the virtual transition through the second excited 2^+ state (see Fig. 3.5c). This has led to large (but discrete) ambiguities in the extracted quadrupole moments, resulting from the different possible sign choices for the interfering amplitude. Within limits, the relative magnitude of the interference term may be reduced by a careful choice of bombarding energy and projectile mass, or its sign may be deduced by requiring consistency of results obtained under different bombarding conditions [La 72, Hä 74].

As a consequence of the technical difficulties mentioned above, early reorientation measurements were plagued by large uncertainties and disturbing discrepancies. A well known example is the nucleus ^{114}Cd for which the largest number of measurements have been performed. The published results cover the full range of possibilities, from $Q(2^+) = 0$ as expected for a harmonic vibrator, up to values appropriate for a strongly deformed rotor. With improved techniques and a better understanding of various effects, however, the quadrupole moment seems to have settled down at an intermediate value [K170, Be72b, La72].

We conclude this discussion of the reorientation effect by noting that great progress has been achieved in recent years, both in the development of precision experimental techniques and in the theoretical analysis of the results. Nevertheless, the interference due to indirect transitions via higher excited states remains a major source of uncertainty in this type of study. The large excitation probabilities typically encountered in heavy-ion collisions not only make new phenomena, like the reorientation effect, accessible to observation; they also enhance competing processes, which may make a model-independent extraction of individual matrix elements difficult, if not impossible.

In the following we consider the population of states above the first excited state by multiple Coulomb excitation. Since the excitation probability depends on the electromagnetic matrix elements connecting all nuclear states involved, a general discussion of this problem is not possible without reference to a nuclear model. In the simplest cases, the level sequence of a pure rotational or vibrational band is populated in successive transitions; the pertinent theory has been developed in great detail by Alder and Winther [Al 60].

From an experimental point of view, multiple excitation within a vibrational band is a somewhat academic problem, since pure vibrational bands do not seem to exist in real nuclei. In some mass regions, states have been

observed in even–even nuclei at approximately twice the excitation energy
of the first 2$^+$ state, with spin and parity 0$^+$, 2$^+$, or 4$^+$ and other properties
suggestive of an interpretation as members of two-phonon triplets. The double
E2 excitation of such states has been studied extensively [see, for example,
St 63, St 65 a]. There are strong indications, however, that these states are not
very well described as pure harmonic surface vibrations, and no higher states
with three or more vibrational quanta have been identified. It appears,
therefore, that the higher vibrational states are dissolved among many levels
of complicated intrinsic structure, due to strong mixing of vibrational and
particle degrees of freedom. The question of multiple vibrational excitation
has, nevertheless, received considerable theoretical attention, as it might
provide a mechanism for the Coulomb-induced fission of heavy nuclei under
heavy-ion bombardment. This point will be discussed in Sect. 3.2.

Rotational bands, on the other hand, are very clearly defined in strongly
deformed nuclei, especially in the rare earth ($A = 150$–180) and actinide
($A = 230$–250) region. The relatively small energy spacings (small ξ_{1f}) and
large $B(E2)$ values (large χ_2) associated with these bands are particularly
favourable for multiple Coulomb excitation. Figure 3.7 shows theoretical
excitation probabilities for the various spin states of a rotational band in
backward scattering ($\Theta = \pi$) as functions of the parameter χ_2, taken for the
transition 0$^+$ → 2$^+$ [Al 60]. The calculations are performed in the "sudden
approximation" which corresponds to zero energy loss ($\xi_{1f} = 0$) and should
be fairly realistic for rotational excitation by heavy projectiles with bombarding energies of the order of 100 MeV. An interesting feature of Fig. 3.7
is that for a given value of χ_2 the excitation probability peaks strongly at a
certain angular momentum, rather than decreasing monotonically with
increasing angular momentum. This behaviour, which is reminiscent of a classical description of the process, has been confirmed experimentally.

Completely classical calculations of multiple Coulomb excitation have
also been performed [Al60, Ho70a, Je70a, Ho72, Ro75]. Such calculations are
useful as they provide a qualitative understanding of various effects. As an
example we quote the classical result for the maximum angular momentum
(L_{max}) which can be transferred in a head-on collision to a deformed nucleus
with quadrupole moment Q_0 [Al 60, Ro 75]:

$$L_{max} = \frac{Z_1 e^2 Q_0}{2\hbar v_\infty a^2} = \frac{Z_1 e}{2\hbar v_\infty a^2} \left\{ \frac{16\pi}{5} B(E2)_{0\to2} \right\}^{1/2} = \frac{3\sqrt{5}}{2} \chi_2. \qquad (3.18)$$

Equation (3.18) implies that the angle between the internuclear axis and the
intrinsic symmetry axis of nucleus 2 is 45°, and does not change significantly
during the encounter ($\xi \to 0$).

Semiclassical calculations of the multiple Coulomb excitation of ^{238}U
by either ^{136}Xe or ^{238}U projectiles were published by Oberacker and Soff
[Ob77b]. These authors studied in detail the coupling between rotational and
vibrational nuclear degrees of freedom and found large effects of the rotation-
vibration interaction on the excitation probability for very high spin states.

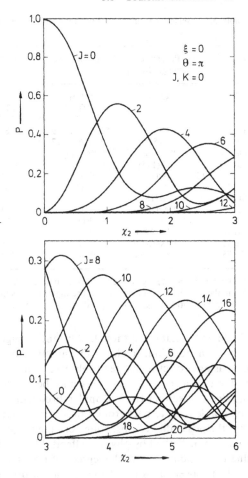

Fig. 3.7. Excitation probabilities for states of spin J by multiple E2 Coulomb excitation of a rotational band in an even–even nucleus at $\theta = \pi$, versus the interaction strength χ_2 for the $0^+ - 2^+$ transition. (From Al 60, Bi 65)

Strong multiple excitation of ground-state rotational bands in deformed nuclei was observed as early as 1959. At that time Stephens et al. reported excitation of ^{232}Th up to the 10^+ level, and of ^{238}U up to the 12^+ level, using 190 MeV argon ions from the Berkeley Hilac [St59]. In the meantime, heavier projectiles have become available in Orsay, Berkeley, Darmstadt, and elsewhere, and spins up to 28^+ have been excited in ^{238}U [Fu 78].

Table 3.1 gives a selection of published work on multiple Coulomb excitation of rotational bands with ^{16}O and heavier projectiles. Figure 3.8 shows a partial level scheme of ^{238}U with rotational transitions induced by argon bombardment. Several features of these results are interesting to note: the maximum spin values are clearly correlated with projectile mass, ranging from 8 for ^{16}O to 28 for ^{208}Pb, in qualitative agreement with the classical estimate of (3.18). The corresponding excitation energies are only moderate, however; it appears, therefore, that energy transfer is not readily achieved by multiple

Table 3.1.   Selected results on multiple Coulomb excitation of ground-state rotational
bands

Reference	Projectile, energy [MeV]	Target nucleus	Highest spin observed[a]	Excitation energy [MeV][b]
Gr 63	$^{16}O(64)$	^{152}Sm	8^+	1.21 (1.55)
		$^{159}Tb(3/2^+)$	$23/2^+$	1.50
		$^{165}Ho(7/2^-)$	$23/2^-$	1.29
Bo 64	$^{16}O(44)$	$150 \lesssim A \lesssim 192$	8^+	0.95 (1.01)
Sa 70b	$^{16}O(52)$	$152 \lesssim A \lesssim 176$	8^+	0.95 (1.61)
Oe 74	$^{16}O(60)$	$^{160,162,164}Dy$	8^+	0.97 (1.44)
Sa 74c	$^{35}Cl(125)$	$^{160,162,164}Dy$	12^+	1.95
St 59, Di 67	$^{40}Ar(190)$	^{232}Th	12^+ (12.0)	1.14
		^{238}U	14^+ (13.4)	1.42
St 68	$^{40}Ar(182)$	$^{235}U(7/2^-)$	$25/2^-$	0.81 (1.05)
Ei 73	$^{40}Ar(145)$	^{232}Th	12^+ (8.0)	1.14
		^{238}U	12^+ (8.9)	1.08
Co 74	$^{84}Kr(371)$	^{232}Th	18^+ (17.6)	2.26
Gr 75	$^{136}Xe(640)$	^{238}U	22^+ (28.2)	3.07
Fu 78	$^{208}Pb(1100)$	^{238}U	28^+ (36.3)	4.6

[a] Numbers in brackets give the classical estimate of L_{max} (3.18).
[b] Numbers without brackets refer to the highest levels observed in the ground-state bands;
numbers in brackets give the highest excitation energy of any other identified level.

Coulomb excitation, although substantial angular momenta may be transferred. In addition, a closer examination of Table 3.1 reveals that with
increasing projectile charge the observed values of L_{max} fall increasingly short
of the classical prediction. This may be partly a consequence of technical
problems, resulting from Doppler broadening of the line shapes and unfavourable peak to background ratios. It is interesting to speculate, however,
that the possibility of observing very high spin states in Coulomb excitation
may also be limited by physical effects, like the damping of collective motion
due to coupling to particle degrees of freedom.

It has been shown that the excitation probabilities of higher levels in a
rotational band may be significantly affected by E4 transitions [Wi 69, Wi 71].
In particular, the direct E4 excitation of the 4^+ level may interfere with the
(usually dominant) double E2 excitation, especially for comparatively light
projectiles like α particles. This effect has been exploited to measure E4 transition matrix elements and to deduce hexadecapole deformations for a number
of nuclei in the deformed mass regions [St 70, St 71, Mc 71]. A problem in
this type of measurement is that usually two values of the E4 matrix element
with different signs are compatible with the data, as a consequence of an
ambiguity in the relative sign of the interfering transition amplitudes. Recently
Eichler et al. [Ei 73] have shown that for heavy projectiles (^{40}Ar) on ^{238}U the
excitation probabilities of higher rotational levels (8^+, 10^+, 12^+) are strongly
affected by E4 contributions; combining their results with α-particle data
for the lower levels, they were able to resolve the existing ambiguity with
respect to the E4 matrix element. This case may serve as another illustration
of the problems and possibilities encountered in spectroscopic studies by

Fig. 3.8. Multiple Coulomb excitation of the ground state rotational band in ^{238}U with ^{40}Ar ions

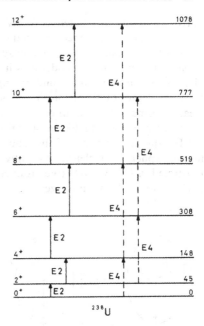

$$^{238}\text{U}$$

multiple Coulomb excitation, similar to those discussed in the context of the reorientation effect. One may hope that also in other cases uncertainties resulting from unknown relative phases of interfering amplitudes can be removed by measurements under different, carefully chosen bombarding conditions.

The discussion of Coulomb excitation in this section has been based— either explicitly or implicitly—on the semiclassical approach. This may be justified by noting that quantum-mecanical corrections to the semiclassical cross sections are proportional, in first order, to n_{if}^{-1} and therefore usually small for heavy projectiles. They, may however, be significant in precise quantitative work, especially if higher-order effects are involved. For details, we refer to the literature [Bi 65, Sm 68, Al 69, Al 72, Ba 74, Al 74].

3.2 Distortion and Break-Up in the Coulomb Field

In this section we consider multiple Coulomb excitation from a somewhat different point of view. We discuss to what extent Coulomb forces will produce a macroscopic distortion of the nuclear surface during a collision. Such a distortion, if it exists in significant measure, would have several important consequences. Firstly, it would affect the Coulomb barrier, and hence the bombarding energies required to reach the domain of nuclear interactions with heavy projectiles. This is clearly a point of concern in the design of heavy-ion accelerators. Secondly, an appreciable increase of the interaction barrier, and the associated collective excitation of the fragments, would strongly reduce our chances of producing superheavy nuclei by fusion reactions. This is because the compound systems could then not be produced in a "cold state", but only at moderately high excitation, and their chance of surviving de-excitation without undergoing fission would be small. Last, but not least, the degree of distortion should depend sensitively on details of the nuclear deformability, and an experimental study of distortion effects should therefore provide valuable information on potential-energy surfaces.

The qualitative effect of the Coulomb interaction on the shape of two initially spherical nuclei approaching each other in a head-on collision is illustrated in Fig. 3.9. At large distances the nuclei tend to assume an oblate deformation, which minimizes the Coulomb potential for a fixed centre

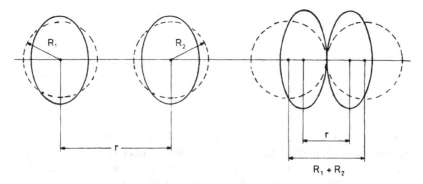

Fig. 3.9. Quadrupole deformation of spherical nuclei due to their Coulomb interaction: a at finite separation, b at the contact point

separation r. This results in a reduction of the value of r at which hard contact between the nuclear surfaces is established, and in an increase of the corresponding Coulomb potential. If the resulting radial force at this point is repulsive, the nuclei may subsequently start separating again, and will then tend to assume a stretched, prolate configuration under the influence of nuclear cohesion.

So far, we have little direct experimental evidence on Coulomb distortion; therefore much of the discussion in this section is necessarily of a theoretical—if not speculative—nature. Early work on the subject is due to Geilikman [Ge 55, Ge 58], who considered the static interaction of two deformed, homogeneously charged liquid drops and derived expressions for the interaction potential. In the late sixties, theoretical predictions concerning the stability of superheavy nuclei and proposals to produce these nuclei in heavy--ion reactions stimulated renewed interest in the problem. Beringer was the first to perform dynamical calculations [Be 67], treating the fragments classically as liquid drops undergoing irrotational flow. He predicted large oblate deformations, which would result typically in an increase of the Coulomb barrier by about 20%. Similar conclusions were reached by Maly and Nix [Ma 67], who also performed dynamical liquid-drop calculations. These latter authors noted, however, that viscous damping of the nuclear motion might reduce the effect. Subsequently Wong pointed out that the stiffness and inertial parameters of the liquid-drop model are not adequate to describe shape oscillations of real nuclei; instead he suggested using "spectroscopic parameters" derived from vibrational excitation energies and transition probabilities [Wo 68 a, Wo 68 b]. This approach yields mass parameters which are about an order of magnitude larger than those of the irrotational flow model, and stiffness parameters which strongly fluctuate around the liquid-drop values as a function of shell structure. The nuclear vibrational periods associated with these parameters are, on the average, much larger than the collision times; therefore at the turning point the dynamical deformation of the fragments is only a small fraction of its static value, and its effect on the Coulomb barrier is correspondingly reduced. Detailed dynamical calculations, including quadrupole and octopole degrees of freedom and the influence of nuclear forces, have been performed by Holm, Greiner, and collaborators [Ho 69, Ho 70 a, Ho 70 b], by Jensen and Wong [Je 70 a, Je 70 b, Je 71], and by Riesenfeldt and Thomas [Ri 70 a].

It is interesting to consider the influence of dynamics not only on the instantaneous deformation, but also on the resulting energy transfer to the deformation mode. One can readily show that no energy is transferred either in the adiabatic limit, where the collision is very slow, or in the "sudden" limit, where the collision is extremely fast compared to the vibrational motion. In the former case, the energy stored temporarily as potential energy of deformation is fed back into the relative motion upon separation of the fragments. In the latter case, on the other hand, the time of interaction is too short to produce significant deformation. The apparent contradiction of this statement to the semiclassical result, that maximum excitation probability occurs for $\xi_{if} \to 0$, may be resolved by noting that $\xi_{if} \to 0$ corresponds to vanishing excitation energy, and therefore a high excitation probability is not in conflict with small energy transfer. A practical example approaching this limit is furnished by the multiple Coulomb excitation of ground-state rotational bands in strongly deformed nuclei, where we have seen in Sect. 3.1 that in spite of high multiple transition probabilities, only moderate excitation energies are reached.

The excitation of quadrupole vibrations represents an intermediate situation between the extreme sudden and adiabatic limits ($\xi_{if} \approx 0.5$). Here energy is fed to the deformation mode during the early stages of the collision in form of kinetic energy; subsequently, as the fragments separate, the deformation amplitudes are built up. The instantaneous transfer of energy is given classically (and in the absence of dissipative forces) by

$$\frac{d}{dt} H_\beta(\beta, \dot{\beta}) = -\sum_{i=1,2} \dot{\beta}_i \frac{\partial}{\partial \beta_i} V_{12}(r, \beta) \tag{3.19}$$

As the time derivatives $\dot{\beta}_i$ are initially zero and increase with time during the

collision, the rate of energy transfer also increases beyond the turning point
and is not symmetric in time with respect to the instant of closest approach
(see Fig. 3.11). It should be noted that this asymmetry is directly related
to the nonadiabatic character of the collision.

We now turn to a discussion of effects due to the nuclear interaction. Here
we have to distinguish two different problems. First we have to define the term
"Coulomb barrier" for spherical nuclei and to show how the radial distribution of the (static) nuclear interaction enters this definition. This question
obviously must be settled before dynamic effects can be discussed. Then we
consider the influence of nuclear forces on the dynamic deformation of the
fragments and hence on the effective Coulomb barrier.

In Chap 1, we introduced a very schematic definition of the Coulomb
barrier [see (1.17)]:

$$E_{\rm C} = \frac{Z_1 Z_2 e^2}{R_{\rm C}}.$$
(3.20)

This definition is based on a classical sharp cut-off model, where one assumes
that for internuclear distances larger than $R_{\rm C}$ the interaction is purely Coulombic, whereas for smaller distances the nuclear interaction dominates.
Clearly in this model the "Coulomb radius" $R_{\rm C}$ cannot be identified with
the sum of effective matter radii of the fragments, but must be taken to include
some effective range of nuclear forces. In this way the details of the nuclear
interaction are absorbed in the empirical parameter $R_{\rm C}$; however, the relationship of the latter to the fragment mass numbers is not obvious and may,
in fact, depend on experimental circumstances. While this approach has been
widely applied, it can and must be refined for the purposes of our present
discussion.

It is known from many analyses of elastic scattering and total reaction
cross sections (see Sects. 1.4 and 3.3) that such data can be well reproduced
with two basic assumptions, namely
i) that all fragment pairs penetrating inside a certain "strong-absorption
distance" $(R_{\rm int})$ are removed from the elastic channel by nuclear reactions;
ii) that for moderately large distances $(r \gtrsim R_{\rm int})$ the inter-nuclear forces can
be derived from a two-body potential, which consists of a Coulomb and
nuclear part.

We prefer the term "interaction distance" $(R_{\rm int})$ to "strong-absorption
distance" in order to make clear that the processes in question include all
types of surface reactions and will not necessarily lead to fusion of the two
fragments. This interaction distance $R_{\rm int}$ differs in principle from the previously introduced Coulomb distance $R_{\rm C}$, due to the fact that it is strictly
defined in configuration space and is not intended to simulate effects of the
neglected tail of the nuclear potential. In practical cases, however, the differences between $R_{\rm int}$ and $R_{\rm C}$ may be small (see Sect. 3.3).

As will be shown in Sect. 3.3, the interaction distance $R_{\rm int}$ is approximately
3 fm larger than the half-density distance in all cases studied so far:

$$R_{\text{int}} \approx R_1 + R_2 + 3 \text{ fm}. \tag{3.21}$$

Here R_1, R_2 are the half-density radii of spherical nuclei with mass numbers A_1, A_2 as given by (3.12). An "interaction barrier" (B_{int}) may now be defined as that bombarding energy in the centre-of-mass system which is needed classically for the fragments to penetrate to the distance R_{int}. This interaction barrier will either correspond to the maximum of the combined Coulomb and nuclear two-body potential, if such a maximum occurs for $r > R_{\text{int}}$, or, otherwise, to the value of the potential at $r = R_{\text{int}}$ (see Fig. 3.10). It should be noted that in the latter case B_{int} is not a potential barrier, but only an energy threshold for nuclear reactions.

In the literature, some authors have identified the "Coulomb barrier" or "interaction barrier" with the maximum value of the two-body potential, regardless at what distance this maximum occurs [see for example Je 70 b, Je 71, Br 74 a]. In the light of our present discussion, this procedure is not generally meaningful for the following reasons: firstly, we have no direct experimental information concerning the existence and magnitude of a two-body potential for $r < R_{\text{int}}$, although it may seem plausible to postulate such a potential (we shall discuss methods to deduce it semiempirically in Chap. 7); secondly, the available evidence on interaction radii R_{int} and two-body potentials indicates that for most havy-ion systems the barrier occurs inside the interaction distance. In this case, the barrier height is irrelevant for elastic scattering or reaction cross sections, but may be identified with the "fusion barrier" (B_{fu}; see Fig. 3.10 and Chap. 7). We conclude that the Coulomb barrier is not a very useful concept for quantitative discussions, unless the nuclear process in question is specified. We therefore prefer to use the terms "interaction barrier" (B_{int}), denoting the threshold energy for nuclear reactions in general, and "fusion barrier" (B_{fu}), denoting the threshold for fusion of the two fragments. The difference between these two "barriers" arises from the fact that for heavy fragments the potential at the interaction distance is normally repulsive, so that additional energy is required to drive the fragments over the fusion barrier. The situation is illustrated qualitatively in Fig. 3.10, which shows potentials for projectiles of different masses incident on a ^{208}Pb target nucleus.

Having clarified our definition of the barriers, we can proceed to review briefly the results of various dynamical calculations on Coulomb distortion. Holm, Greiner et al. have considered quadrupole and octopole vibrations of the nuclear surface, as well as dipole polarization (the latter is, however, of little consequence, as it follows the collision adiabatically). The Coulomb barrier of these authors is based on a contact configuration of sharp nuclear surfaces which have (undistorted) radii equal to $1.2 \, A^{1/3}$ fm, only slightly larger than the effective matter radii. In earlier calculations, neglecting nuclear forces, strong quadrupole distortions were predicted near the barrier [Ho69, Ho70a]. The corresponding increase in the barrier height is of the order of 10 % as compared to about 20 % for static deformation. These results are strikingly modified when the nuclear forces are included [Ho 70 b]: the latter

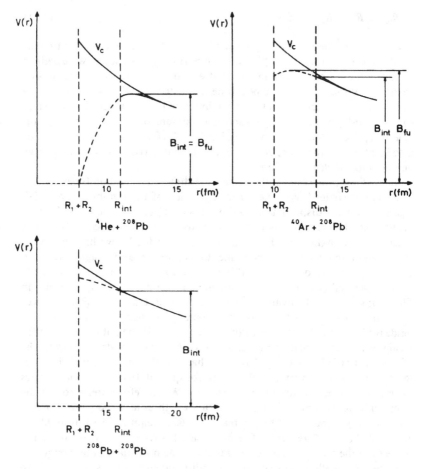

Fig. 3.10. Two-body potentials of various projectiles incident on ²⁰⁸Pb. In order to demonstrate the qualitative differences, each potential is separately normalized for equal magnitude of the Coulomb part at $r = R_1 + R_2$. The classical "interaction barriers" (B_{int}) and "fusion barriers" (B_{fu}) are indicated

strongly counteract the Coulomb forces at small distances and lead to a drastic reduction of the quadrupole deformation; for the octopole deformation, the nuclear effect even overcompensates the Coulomb effect and results in strongly enhanced octopole excitation of essentially nuclear origin. Calculated amplitudes and energies of the quadrupole, octopole, and dipole degrees of freedom in the system ¹⁴⁸Nd + ¹⁴⁸Nd are shown in Fig. 3.11; the deep interference minimum in the quadrupole excitation energy and the rather small quadrupole and octopole deformation amplitudes near the turning point are particularly noteworthy. The calculated barrier heights of various systems are now 2–7 % lower than in the absence of both nuclear interaction and deformation; the corresponding effect of deformation alone

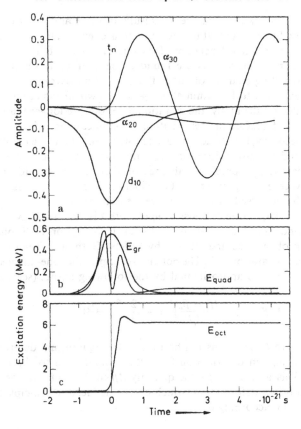

Fig. 3.11. Results of calculations by Holm and Greiner: Amplitudes and excitation energies of quadrupole (α_{20}, E_{quad}), octopole (α_{30}, E_{oct}), and dipole (d_{10}, E_{gr}) vibrations as functions of time in head-on collisions of two ^{146}Nd nuclei at the Coulomb barrier. The time of closest approach is marked as t_n, and the nuclear force is included. (From Ho 70b)

is typically within $\pm 3\%$, as shown by Riesenfeldt and Thomas [Ri 70 b]. It should be noted that the barrier definition of Holm et al. (when nuclear forces are included) corresponds to the fusion barrier (B_{fu}) in our terminology.

Similar calculations have been performed by Jensen and Wong. Without nuclear interaction, but using a large contact distance corresponding roughly to our R_{int}, their interaction barriers are up to 4% higher than in the absence of deformation [Je 70 a]. When nuclear forces are included, and the barrier is defined as fusion barrier, the deformation effects are within $\pm 2\%$ of the undeformed barrier heights [Je 70 b, Je 71]. As in the work of Holm et al., strong octopole deformations are induced by the nuclear interaction and may lead to a slight lowering of the fusion barrier depending on the nuclei considered.

We can summarize the results of these and other calculations [Ri 70 a] by first noting that it is essential to use realistic mass and stiffness parameters and to take full account of the nuclear forces. If this is done, then the dynamical distortions are found to be small up to the turning point, and the resulting changes of either the interaction barrier or the fusion barrier are expected to be within about $\pm 3 \%$. Two effects are mainly responsible for this result: firstly, the vibrational periods are longer than the collision times, and therefore the deformation amplitudes cannot follow the distorting field; secondly, the attractive nuclear force counteracts the repulsive Coulomb force and thereby reduces strongly the quadrupole deformation induced by the latter. The predicted absence of significant distortion effects is in accord with existing experimental evidence on reaction cross sections, which will be discussed in Sect. 3.3 [see also Ri 70 a].

A question of considerable interest is whether heavy nuclei like ^{238}U can undergo Coulomb-induced fission in heavy-ion collisions. This fascinating problem was first studied by Wilets, Guth, and Tenn [Wi 67], following a classical approach. The potential energy in the (one-dimensional) deformation mode was approximated by the following ansatz (see Fig. 3.12):

$$V(\beta) = 3 \, E_b \left(\frac{\beta - \beta_0}{\varDelta\beta} \right)^2 \left(1 - \frac{2}{3} \frac{\beta - \beta_0}{\varDelta\beta} \right). \tag{3.22}$$

Here E_b is the fission barrier, β_0 the equilibrium deformation, and $\varDelta\beta$ the change in deformation from the ground state to the fission saddle point. This ansatz relates the quantity $\varDelta\beta$ to the stiffness parameter C_β at equilibrium deformation (which can be deduced in principle from spectroscopic data) according to

$$C_\beta = \left(\frac{d^2 V}{d\beta^2} \right)_{\beta = \beta_0} = 6 E_b (\varDelta\beta)^2. \tag{3.23}$$

Nevertheless, in actual calculations $\varDelta\beta$ (and thus C_β) was considered a free parameter. Static and dynamical aspects of the problem were discussed in detail, neglecting, however, nuclear forces. Fission cross sections were cal-

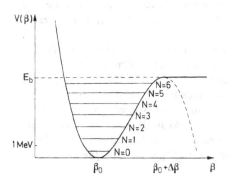

Fig. 3.12. Fission potential used by Wilets et al. (Wi 67; broken line) and Beyer and Winther (Be 69a, Be 69b; solid line) in calculations of Coulomb fission and corresponding energy levels. (From Be 69a)

culated for head-on collisions of various projectiles with ^{238}U by averaging over all possible initial orientations of the nuclear symmetry axis. Large cross sections were predicted for heavy projectiles like Xe or U, of the order of 10–100 mb/sr near the respective barriers. The results were very sensitive to the parameter $\Delta\beta$, however, indicating that information on potential energy surfaces might be obtainable from Coulomb fission studies. Another interesting prediction was that fission fragments should be emitted preferentially at 90° with respect to the beam axis. This is in sharp contrast to angular distributions characteristic of compound nucleus fission, which are approximately proportional to $(\sin\Theta)^{-1}$.

The preferential emission of fission fragments at 90° relative to the beam direction is a typical feature of "instantaneous" fission, which is thought to occur in the presence of the projectile. In this case an orientation of the nuclear symmetry axis perpendicular to the beam axis leads to a lowering of the instantaneous fission barrier in central collisions, and hence temporarily opens a "window" for fission decay. Allternatively, fission may take place only after the projectile has moved to large distances, in which case the ordinary fission barrier must be overcome. Altogether we can distinguish three different mechanisms which may be involved in Coulomb-induced fission:

i) "instantaneous" fission occurring in close proximity to the projectile as discussed obove;

ii) "direct" fission, where sufficient energy is transferred during the collision to the collective fission mode to overcome the (normal) barrier, and fission occurs after the projectile has moved away but before the energy is redistributed among other degrees of freedom;

iii) "thermal" fission where the energy transferred during the collision is thermalized before fission occurs. If the damping of the collective fission mode due to coupling to intrinsic degrees of freedom is sufficiently strong, this may actually limit the amount of energy transferred.

Which of these mechanisms dominates the actual process of Coulomb-induced fission—if it exists at all—should depend on the relative time scales of the orbital motion, the motion in the fission mode, and the damping to other degrees of freedom. Thus interesting information on nuclear dynamics may be expected from studies of Coulomb fission if the dominant mechanism can be established.

The first semiclassical calculations of Coulomb fission were published by Beyer, Winther, and Smilansky [Be 69a, Be 69b]. These authors used the same deformation potential as Wilets et al. [see (3.22) and Fig. 3.12], but with a smaller value of $\Delta\beta$ (0.25), and identified the cross section for fission with that for multiple Coulomb excitation to the highest bound vibrational level, with principal quantum number $N = 6$. The resulting cross sections remain below about 1 mb/sr for head-on collisions, even with uranium projectiles at the Coulomb barrier.

The influence of nuclear forces on Coulomb fission was studied theoretically by Holm and Greiner [Ho 71, Ho 72]. These authors considered backscattering of ^{84}Kr, ^{132}Xe, ^{148}Nd, and ^{222}Rn from ^{238}U and included rotational excitation of the target neglecting, however, the rotation–vibration interaction. Both classical and semiclassical calculations were performed, and found to yield very similar results for bombarding energies above the classical thresholds. The cross sections of Holm and Greiner are intermediate between those of Wilets et al. and of Beyer et al., reaching maxima of the order of 10 mb/sr for heavy projectiles on ^{238}U. A conspicuous feature of the results is a deep minimum of the excitation functions just below the Coulomb barrier, which arises from destructive interference between Coulomb and nuclear excitation (see Fig. 3.14). This phenomenon is well known in the inelastic scattering of heavy projectiles to low-lying collective states and will be discussed in more detail in Sect. 3.4.

A problem of some concern are the large discrepancies which exist between different theoretical predictions of cross sections for Coulomb fission. For some time they were believed to be due to inadequacies of the classical approximation [Be 69a]; later work by Holm and Greiner [Ho 72] and by Oberacker et al. [Ob 77a] suggests, however, that the difficulties are more likely to be associated with approximations in the treatment of nuclear dynamics. We shall return to this point below.

The various theoretical predictions have stimulated considerable experimental activity. Following an earlier, unsuccessful attempt with krypton ions at Orsay [Ng 74], positive evidence for sub-barrier fission was obtained with krypton and xenon ions at Berkeley [Co 76, Bu77], and with xenon ions at Darmstadt [Ha77, Fr77, Fr78c]. The experimental techniques comprise both coincidence measurements between back-scattered projectiles and fission fragments, using solid-state counters [Co76b, Bu77, Ha77], and measurements of induced radioactivity following chemical separation [Fr77, Fr78c]. The two methods are complementary in the sense that the former is more suitable for measuring angular and energy distributions of the reaction products, whereas the latter provides a higher sensitivity and superior resolution in A and Z. The measured cross sections are in all cases smoothly increasing with bombarding energy and reach values of the order of 10–100 mb/sr at the Coulomb barrier.

The Berkeley group studied the dependence of the cross section on the projectile (^{86}Kr, ^{136}Xe) and on the target nucleus (^{232}Th, ^{238}U, ^{244}Pu, ^{248}Cm; see Fig. 3.13). Although larger cross sections were observed with xenon than with krypton projectiles, the absence of a more dramatic dependence on either projectile charge or fission barrier was taken as evidence against fission induced by a purely Coulombic interaction [Bu77]. Similarly the radio-chemical data obtained at Darmstadt suggest that nuclear interactions may be effective down to bombarding energies well below the interaction barrier. In this investigation, one-nucleon transfer cross sections were measured simultaneously with the fission cross sections and found to be of comparable magnitude. The observed dependence of the transfer cross sections on the Q value and bombarding energy and their relationship to the fission cross sections led to the conclusion that a dominant contribution from Coulomb-induced fission

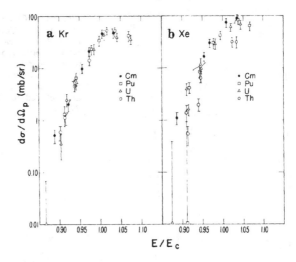

Fig. 3.13. Differential cross sections for fission of different heavy target nuclei, measured in coincidence with back-scattered krypton (**a**) or xenon (**b**) projectiles. The incident energy is expressed as a fraction of the Coulomb barrier $E_C = Z_1 Z_2 e^2/1.16(A_1^{1/3} + A_2^{1/3} + 2)$ fm. (From Bu 77)

may be present at bombarding energies below about 85% of the interaction barrier, but that fission following transfer becomes increasingly important as the bombarding energy is raised [Fr78c]. A similar conclusion was reached also by Habs et al., based on an analysis of energy distributions of back-scattered projectiles [Ha77]. It should be noted that none of the experiments mentioned above permits a direct distinction between Coulomb-induced and transfer-induced fission, since in no case the mass of the scattered particle was unambiguously identified. "Instantaneous" fission seems to be ruled out, however, since the measured angular distributions are approximately proportional to $(\sin \Theta)^{-1}$ [Ha77, Bu77].

The interpretation of the measured low-energy fission cross sections as predominantly due to the Coulomb interaction is supported by recent calculations of Oberacker et al. [Ob77a]. These authors improved the earlier semiclassical calculations of Holm and Greiner [Ho72] by incorporating the rotation–vibration interaction. As a result the sub-Coulomb cross sections are drastically reduced. Figure 3.14 shows a comparison of the calculations of Oberacker et al. and of Holm and Greiner with the experimental data of Habs et al. [Ha77] and of Colombani et al. [Co76b] for the system ^{136}Xe $+ \, ^{238}$U. The calculations of Oberacker et al. are seen to reproduce the magnitude of the cross sections at low energies remarkably well; however, the pronounced minimum predicted theoretically as a consequence of Coulomb–nuclear interference is completely absent in the data. The latter feature could be indicative of transfer-induced fission, or of some other nuclear effect not included in the calculation.

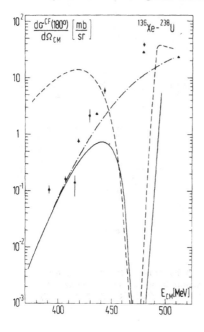

Fig. 3.14. Calculated differential cross sections for Coulomb fission of ^{238}U, induced by back-scattering of ^{136}Xe. Broken curve: Ho 72, full curve: Ob 77a with nuclear interaction, dash-dotted curve: Ob 77a without nuclear interaction. Experimental points from Ha 77 and Co 76b are included for comparison. (From Ob 77a)

Although the existing results provide interesting information concerning the magnitude of the cross sections and their sensitivity to nuclear structure and dynamics, it seems clear that much more work is needed for reaching a detailed understanding of the mechanisms involved. Experimentally one should like to know more about the systematic dependence of the cross sections on various parameters, and the question of transfer should be settled by precise measurements of the mass and energy of the scattered particle. On the theoretical side more realistic nuclear models and mixed excitation mechanisms—involving both nucleon transfer and Coulomb excitation—should be explored.

We close this section with some remarks on the break-up of "light heavy ions" by the Coulomb interaction in collisions with heavy nuclei. The well-known cluster structure of certain light nuclei, such as the lithium isotopes or ^{12}C, adds particular interest to studies of their Coulomb-induced disintegration. The most likely mechanism for such a disintegration is Coulomb excitation of some excited state just above the cluster emission threshold, for example the 3^+ state at 2.18 MeV in ^{6}Li, which can be reached by an E2 transition from the 1^+ ground state. A major difficulty in studies of this kind arises from the fact that nuclear reactions may compete strongly with Coulomb disintegration even at bombarding energies below the classical interaction barrier. Therefore carefully designed coincidence experiments are required in order to separate various reaction mechanisms which can lead to cluster emission.

The break-up of the lithium isotopes has been investigated by various groups [Sp 70, Di 71, Os 72, Pf 73, Os 73, Qu 74]; for details the reader is referred to the literature.

3.3 Elastic Scattering and Reaction Cross Sections

Numerous experimental studies of elastic scattering from medium or heavy target nuclei have been published, and a complete review would be beyond the scope of this book. Instead we shall try to summarize the most important results and conclusions.

The experimental results may be divided roughly into three categories. The first category comprises a large amount of data obtained with electrostatic accelerators using light heavy-ion beams (mostly oxygen, but also lithium, carbon, nitrogen, fluorine, silicon, sulphur, and chlorine) and bombarding energies up to about 8 MeV/u. In the second category we have data obtained with cyclotrons at Berkeley, Oak Ridge, and elsewhere, using projectiles from ^{11}B to ^{20}Ne and typical bombarding energies in the range 8–15 MeV/u. Much of this work was concerned with the doubly magic target nucleus ^{208}Pb [Be 72a, Fo 73, Ba 75, Gr 78, Ol 78, Pi 78]. Finally there is a growing number of experiments on the scattering of argon and heavier projectiles at bombarding energies up to about 10 MeV/u, performed with linear accelerators in Orsay, Berkeley, and Darmstadt and with the heavy-ion cyclotrons in Dubna [Co 72, Le 72, Bi 76, Br 76, Va 76, Og 78, Va 78]. In contrast to most of the results in categories one and two, the results in this last category in general do not represent strictly elastic scattering, but may be contaminated to some extent by inelastic scattering and transfer reactions. This is a consequence of experimental problems associated with the higher bombarding energies and poorer beam qualities encountered in this type of work, and should be taken into account in interpreting the results.

The measured angular distributions typically exhibit a Fresnel-type diffraction pattern with comparatively little structure and a sharp drop in the ratio of elastic scattering to Rutherford scattering beyond the grazing angle. As an example of work in category one, Fig. 3.15 shows results of Becchetti et al. for elastic scattering of 60 MeV oxygen ions from various targets [Be 73]. Further examples are given in Fig. 3.20 (category two) and 3.17 (category three).

Certain conclusions may be drawn from an inspection of scattering data like those shown in Fig. 3.15 even without detailed quantitative analysis. Clearly the dominating feature of the data is strong absorption from the elastic channel within a well-defined geometrical region (see Sect. 1.4). The main piece of information that can be extracted is therefore the critical distance for "strong absorption", or effective interaction distance. In addition, one may hope to learn something about the width of the transition region from no absorption to complete absorption, and about the strength of the nucleus–nucleus interaction just outside the effective interaction distance. One might ask to what extent these quantities are affected by the shell struc-

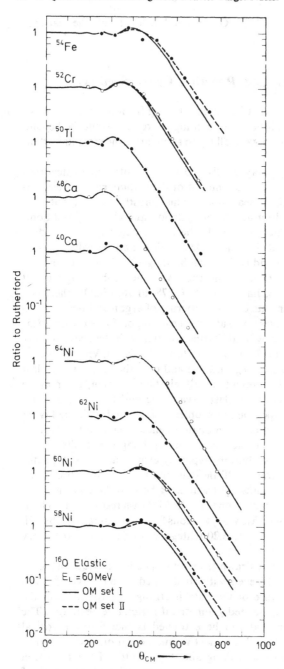

Fig. 3.15. Elastic scattering cross sections (divided by the Rutherford cross section) for ^{16}O at $E_{Lab} = 60$ MeV and different target nuclei. The curves are optical model calculations. (From Be 73)

ture of the fragments; we shall see, however, that elastic scattering among medium and heavy nuclei is rather insensitive to nuclear structure effects.

A particularly simple and fast method to deduce interaction distances (R_C) and grazing angular momenta (L_{gr}) from elastic scattering angular distributions is the so-called "quarter-point recipe" [Bl 54]. This recipe is based on the fact that diffraction by a large totally absorbing object always causes the amplitude at the edge of the geometrical shadow to drop to approximately half of the undisturbed amplitude, almost independently of wavelength and absorber size. In practice one determines the scattering angle at which the differential elastic scattering cross section is one-quarter of the Rutherford cross section ($\Theta_{1/4}$; see Fig. 3.16). Assuming an undisturbed Rutherford trajectory (i.e. neglecting deflection by nuclear forces) the corresponding distance of closest approach and classical angular momentum are then calculated from the following equations [see also (1.11) and (1.16)]:

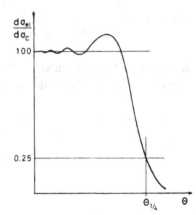

Fig. 3.16. "Quarter-point recipe" for the determination of grazing angular momenta and strong absorption radii

$$R_C = a\left(1 + \csc \frac{1}{2}\,\Theta_{1/4}\right), \tag{3.24}$$

$$L_{gr} = n \cot\left(\frac{1}{2}\,\Theta_{1/4}\right). \tag{3.25}$$

As before we denote by R_C and L_{gr} effective quantities which are based on the simplifying assumptions of total absorption within a sphere of radius R_C and pure Rutherford scattering outside that sphere. It should be noted that these assumptions do not determine R_C and L_{gr} unambiguously for any given experimental situation; instead actual values of R_C and L_{gr} may depend somewhat on the observable (here $\Theta_{1/4}$) from which they are deduced. We shall return to this interesting point later.

The extraction of quarter-point angles from experimental data can be put on a more formal basis by fitting angular distributions with the simple "Fresnel model" [Fr 72a; see Sect. 1.4]. This approach has been used especially in analysing data for very heavy projectiles [Bi 76]. Interaction distances R_C deduced from such analyses for a number of systems are included in Table 3.5. An interesting generalization of the model which removes the unphysical assumption of a sharp cut-off in angular momentum space has been introduced recently by Frahn [Fr 78b].

Another simple and useful parametrization of elastic scattering data is the semiclassical method suggested by Christensen et al. [Ch 73]. These authors write for the ratio of the elastic to the Rutherford cross section

$$\frac{d\sigma}{d\sigma_C} = 1 - P_{abs}(D), \tag{3.26}$$

where $P_{abs}(D)$ denotes the probability that a classical trajectory with turning point D will lead to absorption from the elastic channel. The following ansatz was used for $P_{abs}(D)$:

$$P_{abs}(D) = \begin{cases} 0 & \text{for} \quad D \geq D_0, \\ 1 - \exp\left(\dfrac{D - D_0}{\varDelta}\right) & \text{for} \quad D < D_0, \end{cases} \tag{3.27}$$

where D_0 and $\varDelta$ are parameters related to the interaction distance and the thickness of the transition region, respectively.

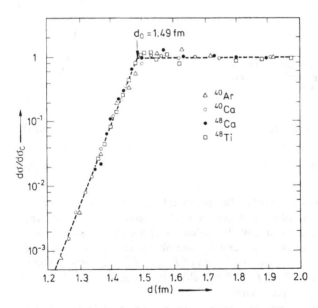

Fig. 3.17. Elastic scattering cross sections for ^{40}Ar, ^{40}Ca, ^{48}Ca, ^{48}Ti + ^{208}Pb, plotted as $d\sigma/d\sigma_C$ versus the distance parameter d_0 (for definition see text). The broken lines represent a fit according to (3.26) and (3.27). (From Og 78)

In fitting experimental data with (3.26) and (3.27) it is customary to translate distances D into scattering angles Θ assuming pure Coulomb scattering, as in (3.24). The method has been applied with remarkable success; an example is given in Fig. 3.17, which shows results of Oganessian et al. for scattering of projectiles with $A = 40–48$ from ^{208}Pb [Og 78]. The parameters $d_0 = D_0$ $(A_1^{1/3} + A_2^{1/3})^{-1}$, D_0, and Δ deduced from a number of such fits are collected in Table 3.2. We note that the values of Δ quoted for the heavier projectiles $(A_1 \geq 40)$ may be affected by contributions of inelastic processes to the "elastic" cross sections.

Table 3.2. Parameters deduced from semiclassical analyses of elastic scattering data

Systems	E_{Lab} [MeV]	n	d_0 [fm]	D_0 [fm]	Δ [fm]	References
^{16}O + ^{90}Zr	45–60	26–30	1.71	12.15	0.57	Ch 73
^{12}C, ^{16}O + ($A = 40$–96)	49–66	13–30	1.68		0.55	
^{40}Ar + ^{209}Bi	286, 340	88, 81	1.495	14.0	0.47 ± 0.05	Bi 76
^{40}Ar + ^{238}U	286, 340	97, 87	1.47	14.1	0.46 ± 0.05	
^{40}Ar + ^{208}Pb	302	85	1.50	14.03 ± 0.02	0.35 ± 0.01	Og 78
^{40}Ca + ^{208}Pb	302	94	1.50	13.98 ± 0.02	0.33 ± 0.01	
^{48}Ca + ^{208}Pb	252	113	1.48	14.17 ± 0.02	0.30 ± 0.01	
^{48}Ti + ^{208}Pb	252	124	1.49	14.21 ± 0.02	0.34 ± 0.01	

We now turn to more sophisticated methods of data analysis, which do not rely on such simplifying assumptions as pure Coulomb scattering outside R_C or the existence of well-defined classical trajectories. The most general and transparent approach is to assume a simple analytical dependence of the scattering phase shifts on angular momentum, as given by the smooth cut-off strong absorption model. This approach has been discussed in detail in Sect. 1.4.

Very good fits to elastic angular distributions have been obtained with only three adjustable parameters [Al 64, Ve 64]. In the phase-shift parametrization of McIntyre et al. [Mc 60; see (1.44), (1.45)] the significant quantities are l_A, Δ_A $(= \Delta_\delta)$ and $\delta [= 2\delta(l_A)]$. A strong absorption radius (R_C) can be derived from l_A by application of (1.48). Note that l_A corresponds to that partial wave for which the scattered intensity is one-quarter of the incident intensity. Nevertheless $l_A(l_A + 1)$ is only approximately comparable to L_{gr}^2, as deduced from the quarter-point recipe (3.25), since the latter implies neglect of nuclear scattering. For the same reason the effective interaction distance deduced from l_A (1.45) and $\Theta_{1/4}$ (3.32) will differ slightly; moreover, both of these distances will be somewhat larger than the true distance of closest approach (R_{int}) which corresponds classically to the angular momentum l_A taking nuclear deflection into account.

The difference between the distances of closest approach with ($D = R_{\text{int}}$) and without ($D_C = R_C$) nuclear interaction at specified angular momentum $l = l_A$ can be estimated simply by requiring energy conservation [see (1.14)]:

$$\frac{R_C - R_{\text{int}}}{R_C} \approx \frac{-V_N(R_{\text{int}})}{2E_{\text{CM}} - V_C(R_C)}, \tag{3.28}$$

where V_C and V_N refer to the Coulomb and nuclear potential, respectively. With typical values of $V_N(R_{\text{int}})$ between 1 and 5 MeV, the differences $R_C - R_{\text{int}}$ amount to a few tenths of a Fermi for medium to heavy target nuclei and bombarding energies between 5 and 10 MeV/u.

In Table 3.3 we have collected parameters deduced from elastic scattering data by phase-shift analyses. The parameters l_A and Δ_A characterize the distribution of absorption in angular momentum space. The diffuseness parameter Δ_A is typically in the range 3 ± 1 for l_A values between 50 and 110; the corresponding diffuseness parameters in configuration space [see (1.49)] are about 0.3 ± 0.1 fm. The quantity $\delta(l_A)$ is the real nuclear phase shift at the angular momentum l_A and reflects scattering by the nuclear potential in the vicinity of the strong absorption radius. Due to the predominance of Coulomb scattering, this quantity is not very accurately determined by the data. Typical values scatter around 0.3 rad.

A point of particular interest in Table 3.3 is our comparison of the interaction distances R_C with the half-density distances $R_{12} = R_1 + R_2$, where R_1 and R_2 are given by (3.12). The difference $R_C - R_{12}$ is found to be remarkably constant over a wide range of masses and energies:

$$R_C = R_1 + R_2 + (3.0 \pm 0.5) \text{ fm}. \tag{3.29}$$

This result implies that a formula based on (3.29) should provide a more useful representation of interaction distances than the conventional parametrization

$$R_C = r_{0C}(A_1^{1/3} + A_2^{1/3}), \tag{3.30}$$

where the radius parameter r_{0C} varies systematically with target and projectile mass. Further since (3.29) presumably reflects a property of the true interaction distance R_{int} [it may be recalled that the latter is only slightly smaller than the effective interaction distance R_C, see (3.28)], it suggests the existence of a critical overlap of nuclear densities in the tail region corresponding to the onset of nuclear reactions. This critical overlap seems to be reached when the density of each nucleus is about 5% of its central density at a point halfway between the nuclear surfaces.

By far the most popular method of analysis of elastic scattering data has been the application of the optical model. In contrast to the methods discussed previously, this model introduces explicit assumptions on the interaction between the fragments. The latter is described by a complex two-body potential as discussed in Sect. 2.2. In most cases a radial dependence of Woods–

Table 3.3. Parameters deduced from phase-shift analyses of elastic scattering data

Projectile	Target	E_{Lab} (MeV)	$l_A{}^a$	$\Delta_A{}^a$	$\delta(l_A)^b$	R_C [fm]	$R_{12}{}^c$ [fm]	R_C-R_{12} [fm]	$r_C{}^d$ [fm]	References[e]
^{12}C	Fe	124	51	2.9	0.3	8.65	6.19	2.46	1.42	Al 64
	Ni	124	56	3.1	0.25	9.32	6.27	3.05	1.51	Al 64
	^{107}Ag	124	60.5	2.5	0.4	10.15	7.28	2.87	1.44	Al 64
	In	124	61.5	2.2	0.4	10.30	7.37	2.93	1.44	Al 64
	^{144}Nd	118	56.9	5.23	0.21	10.30	7.82	2.48	1.37	Fr 72b
	^{146}Nd	118	54.8	6.11	0.26	10.03	7.87	2.16	1.33	Fr 72b
	^{152}Sm	118	53.3	5.38	0.32	9.91	7.95	1.96	1.31	Fr 72b
	^{154}Sm	118	56.5	6.75	0.08	10.31	7.98	2.33	1.34	Fr 72b
	^{181}Ta	124	65.3	3.05	0.18	11.43	8.32	3.11	1.44	Ba 67
	^{197}Au	101	52.5	1.04	0.30	11.81	8.51	3.30	1.46	Ve 64
		118	60.8	1.75	0.24	11.52		3.01	1.42	Ve 64
		121	62.6	2.18	0.29	11.54		3.03	1.42	Ve 64
	^{206}Pb	124	63.9	2.47	0.23	11.64	8.61	3.03	1.42	Ve 64
	^{208}Pb	123	65.5	2.03	0.29	11.91	8.63	3.28	1.45	Ve 64
		125	66.3	4.20	0.34	11.85		3.22	1.44	Ba 67
		118	61.9	3.99	0.08	11.80		3.17	1.44	Fr 72b
	^{209}Bi	118	60.9	2.75	0.18	11.73	8.64	3.09	1.45	Fr 72b
		124	63.9	2.22	0.17	11.69		3.05	1.42	Ve 64
^{16}O	Ni	157	69.5	3.31	0.20	9.56	6.57	2.99	1.49	Ve 64
	^{144}Nd	130	62.4	6.19	0.09	10.56	8.11	2.45	1.35	Fr 72b
	^{146}Nd	130	62.7	10.4	0.36	10.57	8.15	2.42	1.35	Fr 72b
	^{148}Nd	130	64.3	4.77	0.40	10.73	8.18	2.55	1.37	Fr 72b
	^{152}Sm	130	61.1	7.49	0.35	10.51	8.24	2.27	1.32	Fr 72b
	^{154}Sm	130	65.3	5.10	0.28	10.92	8.27	2.65	1.39	Fr 72b
	^{197}Au	157	86.1	2.05	0.41	12.25	8.81	3.44	1.47	Ve 64
		164	88.0	3.62	0.20	12.05		3.24	1.44	Ve 64
	^{206}Pb	130	62.6	4.52	0.30	11.77	8.91	2.86	1.38	Fr 72b
	^{207}Pb	166	86.4	3.46	0.13	11.91	8.92	2.99	1.41	Ve 64
	^{208}Pb	130	63.2	3.69	0.41	11.82	8.93	2.89	1.44	Fr 72b
		166	86.2	2.68	0.26	11.90		2.97	1.41	Ve 64
		169	95.0	3.78	0.24	12.54		3.61	1.48	Ve 64
		170	91.0	4.75	0.33	12.13		3.20	1.44	Ba 67
	^{209}Bi	164	86.4	3.00	0.19	12.06	8.94	3.12	1.43	Ve 64
		170	91.7	4.85	0.28	12.23		3.29	1.45	Ba 67
^{20}Ne	^{197}Au	208	109.7	2.44	0.19	12.15	9.05	3.10	1.42	Ve 64
	^{206}Pb	207	110.0	3.14	0.17	12.29	9.15	3.14	1.43	Ve 64
	^{207}Pb	208	109.2	3.01	0.19	12.19	9.16	3.03	1.41	Ve 64
	^{208}Pb	206	109.4	3.55	0.31	12.27	9.17	3.10	1.42	Ve 64
	^{209}Bi	210	109.8	2.75	0.15	12.21	9.18	3.03	1.41	Ve 64

[a] Defined by (1.44) and (1.45)
[b] Real part of nuclear phase shift for angular momentum l_A [$= \delta/2$ in McIntyre parametrization, (1.44); $= 1/2 \tan^{-1}(\mu/2\Delta)$ in Frahn–Venter parametrization, (1.46) and (1.47)].
[c] $R_{12} = R_1 + R_2$ sum of half-density radii as calculated from (3.12).
[d] Coulomb radius parameter defined by $R_C = r_C (A_1^{1/3} + A_2^{1/3})$.
[e] Quoted references refer to the analysis and not necessarily to the experimental work.

Saxon type has been assumed for both the real and imaginary part of the potential, with up to 6 adjustable parameters [see (2.1) and (2.2)].

In this manner excellent fits to elastic angular distributions have usually been obtained; however, the potential parameters are not determined unambiguously. The existence of "continuous ambiguities" in optical model parameters was first noted for α-particle scattering by Igo [Ig 59], who found families of acceptable potentials connected by the relationship

$$V \exp\left(\frac{R}{a}\right) = \text{const} \qquad (a = \text{const}). \tag{3.31}$$

Equation (3.31) is known as the "Igo ambiguity"; its physical origin is readily understood from the consideration that, because of strong absorption in the interior, only the tail of the nuclear potential at large distances ($r \geq R_C$) should affect elastic scattering. That this is indeed the case is demonstrated rather convincingly by optical model calculations of Becchetti et al. for the system $^{16}O + {}^{96}Zr$ at $E_{Lab} = 60$ MeV [Be 73]. Their results are given in Fig. 3.18 for the real part and in Fig. 3.19 for the imaginary part of the potential. In each case various potentials are shown which produce fits of similar quality to the data; while they are similar in magnitude near the effective interaction distance ($R_C \approx 10.5$ fm), they diverge strongly towards smaller and larger distances. The existence of a crossover of acceptable optical model potentials at a well-defined distance close to R_C has since been confirmed for many scattering systems [Ba 75, Va 76, Sc 78, Wo 78]. Another way of demonstrating the irrelevance of the interior region to optical model analyses for strongly absorbed systems is by truncation of the potential. As shown by Satchler for a number of representative cases, cutting off the potentials near R_C and assuming either vanishing or constant values for smaller distances has a negligible influence on the predicted scattering cross sections [Sa 74b, Ba 75].

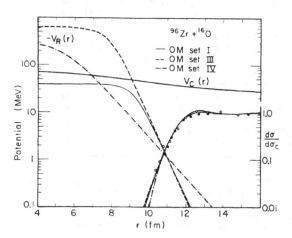

Fig. 3.18. Real part of optical model potentials which fit $^{16}O + {}^{96}Zr$ elastic scattering [$V_R(r)$ nuclear contribution, $V_C(r)$ Coulomb contribution]. Also included are the ratio of the experimental elastic scattering cross section to the Rutherford cross section (full circles: $E_{Lab} = 49$ MeV, open circles: $E_{Lab} = 60$ MeV) as a function of the distance of closest approach (calculated assuming Coulomb trajectories) and the corresponding optical model fits for $E_{Lab} = 60$ MeV. (From Be 73)

Fig. 3.19. Same as Fig. 3.18, but for the imaginary part of the optical model potential. (From Be 73)

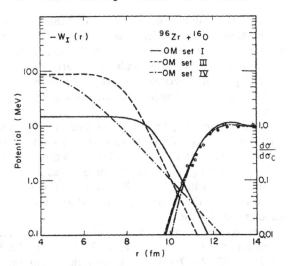

It follows that the usual Woods–Saxon parametrization of the optical model potential contains redundant parameters. The good fits obtained with simple phase-shift parametrizations and semi-classical models indicate that between two and four relevant parameters can be determined by fitting a single angular distribution. This is born out by numerous optical model analyses, where excellent fits to isolated angular distributions were obtained by keeping certain parameters fixed while adjusting the remaining ones. Four-parameter fits, for example, have been performed by assuming equal geometries for the real and imaginary potential ($r_{0V} = r_{0W} = r$, $a_V = a_W = a$) and varying the parameters V, W, r_0, and a [Be 73, Bi 76]. Alternatively, V and W may be fixed, and the geometrical parameters r_{0V}, a_V, r_{0W}, a_W adjusted for optimum fit [Pi 78]. A selection of typical parameter sets obtained by the former procedure is listed in Table 3.4.

Table 3.4. Examples of four-parameter optical model potentials with equal geometry of the real and imaginary part ($r_{0V} = r_{0W} = r_0$, $a_V = a_W = a$).

Systems	E_{Lab} [MeV]	V [MeV]	W [MeV]	r_0 [fm]	a [fm]	References
^{12}C, ^{16}O + (A = 40–96)	38 – 60	40	15	1.30	0.50	Be 73
^{16}O + ^{54}Fe	46	100	18.8	1.154	0.55	Bo 72
^{16}O + ^{63}Cu	40	67.2	9.6	1.22	0.57	Wo 78
	46	63.4	52.8	1.22	0.57	
^{12}C + ^{208}Pb	96	40	25	1.256	0.56	Ba 75
^{16}O + ^{208}Pb	129.5	40	35	1.249	0.615	
	192	40	35	1.226	0.634	
^{40}Ar + ^{209}Bi	286,340	43.2	56.0	1.196	0.529	Bi 76
		68.0	83.9	1.167	0.540	
		214.5	261.1	1.104	0.536	

Certain trends may be recognized in Table 3.4. Larger values of V and W are associated with smaller values of r_0 and (or) a, and vice versa. This obviously reflects the fact that the tail of the potential is determined by the data, in accordance with the Igo ambiguity. Further the ratio of the imaginary to the real potential increases strongly with energy near the Coulomb barrier, and reaches values comparable with or slightly larger than unity at higher energies.

Four parameters, however, are more than adequate to fit an angular distribution. This is demonstrated by successful fits using only two or three adjustable parameters. Figure 3.20 shows a three-parameter fit for the system ^{12}C + ^{208}Pb at $E_{\text{Lab}} = 96$ MeV, where V was kept fixed at 40 MeV and W, r_0, and a were adjusted [Ba 75]. It is interesting to note a pronounced correlation between the imaginary potential and the oscillatory structure in the angular distribution.

Ambiguities in optical model parameters were studied systematically by many groups [see, for example, Bo 72, Be 73, Ba 75, We 75, Bi 76, Co 76a, Cr 76, Pi 78, Wo 78]. In some cases, the constraints imposed by data covering a large energy range were investigated [Cr 77, Pi 78]. It was concluded that the parameters of both the real and imaginary part of the potential should depend smoothly on energy for satisfactory reproduction of the

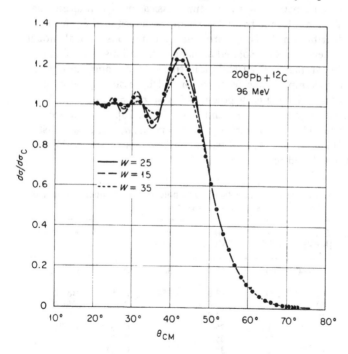

Fig. 3.20. Elastic scattering data for the system ^{12}C + ^{208}Pb at an incident energy of 96 MeV in the laboratory system. The curves represent optical model fits with different imaginary potentials as explained in the text. (From Ba 75)

data [Sa 76, Pi 78]. For further details we refer to two review articles by Satchler, where the problems encountered in optical model analyses of heavy-ion scattering are fully discussed [Sa 74b, Sa 76].

In many cases interaction distances R_C have been deduced from optical model analyses, using the relationship

$$kR_C = n + [n^2 + l_{1/2}(l_{1/2} + 1)]^{1/2}. \tag{3.32}$$

Here $l_{1/2}$ denotes that angular momentum for which the "transmission coefficient" $T_l = 1 - |S_l|^2$ [see (1.30)] is equal to 0.5, and R_C is the classical distance of closest approach for the corresponding trajectory. We note that (3.32) implies semiclassically $d\sigma/d\sigma_C = 0.5$ for the trajectory characterized by R_C. As might be expected from our previous discussion, values of R_C determined in this way are not affected significantly by parameter ambiguities in the optical model. Moreover, it has been shown that these values are quite close to R_C values deduced by Fresnel or quarter-point analyses [Bi 76, Sc 78; see Table 3.5].

Table 3.5 gives a summary of R_C values which were determined by the methods discussed above, and a comparison with the corresponding half-density distances as calculated from (3.12). The differences $R_C - R_{12}$ are again found to be rather constant; excluding data obtained with heavier projectiles at linear accelerators, they are well represented by

$$R_C = R_1 + R_2 + (3.5 \pm 0.3) \text{ fm}. \tag{3.33}$$

For the heavier projectiles, the differences $R_C - R_{12}$ tend to be somewhat smaller than given by (3.33). This could be caused partly by inelastic contributions to the "elastic" data, however. In order to eliminate redundant parameters, it would be desirable to reformulate the optical potential in terms of invariants of the scattering problem, such as the critical distance for strong absorption and the magnitude and derivative of the potential at that distance. A somewhat related procedure consists of imposing an "ingoing-wave boundary condition" at a suitably chosen lower cut-off distance. This method has been discussed by Rawitcher [Ra 64, Ra 66] and by Strutinsky [St 65b]; more recently it was applied successfully by Eisen et al. to the analysis of ^{16,18}O scattering from target nuclei in the Ca–Ni region near the Coulomb barrier [Ei 72a, Ei 72b].

The bewildering variety of potentials in the literature is clearly an unsatisfactory feature, and seems hardly consistent with the spirit of the optical model. This has led Christensen and Winther to deduce a universal nucleus-nucleus potential from a global fit to elastic scattering data [Ch 76]. The analysis of these authors is based on the recognition that each angular distribution determines the corresponding real potential at one particular point. The location of the latter was calculated for a large number of scattering systems, using semiclassical arguments and the empirically determined potentials. Thus a large set of points $V(r)$ was obtained which could be compared

Table 3.5. Effective Coulomb interaction distances R_C deduced from elastic scattering data

System	E_{Lab} [MeV]	R_C[fm] a	b	R_{12} [fm] c	$R_C - R_{12}$ [fm]	References
$^{16}O + {}^{54}Fe$	46 – 52	9.76		6.43	3.33	Bo· 72
$^{16}O + {}^{63}Cu$	42		10.56	6.67	3.89	Wo 78
$^{16}O + {}^{64}Ni$	56		10.0	6.69	3.3	Co 76a
$^{16}O + {}^{74}Ge$	56		10.4	6.93	3.5	
$^{18}O + {}^{74}Ge$	56		10.8	7.05	3.75	
$^{11}B + {}^{208}Pb$	72.2		11.9	8.55	3.35	Ba 75
$^{12}C + {}^{208}Pb$	96		12.2	8.63	3.6	
	116.4		12.1		3.5	
$^{16}O + {}^{208}Pb$	129.5		12.7	8.93	3.8	
	192		12.5		3.6	
$^{20}Ne + {}^{208}Pb$	161.2		13.1	9.17	3.9	
$^{32}S + {}^{24}Mg$	75 – 120		9.4	6.16	3.2	Gu 73b
$^{32}S + {}^{27}Al$	85 – 110		9.5	6.31	3.2	
$^{32}S + {}^{40}Ca$	82.5 – 90		10.55	6.82	3.7	
$^{32}S + {}^{32}S$	91.0		9.86	6.52	3.34	Ri . 77
$^{32}S + {}^{40}Ca$	100		10.2	6.82	3.4	
$^{32}S + {}^{48}Ca$	83.3		10.3	7.07	3.2	
$^{40}Ca + {}^{40}Ca$	121		10.4	7.11	3.3	
	130		10.3		3.2	
$^{35}Cl + {}^{27}Al$	120 – 130	9.93	10.0	6.42	3.55	Sc 78
$^{35}Cl + {}^{58}Ni$	120 – 165	11.04	11.1	7.47	3.6	
$^{35}Cl + {}^{62}Ni$	100 – 170	11.18	11.2	7.57	3.6	
$^{35}Cl + {}^{120}Sn$	165	12.34	12.5	8.71	3.7	
$^{35}Cl + {}^{141}Pr$	160 – 165	12.58	12.8	9.03	3.7	
$^{20}Ne + {}^{238}U$	175	12.8		9.48	3.3	Vi 76b
	252	12.7			3.2	
$^{40}Ar + {}^{109}Ag$	169–236	12.0		8.71	3.3	Br 76
$^{40}Ar + {}^{121}Sb$	282–340	12.0		8.91	3.1	
$^{84}Kr + {}^{65}Cu$	494 – 604	12.1		8.96	3.1	
$^{40}Ar + {}^{209}Bi$	286	13.43	13.35	10.04	3.35	Bi 76
	340	13.21	13.30		3.2	
$^{40}Ar + {}^{238}U$	286–340	13.58	13.69	10.34	3.3	
$^{84}Kr + {}^{209}Bi$	600–712	14.20	14.27	11.18	3.05	
$^{84}Kr + {}^{208}Pb$	494	14.04	13.8–14.2	11.17	2.9	Va 76
	510	14.20			3.0	
	718	14.27			3.1	
	500	13.7			2.5	Co 72
$^{84}Kr + {}^{232}Th$	500	13.5		11.42	2.1	
$^{84}Kr + {}^{238}U$	450	14.3		11.48	2.8	Le 72

ᵃ From quarter-point or Fresnel model analysis (3.24)
ᵇ From optical model analysis
ᶜ Calculated with (3.12).

with suitably chosen universal functions. A good overall reproduction of these points was achieved with the following expression:

$$V_N(r) = 50 \frac{\bar{R}_1 \bar{R}_2}{\bar{R}_1 + \bar{R}_2} \exp\left(-\frac{r - \bar{R}_1 - \bar{R}_2}{0.63 \text{ fm}}\right) \text{ MeV}, \tag{3.34}$$

where the radii $\bar{R}_1$ and $\bar{R}_2$ are given by

$$\bar{R}_i = 1.233 \, A_i^{1/3} - 0.978 \, A_i^{-1/3} \text{ fm}, \tag{3.35}$$

and the functional form of the pre-exponential term is suggested by general geometrical arguments which are discussed in the context of fusion potentials in sect. 7.4. It should be noted that the potential described by (3.34) and (3.35) is meaningful only in the radial region covered by the underlying analysis, i.e. at distances in the vicinity of R_C and beyond. Further, in order to calculate scattering cross sections, a prescription for the imaginary part of the potential must be added.

A question of considerable interest is to what extent the elastic scattering data, or the phenomenological potentials deduced from the data, are consistent with theoretical nucleus–nucleus potentials. Such potentials have been derived from various models, like the single- or double-folding model [see sect. 2.2 and 6.4], the two-centre shell model, and the liquid-drop model. Extensive analyses of experimental data have been performed, in particular, with the folding models [Br 74a, Sa 74b, Ba 75, We 75, Sa 76, Wo 78]. In general the theoretical potentials were shown to be sufficiently flexible to produce agreement with experiment, provided reasonable parameter adjustments were made. The significance of such comparisons appears to be limited, however, by the restricted radial range probed in elastic scattering of strongly absorbed systems, and the lack of reliable information on the absorptive part of the interaction.

We now turn to a dicussion of total reaction cross sections. These are closely related to elastic scattering and can, in fact, be deduced from elastic scattering data in a number of ways. The latter include phase-shift analyses, optical model analyses, and the evaluation of the difference between the experimental scattering cross section and the Rutherford cross section according to (1.51). Direct measurements of total reaction cross sections can be performed in principle by observing the attenuation of a particle beam in a thin target [Wi 63] or by integrating over the yields of all possible reaction products. In practice both these approaches are technically difficult to realize, and therefore only few results are available.

In contrast to elastic scattering, total reaction cross sections for complex projectiles at energies above the interaction barrier can be described adequately in completely classical terms. This is a consequence of summing over all angles and exit channels, which effectively eliminates quantal interference effects. In a sharp cut-off model, where all trajectories with angular momenta up to L_{int} lead to absorption, we can write

$$\frac{d\sigma_R}{dL} = 2\pi \lambda^2 L \tag{3.36}$$

$$\sigma_R = 2\pi \lambda^2 \int_0^{L_{int}} L \; dL = \pi \lambda^2 \, L_{int}^2.$$

(3.37)

In practice there will be a transition region in angular momentum space near L_{int}, where the transmission coefficient varies smoothly from near-unity to practically zero, as indicated by the broken curve in Fig. 3.21. For sufficiently large values of L_{int}, however, this has little effect on the total reaction cross

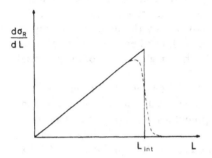

Fig. 3.21. Classical decomposition of the reaction cross section according to angular momentum L [see (3.36) and (3.37)]. Deviations from the sharp cut-off model due to nuclear diffuseness and deformation and quantum-mechanical barrier penetration are indicated schematically by the broken curve.

section. The limiting angular momentum (L_{int}) can be related to the corresponding distance of closest approach (r_{int}), if the interaction between the two fragments is described by a conservative two-body potential $V(r)$:

$$\frac{\hbar^2}{2\mu r_{int}^2} L_{int}^2 = E_{CM} - V(r_{int}).$$

(3.38)

Combining (3.37) and (3.38), we obtain the well-known classical formula for the total reaction cross section:

$$\sigma_R = \pi \, r_{int}^2 \left(1 - \frac{V(r_{int})}{E_{CM}}\right).$$

(3.39)

In the following we assume that the quantity r_{int} is independent of energy ($= R_{int}$) and is characteristic of the system considered. We emphasize, however, that this is not a necessary condition for the derivation of (3.39). It can be shown, for example, that an energy-dependent value of r_{int} should be used in (3.39) if the effective potential $V(r) + (\hbar^2/2\mu r^2)L_{int}^2$ has a barrier at distances beyond the point where nuclear reactions become significant (R_{int}). In this case the location of that barrier, which depends on L_{int} and hence on E_{CM}, must be identified with r_{int} in (3.39). Since we are concerned in this section mainly with moderately heavy systems, where the potential barriers occur close to or inside R_{int} (compare Fig. 3.10), it seems justified to use a constant value $r_{int} = R_{int}$ in (3.39). We note that the situation is different in fusion, where the barrier in general occurs beyond the relevant critical distance (R_{12}). This point is discussed in detail in sect. 7.4.

Equation (3.39) has several important applications. Firstly, it can be used to calculate σ_R if r_{int} and $V(r_{int})$ are known from an analysis of elastic scattering data. Secondly, and more importantly, reaction cross sections can be predicted on the basis of general empirical systematics. A convenient and reliable way to proceed is by writing

$$r_{int} = R_{int} = R_1 + R_2 + 3.2 \text{ fm}, \tag{3.40}$$

$$V(r_{int}) = B_{int} = \frac{Z_1 Z_2 e^2}{R_1 + R_2 + 3.2\text{fm}} - \frac{bR_1 R_2}{(R_1 + R_2)}, \tag{3.41}$$

where R_1 and R_2 are calculated from (3.12) and $b = 1.0$ MeV fm^{-1}. The last term on the right hand side of (3.41), which represents the nuclear contribution to B_{int}, comes from an analysis of fusion cross sections which is discussed in Sect. 7.4.

Finally (3.39) may be used to extract values of R_{int} and B_{int} from experimental data for σ_R as a function of energy. To show this more clearly we rewrite (3.39), with $r_{int} = R_{int}$, as

$$E_{CM}\sigma_R = \pi R_{int}^2 (E_{CM} - B_{int}) \tag{3.42}$$

Thus plotting $E_{CM}\sigma_R$ versus E_{CM} should result in a straight line, which intersects the abscissa at $E_{CM} = B_{int}$ and has a slope equal to πR_{int}^2.

Unfortunately only few measurements of reaction cross sections as a function of energy have been performed which can be analysed in terms of (3.42). The best example available to date is furnished by results of Viola and Sikkeland for various projectiles incident on ^{238}U [Vi 62]. These authors have measured the total cross section for production of fission fragments and identified it with the total reaction cross section, assuming that essentially all nuclear interactions in these heavy systems lead to fission as a consequence of the low fission barriers involved. Supporting evidence for this assumption comes from fission fragment angular correlation studies, which show that appreciable contributions to the fission yield arise from surface reactions between the fragments [Si 63]. Figure 3.22 gives the experimental excitation functions and in Fig. 3.23 they are redrawn as $E_{CM}\sigma_f$ versus E_{CM}. Clearly in the latter representation, the data are very well fitted by straight lines from about 2 to 5 MeV above the classical interaction barrier up to about twice that energy. This means that over the same energy interval, the reaction cross sections are accurately reproduced by the classical formula (3.42) with a constant effective interaction radius R_{int}. In the immediate vicinity of the barriers, deviations occur which can be explained as due to the combined effects of quantum-mechanical barrier penetration and static deformation of one or both fragments. The influence of deformation will be discussed in more detail below.

The interaction radii (R_{int}) and interaction barriers (B_{int}) from the analysis shown in Fig. 3.23 are collected in Table 3.6. Also included are parameters which have been derived by the same method from results of Gutbrod et al.

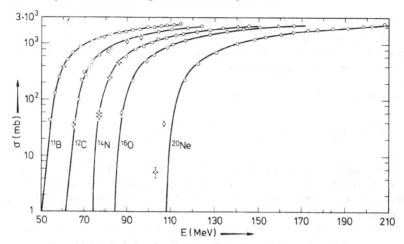

Fig. 3.22. Experimental total cross sections for fission of ^{238}U induced by heavy ions. The curves represent optical model calculations of the total reaction cross section. (From Vi 62)

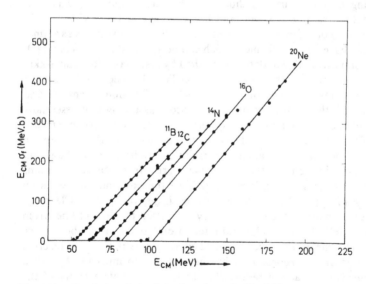

Fig. 3.23. Classical analysis of the data shown in Fig. 3.22

on fusion cross sections of ^{32}S projectiles with different targets [Gu 73 a]. In these cases the total reaction cross section is approximately equal to the fusion cross section (determined by observation of heavy recoil nuclei), as has been ascertained by the authors by comparison with elastic scattering data [Gu 73 b].

Table 3.6. Parameters deduced from total reaction cross sections

Pro-jectile	Target	E_{CM} [MeV]	R_{int}[a] [fm]	R_{12}[b] [fm]	R_{int}-R_{12} [fm]	r_{int}[c] [fm]	$V(R_{int})$ [MeV]	$V_C(R_{int})$ [MeV]	$V_N(R_{int})$ [MeV]	References[d]
^{11}B	^{238}U	52–110	11.68	8.86	2.82	1.39	52.5	56.7	−4.2	Vi 62[e]
^{12}C		63–118	11.68	8.93	2.75	1.38	63.0	68.0	−5.0	
^{14}N		74–137	12.04	9.09	2.95	1.40	73.6	77.0	−3.4	
^{16}O		83–156	12.31	9.23	3.08	1.41	82.8	86.1	−3.3	
^{20}Ne		103–192	12.51	9.47	3.04	1.41	102.7	106.0	−3.3	
^{32}S	^{24}Mg	30–39	8.76	6.16	2.60	1.44	28.3	31.5	−3.2	Gu 73 a[f]
	^{27}Al	31–41	8.80	6.31	2.49	1.42	29.9	34.0	−4.1	
	^{40}Ca	44–54	9.35	6.82	2.53	1.425	43.6	49.3	−5.7	

[a] From (3.42).
[b] $R_{12} = R_1 + R_2$, where R_1, R_2 are calculated from (3.12).
[c] $r_{int} = R_{int} (A_1^{1/3} + A_2^{1/3})^{-1}$
[d] Sources of experimental data.
[e] The total reaction cross section was assumed equal to the measured fission cross section.
[f] The total reaction cross section was assumed equal to the measured fusion cross section.

The results given in Table 3.6 confirm our earlier conclusions from elastic scattering data, since differences $R_{int} - R_{12}$ of about 3 fm are consistently observed. The fact that these differences tend to be somewhat smaller than the values of $R_C - R_{12}$ deduced from elastic scattering may be due partly to real differences between R_C and R_{int} [see (3.28)]; in addition, small fractions of the total reaction cross section may have been missed in the measurements, and such losses would also tend to lower the deduced values of R_{int}.

It should be noted that our classical analysis of reaction cross sections so far disregards nuclear deformation. Possible consequences of a dynamic distortion of the fragments have been treated in Sect. 3.2 and are expected to be negligible. The strong static deformation of nuclei like ^{20}Ne, ^{24}Mg, and ^{238}U, on the other hand, cannot be ignored. Nevertheless the results given in Tables 3.4 and 3.5 show that systematic differences between deformed and undeformed nuclei with respect to reaction cross sections are not observed at energies significantly above the classical barriers. This result may be understood qualitatively as follows: if one of the fragments is deformed, then the interaction potential V depends not only on the centre distance r but also on the angle ϑ between the symmetry axis of the deformed fragment and the line connecting the fragment centres. In an approximate manner, i.e. disregarding changes of ϑ during the relevant part of the collision, the (classical) reaction cross section can be calculated for fixed ϑ from (3.42) [using the appropriate $V(r, \vartheta)$] and then averaged over all orientations ϑ. The variations of the Coulomb and nuclear potential with ϑ are of different sign and therefore the resulting fractional variation of V with ϑ is less than would be expected for each component separately. The resulting averaged potential (and hence the averaged reaction cross section) will practically coincide with that for spherical fragments except for energies close to the barrier. The latter will be shifted downward and will now correspond to the most favourable orientation of the deformed fragment, i.e. that angle ϑ which yields the minimum value of $V(R_{\text{int}}, \vartheta)$. In practice this effect is modified by quantum-mechanical barrier penetration and therefore is difficult to extract quantitatively; moreover, a classical analysis would of course not be meaningful at near-barrier energies.

A number of authors have discussed heavy-ion reaction cross sections in terms of different models [Th 59, Th 68, Ri 70a, Wo 73, Va 74]. Both classical and quantum-mechanical calculations (the latter based on the optical model) have been presented and compared with the data, and possible effects of deformation and barrier penetration have been estimated. In our present discussion we have attempted to incorporate the results of these studies and their physical implications; for further details the reader is referred to the references given above. We have emphasized here the classical approach which successfully accounts for the bulk of the existing evidence on total reaction cross sections. There are situations, however, where such a simple picture must be modified, as for example in the barrier region. Also at high energies, complications may arise in the scattering of deformed nuclei due to strong Coulomb excitation of low-lying collective states. In such cases the concept of a well-defined interaction distance breaks down as a consequence of the long range of the Coulomb interaction. An example is provided by recent work of Thorn et al. who succeeded in resolving elastic and inelastic scattering of ^{16}O from ^{184}W at $E_{\text{Lab}} = 90$ MeV (Th 77). It was demonstrated that strong coupling to low-lying inelastic channels causes an appreciable reduction in elastic intensity at angles well below the grazing angle, corresponding to distances much larger than the classical interaction distance. The results of this and other studies suggest, on the other hand, that the simplicity of the classical description may be largely restored by including the low-lying collective states in the "elastic" channel.

To summarize the results of this section, we note that elastic scattering and reaction cross sections for systems where at least one of the fragments is heavy appear to follow simple systematics and do not exhibit the pronounced fluctuations which are characteristic of lighter systems. Two dominant effects can be recognized in the data: the long-range Coulomb interaction, and the existence of a fairly well-defined interaction distance R_{int} where the system is absorbed from the incident channel. The magnitude of the latter can be deduced quite accurately from the data; it turns out to correspond consistently to a distance of about 3 fm between the half-density surfaces. At this point the nuclear part of the nucleus–nucleus potential is typically of the order of 1 MeV and hence much smaller than the Coulomb potential.

3.4   Inelastic Scattering Near and Above the Coulomb Barrier

In this section we return once more to inelastic scattering. The term inelastic here implies excitation of discrete low-lying states of either target nucleus or projectile. Processes involving massive transfers of energy and (or) mass will be discussed in Chap. 6. Whereas in Sect. 3.1 and 3.2 we were concerned with situations where inelastic scattering is induced exclusively or predominantly by the Coulomb interaction, we now consider higher bombarding energies where nuclear forces play a significant part.

From the theoretical discussions of Coulomb distortion and break-up presented in Sect. 3.2 we expect that nuclear excitation will interfere strongly and destructively with Coulomb excitation at energies near the barrier. This is indeed born out by the experimental evidence on inelastic scattering, as will be shown below. The interplay between Coulomb and nuclear effects is a specific feature of heavy-ion scattering, which—in contrast to the scattering of light projectiles—persists to high bombarding energies well above the Coulomb barrier.

In quantum mechanics, different mechanisms leading to identical final states contribute coherently to the cross section. In the case of inelastic scattering, this means that the Coulomb and nuclear amplitudes have to be added; since the former is well known, the cross section may be expected to serve as a sensitive probe of details of the nuclear interaction, especially under conditions of destructive interference. Another interesting consequence of the coherent nature of Coulomb and nuclear excitation is that the total (nuclear) reaction cross section can no longer be separated from the Coulomb excitation cross section in a rigorous quantum-mechanical description. At energies well above the barrier, where classical or semiclassical methods become applicable, the total (nuclear) reaction cross section nevertheless remains a useful concept.

Before entering a quantitative discussion of nuclear inelastic scattering, we have to ask what states are preferentially excited and what are the relevant properties of the nuclear interaction. Experimentally one finds a pronounced selectivity for excitation of collective states of either rotational or vibrational character, just as in Coulomb excitation. This feature appears to be a general property of the scattering of "strongly absorbed" particles, and has been extensively exploited in inelastic scattering studies with light complex projectiles, such as deuterons and α particles. It is qualitatively clear that the excitation in all these cases must be predominantly a surface phenomenon, since strong overlap of the fragments would lead to complicated reactions rather than re-emergence of one fragment in a low-lying collective state. In contrast, Coulomb excitation represents a coherent excitation of nucleons throughout the nuclear volume by virtue of the long range of the Coulomb force, and may therefore be considered a volume effect (the same holds, to a lesser extent, for nuclear excitation of collective states by high energy nucleons as pointed out by Pinkston and Satchler [Pi 61]). These considerations imply that different regions of the nucleus are probed in Coulomb and nuclear excitation, and that the ratio of the corresponding strengths should in principle depend on details of nuclear structure. Nevertheless in practice both mechanisms are observed to have very similar selectivity, presumably due to the fact that the strongly excited states are of rather uniform character and well described as surface vibrations or rotations.

The nuclear part of the interaction responsible for excitations of the nuclear surface has been customarily derived by a simple generalization of the optical model potential $U(r)$ which is used to analyse elastic scattering. The excited states are described in the collective model by writing for the radius of the nucleus in question ($i = 1,2$):

$$R_i(\vartheta_i, \varphi_i) = R_{0i} [1 + \sum_{\lambda\mu} \beta_{\lambda\mu} Y_{\lambda\mu}(\vartheta_i, \varphi_i)], \tag{3.43}$$

where the angles ϑ_i, φ_i refer to a body-fixed coordinate system. For simplicity we assume in the following that only one term $\lambda\mu$ is significant and drop the indices $\lambda\mu$. Further we consider vibrational excitation of a target nucleus ($i = 2$) with spherical equilibrium shape, simply to deal with a definite situation. More general cases can be treated in an analogous fashion. The potential for undeformed fragments U_0 (corresponding to $\beta = 0$) is taken to depend only on the difference $r - R_U$, where r is the centre separation and R_U the sum of the potential radii of the undeformed fragments:

$$U_0 \equiv U_0(r - R_U) \text{ with } R_U = R_{U1} + R_{U2}. \tag{3.44}$$

Here we have used the subscript U to indicate potential radii as distinct from matter or charge radii. The effect of target deformation on the potential is obtained by substituting in (3.44) for R_{U2} an expression analogous to the right-hand side of (3.43). To first order in β_{U2}, this yields

$$U_\beta(r, \vartheta_2, \varphi_2) = U_0(r - R_U) - R_{U2}\beta_{U2}\frac{dU_0(r - R_U)}{dr} Y(\vartheta_2\varphi_2) \tag{3.45}$$

Thus we have decomposed the interaction into a term independent of deformation, which causes elastic scattering, and a deformation-dependent term, which causes inelastic scattering. The latter is seen to be proportional to the derivative of the central potential and to the product $R_{U2}\beta_{U2}$. It is important to realize that neither of the quantities R_{U2} and β_{U2} has an unambiguous physical significance, and that only their product enters the interaction. The angles ϑ_2 and φ_2 now specify the orientation of fragment 2 with respect to the line connecting the fragment centres.

It should be noted that (3.45) is based not only on the collective model for the excited state, but also on the assumption that the interaction is completely determined by the local separation of the fragment surfaces along the centre-to--centre line. However, in view of the finite range of the nuclear interaction and the diffuse nature of the nuclear surface, the possibility cannot be excluded— and is indeed indicated by more elaborate calculations of the nucleus–nucleus potential—that the interaction depends on additional parameters such as the local curvature of the surfaces. Nevertheless the dominant part of the interaction seems to be correctly described by the simple arguments outlined above. This will become clear from quantitative analyses of experimental data to be discussed below.

So far we have not specified the real or imaginary nature of the potential

U. In heavy-ion scattering the nuclear potential must be complex, and both the real and the imaginary part of the potential will be represented by equations of type (3.44) and (3.45) with, in general, different radius and deformation parameters:

$$U_0(r - R_U) \rightarrow -Vf(r - R_V) - iW g(r - R_W). \tag{3.46}$$

Further we have to add the (real) Coulomb potential V_C. Confining our attention to that part of the total potential which depends on target deformation, we obtain, again in first order,

$$U_\beta - U_0 = \left[\beta_{C2}\frac{\partial V_C}{\partial \beta_{C2}} + (R_{V2}\,\beta_{V2})V\frac{df}{dr} + i(R_{W2}\,\beta_{W2})W\frac{dg}{dr}\right]Y(\vartheta_2\varphi_2) \tag{3.47}$$

with

$$\frac{\partial V_C}{\partial \beta_{C2}} = \frac{3R_{C2}^\lambda}{2\lambda+1}\frac{Z_1Z_2e^2}{r^{\lambda+1}} \qquad (r \geq R_{C1} + R_{C2}), \tag{3.48}$$

for a homogeneous charge distribution of radius R_{C2} with deformation β_{C2} of multipole order λ. It is easy to see that the first two terms on the right-hand side of (3.47) are of opposite sign and will therefore interfere destructively in the cross section as discussed earlier. The imaginary part of the potential will in general also contribute to the cross section; the relative effects of the imaginary and real nuclear potentials in inelastic scattering have been discussed by Satchler [Sa 70a, Sa 71].

The theoretical analysis of inelastic scattering data has been performed by a variety of different methods. In the following we review briefly the main lines of approach: the distorted wave Born approximation (DWBA), the "adiabatic theory" of Blair and Austern, and the semiclassical and the quantum mechanical version of the coupled channels method (CC).

In the DWBA, the differential cross section and transition amplitude can be written as follows [see, for example, Ro 60, Ba 62]:

$$\frac{d\sigma}{d\Omega} = \left(\frac{\mu}{2\pi\hbar^2}\right)^2 \left(\frac{k_f}{k_i}\right) \sum |T_{fi}|^2, \tag{3.49}$$

$$T_{fi} = \int d^3r \, \chi_f^{(-)*}(k_f, r) \, \langle \psi_f^* | V_{if} | \psi_i \rangle \, \chi_i^{(+)}(k_i, r). \tag{3.50}$$

The sum on the right hand side of (3.49) is meant to imply summation over final state quantum numbers and averaging over initial state quantum numbers. In (3.50), $\chi_i^{(+)}$ and $\chi_f^{(-)}$ are the "distorted waves", describing the relative motion of the fragments in the initial (i) and final (f) channels under the influence of the deformation-independent part of the two-body interaction. These functions are obtained by numerical solution of the Schrödinger equation with appropriate optical model potentials. The matrix element under the integral sign connects the intrinsic wave functions in the initial and final states via the interaction V_{if} which induces the transition [see (3.47)]. The radial part of this matrix element is usually called the "form factor" $F_\lambda(r)$. In the present case it is essentially equal to the right-hand side of (3.47), where the deformation variables $R\beta$ are replaced by appropriate matrix elements $\langle R\beta \rangle$ taken between the states involved in the transition. For a collective transition of multipole order λ, the form factor is thus given by [Ba 62]

$$F_\lambda(r) = \frac{4\pi[B(E\lambda)]^{1/2}}{2\lambda + 1} \frac{Z_1 e}{r^{\lambda+1}} + \langle R_{V2}\beta_{V2}\rangle V \frac{df}{dr} + i\langle R_{W2}\beta_{W2}\rangle W \frac{dg}{dr}, \quad (3.51)$$

where the reduced electromagnetic transition probability $B(E\lambda)$ can also be written in terms of the collective model as

$$[B(E\lambda)]^{1/2} = \frac{3}{4\pi} Z_2 e \langle R_{C2}^\lambda \beta_{C2}\rangle. \quad (3.52)$$

We emphasize that the strength of the nuclear excitation is determined by the matrix elements $\langle R\beta \rangle$, which are usually referred to as "deformation lengths". We further note that (3.51) and (3.52) are applicable both to vibrational excitation of spherical nuclei and to rotational excitation of deformed nuclei, if in each case the quantities $\langle R\beta \rangle$ are interpreted as root-mean-square displacements of the associated nuclear surfaces.

The "adiabatic method" in the theory of inelastic scattering of strongly absorbed particles is in some respects related to the DWBA. It has, in fact, been shown to be formally identical with the latter in the limit of small deformation [Ro 60]. In contrast to the DWBA, however, the method is not restricted to one-step excitation in general. The term "adiabatic" here implies a small ratio of excitation energy to incident energy, and hence a small relative energy transfer per collision; nevertheless, the transition probability may be appreciable and multiple excitations are readily included (we note that in Coulomb excitation theory the same situation would be characterized by a small "adiabaticity parameter" ξ).

The adiabatic calculation is performed in two steps. In the first step an elastic scattering amplitude is calculated for a fixed value of the deformation coordinate β, e.g. for a fixed deformation of the nuclear surface. Subsequently, in the second step, the matrix element of this scattering amplitude between the initial and final intrinsic states is evaluated by integration over β. A detailed discussion of the method has been given by Austern and Blair [Au 65].

In early work, using the adiabatic method in the "Fraunhofer approximation", Blair considered inelastic scattering of high-energy α particles [Bl 59]. He derived the famous "Blair phase rule" which states that the maxima and minima of inelastic angular distributions are in phase with the corresponding structures of the elastic angular distribution for excitations of odd parity, and out of phase for excitations of even parity. This rule has met with considerable success in situations where the underlying approximations are applicable.

More recently Austern and Blair have shown that—provided the scattering is well localized in angular momentum space—the radial integrals of the theory may be replaced by derivatives of the scattering coefficients S_l with respect to the nuclear radius [Au 65]. Thus a direct link is established between elastic and inelastic scattering, and the optical model potential no longer appears explicitly in the formulae describing inelastic scattering. Using a smooth cut-off parametrization for S_l [Fr 63], Potgieter and Frahn derived simple closed expressions for nuclear inelastic scattering cross sections, and applied them successfully in analyses of α-particle and heavy-ion scattering data [Po 66a, Po 66b, Po 67]. In later work, Frahn developed a closed formalism which explicitly includes Coulomb excitation and has been used to analyse Coulomb–nuclear interference effects in inelastic scattering [Fr 76, Fr 78a].

Next we consider semiclassical methods in the theory of inelastic scattering which are based on the concept of classical trajectories. Many of the basic ideas were worked out and applied to elastic scattering by Ford and Wheeler in 1959 [Fo 59]; summaries of more recent developments, as applied to inelastic scattering, have been given by Broglia, Winther, and collaborators [Br 72, 74b], and by Glendenning [Gl 74]. In the semiclassical coupled-channels method, the elastic and inelastic amplitudes are obtained by solving a system of coupled equations of motion. The method is closely analogous to that employed in the theory of Coulomb excitation, although the inclusion of the nuclear interaction introduces qualitatively new phenomena to be discussed below.

The semiclassical treatment usually starts by introducing the classical deflection function $\Theta(L)$ which describes, in the classical limit, the relationship between scattering angle and orbital angular momentum (or impact parameter $b = \lambda L$). To be consistent with the literature, and in contrast to other parts of this book, we use here the symbol Θ for the deflection function and the symbol ϑ_{CM} for the physical scattering angle in the centre-of-mass system. Note that ϑ_{CM} is positive by definition, whereas Θ may assume either positive or negative values, depending on whether the classical trajectory crosses the z axis or not.

Adding the attractive nuclear field to the repulsive Coulomb field leads to the consequence that different classical trajectories, corresponding to different angular momenta L, may be associated with the same scattering angle. This is demonstrated qualitatively in Fig. 3.24 for trajectories 1 and 3. The former passes the scattering centre at a relatively large distance and is essentially determined by the Coulomb interaction, whereas the latter approaches to a significantly smaller distance where the additional Coulomb deflection is compensated by nuclear deflection. At some intermediate value of the angular momentum the scattering angle passes through a maximum (trajectory 2). If the angular momentum is decreased further, a situation may arise where the nuclear force just balances the Coulomb and centrifugal forces and the two fragments start to rotate around each other as indicated by trajectory 4; this phenomenon is called "orbiting". At this point, however, the nuclear interaction will usually be strong enough to remove the system from the elastic channel within a time small compared to the rotational period.

The deflection function which corresponds to Fig. 3.24 is shown schematically in Fig. 3.25. The angular momenta associated with trajectories 1 to 4 are denoted by L_1 to L_4. By definition we have $\Theta(L_1) = \Theta(L_3)$; further for $L = L_2$, the function $\Theta(L)$ goes through a maximum. By analogy with the

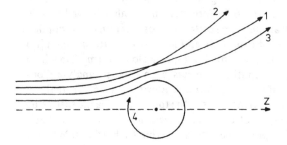

Fig. 3.24. Influence of the nuclear interaction on classical trajectories in heavy-ion elastic scattering (for explanation see text)

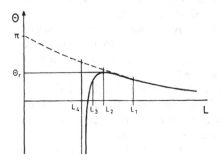

Fig. 3.25. Classical deflection function $\Theta(L)$ for heavy-ion elastic scattering. The broken curve represents deflection by the Coulomb field only, and the full curve deflection by the combined Coulomb and nuclear field. The angular momenta L corresponding to the trajectories shown in Fig. 3.24 are indicated

scattering of light by rain drops this value of Θ is called the "rainbow angle" (Θ_r). Whenever $\Theta(L)$ has an extremum, e.g. at rainbow angles, the classical scattering cross section $d\sigma_{CL}/d\omega_{CM}$ goes to infinity as can be seen from the relationship

$$\frac{d\sigma_{Cl}}{d\omega_{CM}} = 2\pi\hbar^2 L\left(\frac{d\omega_{CM}}{dL}\right)^{-1} = \hbar^2 L\left(\sin\Theta\,\frac{d\Theta}{dL}\right)^{-1}. \tag{3.53}$$

The classical deflection function $\Theta(L)$ is closely related to the quantum-mechanical scattering phase shift as calculated in the JWKB approximation. The latter approximation [see, for example, Be 72c] is applicable whenever the local wave number

$$K_L(r) = \left\{\frac{2\mu}{\hbar^2}[E_{CM} - U(r)] - \frac{L^2}{r^2}\right\}^{1/2} \tag{3.54}$$

varies slowly with distance, i.e. if the following inequality is satisfied:

$$\frac{d}{dr} K_L(r) \ll K_L^2(r).$$

(3.55)

Denoting by $\alpha_L = (\delta_L + \sigma_L)_{\text{JWKB}}$ the JWKB phase shift, which describes the combined effects of Coulomb and nuclear scattering, we have

$$\alpha_L = \frac{\pi}{2} L - kr_L + \int_{r_L}^{\infty} \{K_L(r) - k\} \, dr,$$

(3.56)

where $L = l + 1/2$ and the quantity r_L is defined by

$$K_L(r_L) = 0.$$

(3.57)

For real potentials $U(r)$, the quantity r_L is identical with the classical turning point for angular momentum L. In this case the classical deflection function is obtained by interpreting L as a continuous variable and writing

$$\Theta(L) = 2 \frac{d\alpha_L}{dL}.$$

(3.58)

In the more general case of a complex potential $U(r)$ [or in clasically forbidden regions, where $K_L^2(r) < 0$], one sees from (3.54), (3.56), and (3.57) that K_L, α_L, and r_L become complex quantities. The deflection function is then usually defined as the real part of (3.58). The concepts of complex turning points and complex trajectories have been applied successfully in generalized semiclassical scattering theories, in order to treat situations where diffraction effects caused by strong absorption are important [Kn 74, Ma 74, Ko 75, Kn 76, La 76].

We return now to the specific case of inelastic excitation by combined Coulomb and nuclear forces. We note that, from a semiclassical point of view, the phenomenon known as "Coulomb–nuclear interference" is of a twofold nature. Firstly, for a given classical trajectory the two interactions contribute destructively to the inelastic amplitude and may, if of comparable strength, practically cancel each other. Secondly, we have interference between different trajectories arriving at the same scattering angle, which again may lead to cancellation if one of the trajectories is dominated by Coulomb interaction and the other by nuclear interaction. Malfliet et al. have emphasized the latter mechanism [Ma 73] and deduced that inelastic angular distributions should be out of phase compared to the corresponding elastic angular distribution with respect to the position of maxima and minima, in agreement with observation. It is interesting to note that this "phase rule" is expected to hold irrespective of the parity of the excitation in contrast to the Blair phase rule, which predicts opposite phases for excitations of different parity (under conditions of purely nuclear excitation in the adiabatic Fraunhofer approximation). Another important result of both semiclassical and quantal calculations is that the Coulomb–nuclear interference pattern as a function of angle or energy may be significantly shifted due to the reorientation effect (see Sect. 3.1) and may thus be utilized to determine static quadrupole moments of excited states. Examples where this effect has been exploited in experimental work are discussed below.

Semiclassical analyses of inelastic scattering have been performed by a number of authors with remarkable qualitative success. It must be realized, however, that the semiclassical method relies upon approximations which cannot be expected to yield quantitatively reliable results under conditions of strong nuclear scattering and absorption. Moreover, the strong absorption models show that quantum-mechanical diffraction plays an essential role in producing certain features, like the rise of the elastic cross section above the Rutherford value just below the grazing angle. Therefore attempts to explain such features semiclassically as due to interfering trajectories, or classically as due to "rainbow scattering", may give a misleading or incomplete picture of their physical origin. This point has been emphasized, in particular, by Frahn [Fr 76, Fr 78b]. The problem may be resolved by performing refined semiclassical calculations with complex trajectories (see above), but only at the expense of losing much of the simplicity and transparency of the semiclassical approach.

In recent years, the quantum-mechanical coupled-channels treatment has emerged as the most powerful method of analysis for inelastic scattering data. Here one avoids the limitations of the DWBA (being restricted to one-step excitation) and of the semiclassical treatment (being based on the trajectory concept). Large computer codes have been developed for performing the necessary numerical calculations, and impressive fits to experimental data have been achieved (see Fig. 3.31).

In Fig. 3.26 and 3.27 we show experimental data from the work of Christensen et al. at Copenhagen [Ch 73]. These authors studied elastic and inelastic scattering of 35 to 60 MeV oxygen ions from ^{58}Ni, ^{88}Sr, and ^{142}Nd. Figure 3.26 shows angular distributions and Fig. 3.27 excitation functions for the ^{58}Ni target. In both cases the inelastic excitation of the first 2^+ state exhibits a marked interference pattern, with a pronounced dip which appears to be correlated with a maximum in elastic scattering. As mentioned above, an explanation for the latter feature in terms of interfering trajectories has been suggested by Malfliet et al. [Ma 73]. The curves in Figs. 3.26 and 3.27 are optical model and DWBA fits which are seen to reproduce the data quite well. Optical model potentials with 6 free parameters, i.e. with independent geometrical parameters for the real and imaginary parts of the potential, were used. Further the deformation parameters $\langle \beta R \rangle$ for the Coulomb and nuclear excitation were independently adjusted. With 4 parameter optical model potentials, where the real and imaginary geometries were assumed equal, fits of somewhat inferior quality were obtained. From the six parameter fits a preference for imaginary potentials of smaller diffuseness (about 0.2 to 0.3 fm) than that of the real potential (0.5 to 0.6 fm) can be deduced. The same conclusion holds for the other nuclei investigated.

Elastic and inelastic scattering from the doubly magic nucleus ^{208}Pb has

Fig. 3.26. Experimental angular distributions for elastic and inelastic scattering of 60 MeV ^{16}O ions from ^{58}Ni. The curves are optical model and DWBA calculations. (From Ch 73)

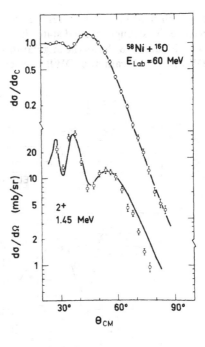

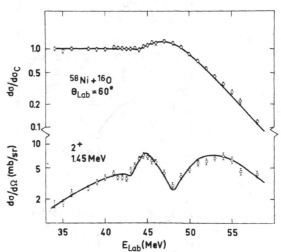

Fig. 3.27. Excitation functions for elastic and inelastic scattering of ^{16}O ions from ^{58}Ni at a laboratory angle of 60°. The curves are optical model and DWBA calculations. (From Ch 73)

been studied by Becchetti et al. at Berkeley with 104 MeV ^{16}O ions [Be 72a] and by Ford et al. at Oak Ridge with 72 MeV ^{11}B ions [Fo 73]. Both groups have used heavy-ion beams from isochronous cyclotrons and magnetic spectrograph detection techniques for the scattered particles, and were able to resolve the strongly excited states in ^{208}Pb up to about 4 MeV excitation energy. Figure 3.28 shows the angular distributions of Ford et al. together with DWBA fits. Again the DWBA calculations (based on a four-parameter

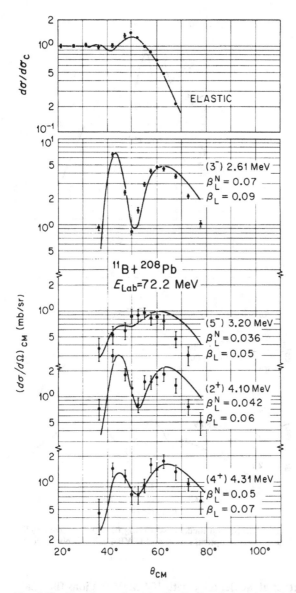

Fig. 3.28. Experimental angular distributions for elastic and inelastic scattering of 72 MeV ^{11}B ions from ^{208}Pb to different final states as indicated. The curves are optical model and DWBA calculations. (From Fo 73)

optical model potential) reproduce well the measured angular distributions.

First discrepancies between DWBA calculations and inelastic scattering data were noted in later work, which included detailed comparisons between projectile excitation and target excitation in the systems ^{22}Ne + ^{88}Sr [Gr 74], ^{18}O + ^{58,64}Ni [Re 75, Vi 76a], ^{18}O + ^{92}Mo, and ^{18}O + ^{120}Sn[Re 75]. Projectile excitation was generally found to produce more pronounced Coulomb--nuclear interference patterns, in qualitative accord with the DWBA calculations. The latter failed, however, to reproduce the angular position of the interference dip, which was shifted towards forward angles in projectile excitation compared to target excitation. A comparison with coupled channels calculations led to the conclusion that this effect results from quadrupole reorientation in the final 2^+ states [Vi 76a].

In Fig. 3.29–31 we present results of a systematic study by Hillis et al. [Hi 77] of inelastic scattering of 70 MeV ^{12}C ions from different Nd isotopes, ranging in shape from nearly spherical (^{142}Nd, $\beta_2 \approx 0.1$) to strongly deformed (^{150}Nd, $\beta_2 \approx 0.3$). Figures 3.29 and 3.30 show experimental data for excitation of a 2^+ state and a 3^- state, respectively, together with DWBA calculations. Theoretical cross sections representing Coulomb excitation only, nuclear excitation only, and their coherent superposition are given separately. Clearly the cross section is dominated over most of the angular range by Coulomb excitation in the case of Fig. 3.29, where a low-lying rotational state is excited via a strong E2 transition ($\lambda = 2$). In contrast the excitation of an octopole vibrational state ($\lambda = 3$) shown in Fig. 3.30 is characterized by Coulomb and nuclear amplitudes of similar magnitude, giving rise to a much more pronounced interference pattern. It is interesting to note that in both cases the oscillations of the resulting inelastic cross section are out of phase with those

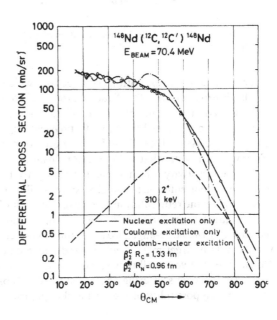

Fig. 3.29. Angular distribution for inelastic scattering of 70 MeV ^{12}C ions exciting a low-lying 2^+ state in ^{148}Nd. The curves represent DWBA calculations as explained in the figure. (From Hi 77).

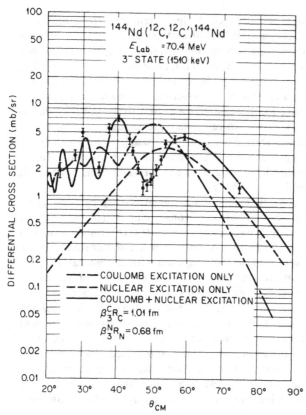

Fig. 3.30. Same as Fig. 3.29, but for excitation of a 3⁻ state at 1.5 MeV in ^{144}Nd. (From Hi 77)

of the pure Coulomb excitation cross section, which in turn are in phase with the elastic cross section. This "180° phase rule" connecting elastic and inelastic cross sections is a general feature of heavy-ion-induced inelastic scattering in the regime of Fresnel diffraction. Its physical origin has been discussed both from a semiclassical [Ma 73] and from a quantum-mechanical [Fr 78a] point of view, with somewhat different conclusions.

The sensitivity of inelastic scattering to higher-order effects—here quadrupole reorientation in a final 3⁻ state—is illustrated in Fig. 3.31. As mentioned above, the reorientation effect causes a shift of the principal Coulomb–nuclear interference dip towards smaller angles and, in addition, a significant reduction of the large-angle cross section. These effects and their systematic variation with target mass number—and hence deformation—for a series of Nd isotopes are very well reproduced by the coupled channels fits shown in Fig. 3.31. Inelastic scattering above the Coulomb barrier therefore provides an interesting possibility of measuring the quadrupole moment of an excited

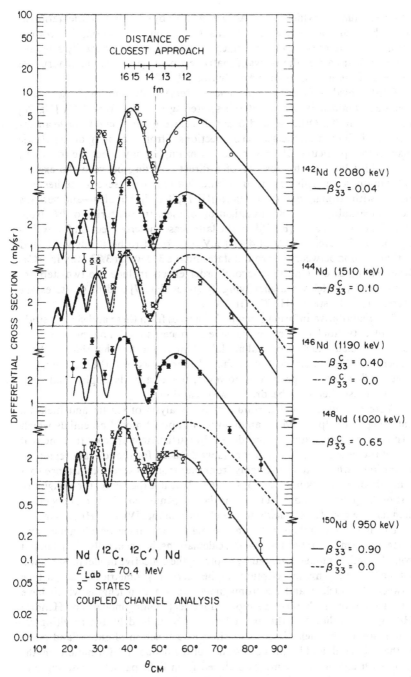

Fig. 3.31. Influence of quadrupole reorientation in final 3⁻ states on inelastic scattering angular distributions. The full curves are coupled channels fits including reorientation, and the broken curves result when the reorientation effect is neglected. (From Hi 77)

3^- state, a quantity which is not easily accessible by Coulomb reorientation at "safe" bombarding energies. In contrast to Coulomb excitation, the reorientation effect at these higher energies was found to be primarily sensitive to the nuclear part of the relevant matrix element $\langle \beta R \rangle$, i.e. the quadrupole moment of the nuclear matter distribution [Hi 77].

Further evidence on the importance of reorientation in projectile excitation was obtained in studies of the scattering of ^{20}Ne from ^{40}Ca [Ng 78] and ^{208}Pb [Gr 78]. Other second-order effects were also found to have an appreciable influence on inelastic cross sections. Strong coupling of the elastic channel to a particular 2^+ or 3^- state, for example, may affect the transition into a third level in the absence of a direct coupling between the excited states in question. This conclusion results not only from several studies performed with strongly deformed nuclei [Hi 77, Th 77], but also—and perhaps more remarkably—from an investigation of inelastic scattering of ^{16}O from ^{40}Ca due to Rehm et al. [Re 78]. In the latter case the cross section for excitation of the 5^- level in ^{40}Ca at 4.49 MeV was found to be strongly affected by the coupling between the ground state and the 3^- level at 3.73 MeV in ^{40}Ca. Other examples of important second-order effects include the two-step excitation of 2^+_1 states via higher lying 2^+_2 states [Hi 77] and of 4^+ states via intermediate 2^+ states [Hi 77, Gr 78].

The quantitative information which emerges from these and similar studies is of a twofold nature: on one side it concerns the "monopole interaction", or nucleus–nucleus optical model potential; on the other side it reflects properties of the inelastic "form factor" and especially the deformation lengths $\langle \beta R \rangle$. Both aspects are, however, not clearly decoupled and conclusions must therefore be drawn with caution.

There is evidence that a simultaneous analysis of elastic and inelastic scattering can help to reduce ambiguities in optical model potentials which exist if only elastic data are considered. In particular it appears that the real part of the potential can be fairly reliably deduced for the relevant region of separations, which extends roughly from the strong absorption distance to a point about 3 fm beyond [Ch 73, Co 76a]. Not unexpectedly, the imaginary part of the potential required to fit the data depends sensitively on how many channels are coupled explicitly to the elastic channel [Vi 76a, Hi 77, Ng 78].

The nuclear coupling between the elastic and inelastic channels, and hence the form factor used in DWBA calculations, has generally been derived from the optical model potential by applying the collective model prescription outlined above. This prescription implies certain geometrical relationships between the nuclear and electromagnetic deformation lengths which have been derived on the basis of a simplified model ("rolling model") by Hendrie [He 73]. The validity of this procedure can be judged by the resulting reproduction of the inelastic cross sections as a function of energy and angle on the one hand, and by the consistency of the deduced deformation lengths $\langle R \beta \rangle$ with analogous quantities deduced from light particle scattering and Coulomb excitation on the other hand. On both counts we can conclude that the simple collective model yields a correct first-order description of nuclear inelastic scattering, at least for strong transitions. Improved calculations of the coupling potentials, using double-folding techniques and microscopic or semimicroscopic nuclear density distributions, have also been performed with similar success [Sa 76].

We conclude that the mechanism of heavy-ion inelastic scattering is basically well understood, and provides a promising tool for the study of static and dynamic multipole moments of collective nuclear states. Problems of quantitative precision still exist, however, which are due to interference from higher-order effects of unknown strength, and to the approximations made in calculating the coupling potentials.

References Chapter 3
(For AR 76, AS 63, HE 69, MU 73, NA 74 see Appendix D)

Al 56   Alder, K. et al.: Rev. Mod. Phys. 28, 432 (1956)
         [Erratum: Rev. Mod. Phys. 30, 353 (1958)]
Al 60   Alder, K.; Winther, A.: Mat. Fys. Medd. Dan. Vid. Selsk. 32, no. 8 (1960)
Al 64   Alster, J.; Conzett, H.E.: Phys. Rev. 136, B 1023 (1964)
Al 66   Alder, K.; Winter, A. (eds.): Coulomb Exitation. New York: Academic Press
         1966
Al 69   Alder, K.; Pauli, H.: Nucl. Phys. 128, 193 (1969)
Al 72   Alder, K.; Roesel, F.; Morf, R.: Nucl. Phys. A186, 449 (1972)
Al 74   Alder, K.; Roesel, F.; Saladin, J.X.: NA 74 Vol. 1 p. 28
Al 75   Alder, K.; Winter, A.: Electromagnetic Excitation: Theory of Coulomb Excitation
         with Heavy Ions. Amsterdam: North-Holland American Elsevier 1975
Au 65   Austern, N.; Blair, J.S.: Ann. Phys. (N.Y.) 33, 15 (1965)
Ba 62   Bassel, R.H.: Phys. Rev. 128, 2693 (1962)
Ba 67   Baker, S.D.; McIntyre, J.A.: Phys. Rev. 161, 1200 (1967)
Ba 74   Barnett, A.R.; Feng, D.H.; Goldfarb, L.J.B.: Phys. Lett. 48B, 290 (1974)
Ba 75   Ball, J.B. et al.: Nucl. Phys. A252, 208 (1975)
Be 67   Beringer, R.: Phys. Rev. Lett. 18, 1006 (1967)
Be 69a Beyer, K.; Winther, A.: Phys. Lett. 30B, 296 (1969)
Be 69b Beyer, K.; Winther, A.; Smilansky, U.: HE 69 p. 804
Be 72a Becchetti, F.D. et al.: Phys. Rev. C6, 2215 (1972)
Be 72b Berant, Z. et al.: Nucl. Phys. A196, 312 (1972)
Be 72c Berry, M.V.; Mount, K.E.: Rep. Progr. Phys. 35, 315 (1972)
Be 73   Becchetti, F.D. et al.: Nucl. Phys. A203, 1 (1973)
Bi 65   Biedenharn, L. C.; Brussaard, P. J.: Coulomb Excitation. Oxford: Clarendon Press
         1965
Bi 76   Birkelund, J.R. et al.: Phys. Rev. C13 ,133 (1976)
Bl 54   Blair, J.S.: Phys. Rev. 95, 1218 (1954)
Bl 59   Blair, J.S.: Phys. Rev. 115, 928 (1959)
Bo 64   de Boer, J.; Goldring, G.; Winkler, H.: Phys. Rev. 134, B 1032 (1964)
Bo 68   de Boer, J.; Eichler, J.: In Advances in Nuclear Physics, ed. by Baranger, M.;
         Vogt, E.: Vol. 1, 1 (1968)
Bo 72   Bonche, P. et al.: Phys. Rev. C6, 577 (1972)
Br 56   Breit, G.; Gluckstern, R.L.; Russel, J.E.: Phys. Rev. 103, 727 (1956)
Br 72   Broglia, R.A.; Winther, A.: Phys. Rep. 4, 153 (1972)
Br 74a Brink, D.M.; Rowley, N.: Nucl. Phys. A219, 79 (1974)
Br 74b Broglia, R.A. et al.: Phys. Rep. 11, 1 (1974)
Br 76   Britt, H.C. et al.: Phys. Rev. C13, 1483 (1976)
Bu 77   Butler, P.A. et al.: Phys. Lett. 68B, 122 (1977)
Ch 72   Christy, A.; Häusser, O.: Nucl. Data Tables 11, 281 (1972)
Ch 73   Christensen, P.R. et al.: Nucl. Phys. A207, 433 (1973)
Ch 76   Christensen, R.; Winther, A.: Phys. Lett. 65B, 19 (1976)
Cl 69   Cline, D. et al.: Nucl. Phys. A133, 445 (1969)

Co 72   Colombani, P. et al.: Phys. Lett. *42B*, 197 (1972)
Co 74   Colombani, P. et al.: Phys. Lett. *48B*, 315 (1974)
Co 76a Cobern, M.E.; Lisbona, N.; Mermaz, M.C.: Phys. Rev. *C13*, 674 (1976)
Co 76b Colombani, P. et al.: Phys. Lett. *65B*, 39 (1976)
Cr 76   Cramer, J.G. et al.: Phys. Rev. *C14*, 2158 (1976)
Di 67   Diamond, R.M.; Stephens, F.S.: Ark. Fys. *36*, 221 (1967)
Di 71   Disdier, D.L. et al.: Phys. Rev. Lett. *27*, 1391 (1971)
Ei 72a Eisen, Y.; Vager, Z.: Nucl. Phys. *A187*, 219 (1972)
Ei 72b Eisen, Y. et al.: Nucl. Phys. *A195*, 513 (1972)
Ei 73   Eichler, J. et al.: Phys. Rev. Lett. *30*, 568 (1973)
El 65   Elton, L.R.B.: *Introductory Nuclear Theory*, 2nd ed. p.30. London: Pitman 1965
Fo 59   Ford, K.W.; Wheeler, J.A.: Ann. Phys. (N.Y.) *7*, 259 (1959)
Fo 73   Ford, J.L.C. et al.: Phys. Rev. *C8*, 1912 (1973)
Fr 63   Frahn, W.E.; Venter, R.H.: Ann. Phys. (N.Y.) *24*, 243 (1963)
Fr 72a Frahn, W.E.: Ann. Phys. (N.Y.) *72*, 524 (1972)
Fr 72b Friedman, A.M.; Siemssen, R.H.; Cuninghame, J.G.: Phys. Rev. *C6*, 2219 (1972)
Fr 76   Frahn, W.E.: Nucl. Phys. *272*, 413 (1976)
Fr 77   Franz, G. et al.: GSI-Report J–1–77, 41 (1978)
Fr 78a Frahn, W.E.; Rehm, K.E.: Phys. Rep. *37*, 1 (1978)
Fr 78b Frahn, W.E.: Nucl. Phys. *A302*, 267 (1978)
Fr 78c Franz, G. et al.: GSI-Report J–1–78, 25 (1978)
Fu 78   Fuchs, P. et al.: GSI-Report J–1–78, 65 (1978)
Ge 55   Geilikman, B. T.: *Proc. of the Intern. Conf. on Peaceful Uses of Atomic Energy*,
         Geneva 1955 (United Nations, New York), Vol. 2, p. 201
Ge 58   Geilikman, B. T.: *Proc. 2nd. Intern. Conf. on Peaceful Uses of Atomic Energy*,
         Geneva 1958 (United Nations, New York), Vol. 15, p. 273
Gl 74   Glendenning, N.K.: NA 74 Vol. 2 p. 137
Gr 63   Greenberg, J.S. et al.: AS 63 p. 295
Gr 74   Gross, E.E. et al.: Phys. Rev. *C10*, 45 (1974)
Gr 75   Grosse, E. et al.: Phys. Rev. Lett. *35*, 565 (1975)
Gr 78   Gross, E.E. et al.: Phys. Rev. *C17*, 1665 (1978)
Gu 73a Gutbrod, H.H.; Winn, W.G.; Blann, M.: Nucl. Phys. *A213*, 267 (1973)
Gu 73b Gutbrod, H.H.; Blann, M.; Winn, W.G.: Nucl. Phys. *A213*, 285 (1973)
Hä 74   Häusser, O.: In *Nuclear Spectroscopy and Reactions*, Part C, ed. by J. Cerny. New
         York, London: Academic Press 1974
Ha 77   Habs, D. et al.: Z. Phys. *A283*, 261 (1977)
He 73   Hendrie, D.L.: Phys. Rev. Lett. *31*, 478 (1973)
Hi 77   Hillis, D.L. et al.: Phys. Rev. *C16*, 1467 (1977)
Ho 69   Holm, H.; Scheid, W.; Greiner, W.: Phys. Lett. *29B*, 473 (1969)
Ho 70a Holm, H. et al.: Z. Phys. *231*, 450 (1970)
Ho 70b Holm, H.; Greiner, W.: Phys. Rev. Lett. *24*, 404 (1970)
Ho 71   Holm, H.; Greiner, W.: Phys. Rev. Lett. *26*, 1647 (1971)
Ho 72   Holm, H.; Greiner, W.: Nucl. Phys. *A195*, 333 (1972)
Ig 59   Igo, G.: Phys. Rev. *115*, 1665 (1959)
Je 70a Jensen, A.S.; Wong, C.Y.: Phys. Rev. *C1*, 1321 (1970)
Je 70b Jensen, A.S.; Wong, C.Y.: Phys. Lett. *32B*, 567 (1970)
Je 71   Jensen, A.S.; Wong, C.Y.: Nucl. Phys. *A171*, 1 (1971)
Kl 70   Kleinfeld, A.M. et al.: Nucl. Phys. *A158*, 81 (1970)
Kn 74   Knoll, J.; Schaeffer, R.: Phys. Lett. *52B*, 131 (1974)
Kn 76   Knoll, J.; Schaeffer, R.: Ann. Phys. (N.Y.) *97*, 307 (1976)
Ko 75   Koeling, T.; Malfliet, R.A.: Phys. Rep. *22*, 181 (1975)
La 72   Larsen, R.D. et al.: Nucl. Phys. *A195*, 119 (1972)
La 76   Landowne, S. et al.: Nucl. Phys. *A259*, 99 (1976)
Le 72   Lefort, M. et al.: Nucl. Phys. *A197*, 485 (1972)
Ma 67   Maly, J.; Nix, J.R.: Proc. Intern. Conf. Nucl. Struct. Tokyo 1967 p. 678
Ma 73   Malfliet, R.; Landowne, S.; Rostokin, V.: Phys. Lett. *44B*, 238 (1973)
Ma 74   Malfliet, R.A.: HE 74 p. 86
Mc 60   McIntyre, J.A.; Wang, K.H.; Becker, L.C.: Phys. Rev. *117*, 1337 (1960)
Mc 71   McGowan, F.K. et al.: Phys. Rev. Lett. *27*, 1741 (1971)
Mc 74   McGowan, F.K.; Stelson, P.H.: In *Nuclear Spectroscopy and Reactions*, Part C, ed.
         by J. Cerny New York, London: Academic Press 1974

Ng 74 Ngô, C.; Péter, J.; Tamain, B.: Nucl. Phys. *A221*, 37 (1974)
Ng 78 Nguyen Van Sen, N. et al.: Phys. Rev. *C17*, 639 (1978)
Ob 77a Oberacker, V.; Soff, G.; Greiner, W.: J. Phys. *G3*, L 271 (1977)
Ob 77b Oberacker, V.; Soff, G.: Z. Naturforsch. *32a*, 1465 (1977)
Oe 74 Oehlberg, R.N. et al.: Nucl. Phys. *A219*, 543 (1974)
Og 78 Oganessian, Yu. Ts. et al.: Nucl. Phys. *A303*, 259 (1978)
Ol 78 Olmer, C. et al.: Phys. Rev. *C18*, 205 (1978)
Os 72 Ost, R. et al.: Phys. Rev. *C5*, 1835 (1972)
Os 73 Ost, R. et al.: MU 73 Vol. 1 p. 402
Pf 73 Pfeiffer, K.O.; Speth, E.; Bethge, K.: Nucl. Phys. *A206*, 545 (1973)
Pi 61 Pinkston, W.T.; Satchler, R.: Nucl. Phys. *27*, 270 (1961)
Pi 78 Pieper, S.C. et al.: Phys. Rev. *C18*, 180 (1978)
Po 66a Potgieter, J.M.; Frahn, W.E.: Nucl. Phys. *80*, 434 (1966)
Po 66b Potgieter, J.M.; Frahn, W.E.: Phys. Lett. *21*, 211 (1966)
Po 67 Potgieter, J.M.; Frahn, W.E.: Nucl. Phys. *A92*, 84 (1967)
Pr 69 Pryor, R.S. et al.: HE 69 p. 450
Qu 74 Quebert, J.L. et al.: NA 74 Vol. 1 p. 80
Ra 64 Rawitscher, G.H.: Phys. Rev. *135*, B 605 (1964)
Ra 66 Rawitscher, G.N.: Nucl. Phys. *85*, 337 (1966)
Re 75 Rehm, K.E. et al.: Phys. Rev. *C12*, 1945 (1975)
Re 78 Rehm, K.E. et al.: Phys. Rev. Lett. *40*, 1479 (1978)
Ri 70a Riesenfeldt, P.W.; Thomas, T.D.: Phys. Rev. *C2*, 711 (1970)
Ri 70b Riesenfeldt, P.W.; Thomas, T.D.: Phys. Rev. *C2*, 2448 (1970)
Ri 77 Richter, M. et al.: Nucl. Phys. *A278*, 163 (1977)
Ro 60 Rost, E.; Austern, N.: Phys. Rev. *120*, 1375 (1960)
Ro 75 Rowley, N.; Colombani, P.: Phys. Rev. *C11*, 648 (1975)
Sa 70a Satchler, G.R.: Phys. Lett. *33B*, 385 (1970)
Sa 70b Sayer, R.O. et al.: Phys. Rev. *C1*, 1525 (1970)
Sa 71 Satchler, G. R.: Part. Nucl. *2*, 265 (1971)
Sa 74a Saladin, J.X. et al.: NA 74 Vol. 1 p. 15
Sa 74b Satchler, G.R.: NA 74 Vol. 2 p. 171
Sa 76 Satchler, G.R.: AR 76 p. 33
Sc 78 Scobel, W. et al.: Z. Phys. *A284*, 343 (1978)
Si 63 Sikkeland, T.; Viola, V.E.: AS 63 p. 232
Sm 68 Smilanski, U.: Nucl. Phys. *A122*, 185 (1968)
Sm 69 Smilanski, U.: HE 69 p. 392
Sp 70 Speth, E.; Pfeiffer, K.O.; Bethge, K.: Phys. Rev. Lett. *24*, 1493 (1970)
St 59 Stephens, F.S.; Diamond, R.M.; Perlman, I.: Phys. Rev. Lett. *3*, 435 (1959)
St 63 Stelson, P.H.; McGowan, F.K.: Ann. Rev. Nucl. Sci. *13*, 163 (1963)
St 65a Stelson, P.H.: In *Proc. of the Summer Study Group on the Physics of the Emperor Tandem Van de Graaff Region*, Report BNL 948, Vol. III p. 1005 (1965)
St 65b Strutinsky, V.M.: Nucl. Phys. *68*, 221 (1965)
St 68 Stephens, F.S. et al.: Nucl. Phys. *A115*, 129 (1968)
St 70 Stephens, F.S. et al.: Phys. Rev. Lett. *24*, 1137 (1970)
St 71 Stephens, F.S.; Diamond, R.M.; de Boer, J.: Phys. Rev. Lett. *27*, 1151 (1971)
Th 59 Thomas, T.D.: Phys. Rev. *116*, 703 (1959)
Th 68 Thomas, T.D.: Ann. Rev. Nucl. Sci. *18*, 370 (1968)
Th 77 Thorn, C.E. et al.: Phys. Rev. Lett. *38*, 384 (1977)
Va 74 Vaz, L.C.; Alexander, J.M.: Phys. Rev. *C10*, 464 (1974)
Va 76 Vandenbosch, R. et al.: Phys. Rev. *C13*, 1893 (1976)
Va 78 Vandenbosch, R. et al.: Phys. Rev. *C17*, 1672 (1978)
Ve 64 Venter, R.H.; Frahn, W.E.: Ann. Phys. (N.Y.) *27*, 401 (1964)
Vi 62 Viola, V.E.; Sikkeland, T.: Phys. Rev. *128*, 767 (1962)
Vi 76a Videbaek, F. et al.: Nucl. Phys. *A256*, 301 (1976)
Vi 76b Viola, V.E. et al.: Nucl. Phys. *A261*, 174 (1976)
We 75 West, Jr. L.; Kemper, K.W.; Fletcher, N.R.: Phys. Rev. *C11*, 859 (1975)
Wi 63 Wilkins, B.; Igo, G.: AS 63 p. 241
Wi 65 Winther, A.; de Boer, J.: *California Institute of Technology Technical Report* (1965); reprinted in Al 66 (p. 303)
Wi 67 Wilets, L.; Guth, E.; Tenn, J.S.: Phys. Rev. *156*, 1349 (1967)
Wi 69 Winkler, P.: HE 69 p. 464

Wi 71   Winkler, P.: Nucl. Phys *A168*, 139 (1971)
Wo 68a Wong, C.Y.: Phys. Lett. *26B*, 120 (1968)
Wo 68b Wong, C.Y.: Nucl. Data *A4*, 271 (1968)
Wo 73   Wong, C.Y.: Phys. Rev. Lett. *31*, 766 (1973)
Wo 78   Wojciechowski, H. et al.: Phys. Rev. *C17*, 2126 (1978)

4. General Aspects of Nucleon Transfer

4.1 Introduction

A rather difficult point in discussing transfer reactions is that of notation. The notation perhaps most widely used in the literature is that originally introduced by the Oak Ridge group for deuteron stripping and other light-projectile-induced reactions [Au 64, Sa 73; see Table 4.1]. Unfortunately this notation is not well adapted to the symmetry properties of heavy-ion-induced transfer, where the two cores may be of comparable size. Another well-known notation, which avoids this disadvantage but is somewhat cumbersome due to an extensive use of indices, is that used by Buttle and Goldfarb [Bu 66, Bu 68, Bu 71]. Since there is no generally accepted notation, we have in this book— at the risk of adding to the existing confusion—tried to combine what we consider the advantages of the two notations mentioned above. In order to facilitate comparison with the literature, we give in Table 4.1 a survey of all three notations.

Table 4.1. Notations used in the description of transfer reactions

Reaction formula	Core 1	Core 2	Transferred particle	Donor nucleus	Acceptor nucleus	Reference
$A(a,b)B$	b	A	x	$a = b + x$	$B = a + x$	Oak Ridge (Au 64, Sa 73)
$c_2(a_1,c_1)a_2$	c_1	c_2	t	$a_1 = c_1 + t$	$a_2 = c_2 + t$	Bu 66, Bu 68, Bu 71
$C_2(A,C_1)B$	C_1	C_2	x	$A = C_1 + x$	$B = C_2 + x$	This book

A transfer reaction may be defined as a nuclear collision, during which a nucleon or a group of nucleons, denoted by x, dissociates from a core C_1 and attaches itself to a core C_2. Symbolically we may write

$$A + C_2 \rightarrow C_1 + B + Q, \tag{4.1}$$

where we have put $A = C_1 + x$ and $B = C_2 + x$, and Q denotes the gain in relative kinetic energy. Neglecting the internal structure of x, C_1, and C_2 we can represent the relative positions of the various fragments during the reaction by the vector diagram shown in Fig. 4.1. Each of the five relative vectors $r_{1x}, r_{2x}, r_{12}, r_i, r_f$ can be expressed in terms of two independent basis vectors. Using for the latter either r_{12} and r_{1x} or r_{12} and r_{2x}, we obtain

$$r_i = r_{12} + \frac{m_x}{M_1 + m_x} r_{1x} = \frac{M_1}{M_1 + m_x} r_{12} + \frac{m_x}{M_1 + m_x} r_{2x}, \qquad (4.2)$$

$$r_f = \frac{M_2}{M_2 + m_x} r_{12} - \frac{m_x}{M_2 + m_x} r_{1x} = r_{12} - \frac{m_x}{M_2 + m_x} r_{2x}. \qquad (4.3)$$

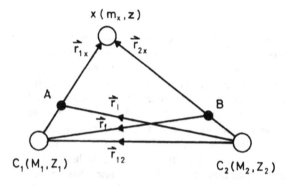

Fig. 4.1. Vector diagram of transfer reaction $A + C_2 \rightarrow C_1 + B$ (C_1, C_2 cores; x transferred particle; $A = C_1 + x$, $B = C_2 + x$)

We shall not immediately, however, enter a discussion of quantitative approaches, but rather begin with the most important qualitative features of transfer reactions.

At the outset it should be realized that the detailed mechanism of a transfer reaction—as the term is understood here—can be of very different complexity. In the simplest case the transferred particle x is a single nucleon and the cores C_1 and C_2 remain completely inert during the reaction, playing merely the part of "spectators". This is the classical case of a "direct", one-step reaction, where the term "direct" implies that only one or a few internal degrees of freedom are involved and the reaction proceeds within a time of the order of the transit time of the fragments through the interaction region. The most prominent example of this type of reaction is probably the (d,p) "stripping reaction", although completely analogous reactions induced by heavy projectiles are well known and will be discussed in this and the following chapter.

A somewhat more complicated situation arises if two or more nucleons are transferred (or exchanged) simultaneously. Such reactions may still be classified as "direct", if the internal states of the cores C_1, C_2 and the transferred group of nucleons x do not change upon transfer. The assumption that C_1,

C_2, and x remain well defined, distinguishable entities throughout the reaction is, of course, not compatible with a rigorous quantum-mechanical description of the reaction; its apparent success in the qualitative and quantitative analysis of a large body of experimental data proves convincingly, however, the validity of the concept of a "direct transfer reaction". The distinguishing feature of this type of reaction is that it populates well-defined quantum states—usually at rather low excitation—with a high degree of selectivity. In order to emphasize the simplicity of these reactions we refer to them in the following as "quasi-elastic" transfer reactions.

In contrast, the opposite extreme of a strongly "inelastic" transfer reaction will in general proceed through a sequence of direct transfers and complicated internal rearrangements of the participating fragments. Under these circumstances, the subgroups C_1, C_2, and x can no longer be defined in a unique way during the reaction, and will retain only an integral significance. The emerging products may be highly excited and may subsequently decay by particle emission. As a consequence of high-level densities, individual quantum states are neither populated selectively, nor can they be resolved experimentally—at best one can hope to identify the masses and charges of the products, as well as their approximate excitation energies and angular momenta. Any selectivity on the basis of microscopic nuclear structure will be lost, or swamped by statistical and kinematic effects.

One may ask at this point what defines the borderline between strongly inelastic transfer reactions and more complex rearrangement processes, such as compound nucleus formation and decay. Intuitively, one would like to associate with a transfer reaction the fact that the two fragments retain their identity at least partially during the collision. This implies—as a necessary, but not sufficient criterion—that the mass fragmentation in the exit channel should be similar to that in the entrance channel, or the transferred mass should be small compared to the total mass of the system. Obviously, this simple criterion is not very useful, if the masses of projectile and target nucleus are similar. Other, more refined criteria refer to the time scale of the reaction and the related question, whether the composite system loses its memory concerning its mode of formation. The relevant observables are the symmetry properties of angular distributions, and the dependence of the branching ratios for different exit channels on entrance channel for a given composite system, at a given excitation energy and angular momentum. Again, however, these criteria are not always applicable or conclusive, either due to experimental problems or for more fundamental reasons. For example, it is conceivable that a transfer reaction (as defined above) proceeds through a long-lived intermediate state of quasi-molecular structure, with an angular distribution characteristic of compound nucleus decay. Alternatively, complete amalgamation of target nucleus and projectile may occur, and yet the dynamical evolution of the compound system may favour decay modes specifically related to the entrance channel, with obvious symptoms of a direct transfer reaction. These examples, although admittedly somewhat hypothetical, may serve to illustrate two basic problems. Firstly, there is a continuous transition from one reaction mechanism to another, and therefore no unambiguous definition of what a transfer reaction (or any other type of reaction) is can be given. Secondly, there is no completely unambiguous way to infer details of a reaction mechanism from experimental data, especially if one deals with reactions between complex nuclei. With these reservations in mind, we nevertheless wish to emphasize that the term transfer reactions stands for a well-established and relatively distinct class of reaction.

The existence of indistinguishable reaction mechanisms is an essential feature of heavy-ion-induced transfer reactions, both of the inelastic and the quasi-elastic type. The transition from a given initial channel into a given final channel may, in general, proceed through many different sequences of inter-mediate, unobserved configurations, and each of these different reaction paths will contribute coherently to the transfer amplitude. Even if a one-step mechanism is assumed, the observation of definite initial and final states does not determine unambiguously the nature and internal state of the transferred particle. This is demonstrated graphically in Fig. 4.2, which relates to the reaction described by (4.1). One can see that the transfer of particle x or that of particle $y = C_2 - C_1$ lead to identical final nuclei and are therefore not distinguishable. (Note that the transfer of y can proceed either way, depending on which of the two "cores" C_1 or C_2 is heavier.) An important conceptual difference between the two mechanisms is that they associate a given product with different "parent nuclei"; hence if our assumption of a direct, one-step process is valid, the transfer amplitudes describing x and y transfer yield very different angular distributions and the two mechanisms are "kinematically separated". In simpler terms, one expects that x transfer leads to forward emission of C_1, and y-transfer to backward emission of C_1, with respect to the initial momentum of A.

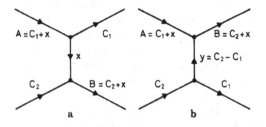

Fig. 4.2. Graphical representation of indistinguishable transfer mechanisms, contributing to the reaction $A + C_2 \rightarrow C_1 + B$. **a** Transfer of $x = A - C_1$; **b** Transfer of $y = C_2 - C_1$

Evidently the coherent mixing of the two transfer mechanisms will be most important if the two cores are of equal or similar mass. In the limiting case $C_1 = C_2$, y transfer corresponds to elastic or inelastic scattering, and x transfer to "elastic" or "inelastic" transfer. A detailed discussion of elastic transfer among light nuclei has been given in Sect. 2.4; inelastic transfer, populating an excited state in at least one of the outgoing fragments, has also been identified [see, for example, Ge 74, Go 74, Ka 75, Re 75]. An interesting example involving different cores C_1, C_2 is given in Fig. 4.3, which

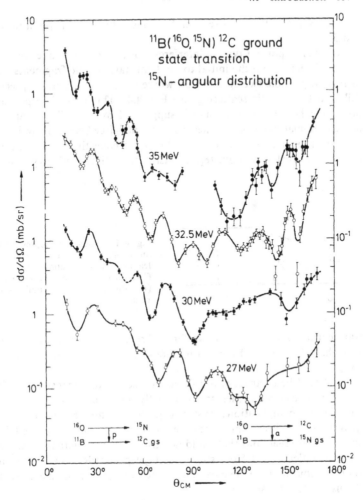

Fig. 4.3. Angular distributions of ^{15}N nuclei from the reaction ^{11}B(^{16}O, ^{15}N)^{12}C, populating the ground states of the final nuclei, at different laboratory energies of the ^{16}O projectile. The cross section is attributed at forward angles mainly to p transfer, and at backward angles mainly to α transfer. (From Bo 66)

shows angular distributions of ^{15}N nuclei from the reaction ^{11}B(^{16}O, ^{15}N)^{12}C obtained by Bock et al. at Heidelberg [Bo 66]. In this case one has interference between proton and α-particle transfer, with the former dominating in the forward hemisphere, and the latter dominating in the backward hemisphere.

We note in passing that coherent mixing of different transfer modes also occurs in the transfer of two or more nucleons, if the transferred group can occupy different internal states. Here, however, the different modes are not localized in different angular regions, and their interference mainly affects the absolute value of the cross section. We shall return to this point in Sect. 5.1 and 5.5.

Another important type of interference is that between one-step transfer, leading directly from the initial to the final states of the fragments, and one or several possible two-step routes, where the transfer is either preceded or followed by inelastic scattering [see Fig. 4.4]. This effect is especially pronounced where the direct route is suppressed by a small overlap between initial and final states or by kinematic mismatch [see Sect. 4.3], and where the indirect transitions can proceed via low-lying collective states. A review of theoretical work on multistep inelastic transitions in heavy-ion-induced

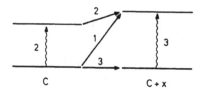

Fig. 4.4. Schematic representation of interfering routes in a transfer reaction leading from the ground state of nucleus C to an excited state of nucleus C + x (1 = direct route; 2,3 = indirect, two-step routes)

transfer (mainly two-neutron transfer) and references to relevant experimental work have been given by Ascuitto and Vaagen [As 74; see also As 73a, As 73b].

It should be clear from our discussion so far that transfer reactions induced by heavy projectiles are potentially quite complex processes. Hence considerable complications must be expected in any serious attempt at their quantitative analysis. This raises the question what can be learned from the study of these reactions that could not also be learned with less experimental and theoretical effort from studies of light-projectile-induced transfer. We try to summarize here the most important points, postponing a more detailed discussion to later sections.

One important simplifying feature with heavy projectiles is the applicability of classical and semiclassical concepts, especially at bombarding energies below or near the Coulomb barrier. Moreover, at sufficiently low energies the trajectories are unaffected by the nucleus–nucleus potential, which therefore does not enter the transfer analysis. As a result, uncertainties due to an incomplete knowledge of the nuclear potential may be reduced or eliminated, and an intuitive understanding of qualitative trends may be facilitated.

The classical nature of the relative motion, together with "strong absorption" at small distances, leads to striking kinematic selectivities in heavy-projectile-induced transfer. Final states with high angular momentum and (or) high excitation energy may be preferentially populated in contrast to light-projectile-induced transfer.

Further a great variety of multinucleon transfer reactions become accessible to investigation with heavy projectiles. Two-, three-, and four-nucleon correlations can be probed in various ways, using different target–projectile combinations. Transfers of even larger numbers of nucleons may serve as a tool to produce and study new and exotic nuclear species.

We conclude that the study of transfer reactions with complex projectiles should extend our knowledge both quantitatively and qualitatively in such diverse fields as nuclear spectroscopy, nuclear reaction dynamics and nuclear chemistry. In the following sections, as well as in Chap. 5 and 6, we shall try to show this in more detail, concentrating mainly on those aspects of transfer reactions which are specific to the use of heavy projectiles.

4.2 Semiclassical Considerations and the Structure of Quasi-Elastic Angular Distributions

Angular distributions of tansfer reactions often exhibit a single peak at the grazing angle. This feature may be understood qualitatively—and to some extent quantitatively—from semiclassical considerations, where one assumes the cores to move along classical trajectories. Neglecting differences in the kinematic variables between the initial (i) and final (f) channel, and treating the transfer reaction in first-order perturbation theory, we can write for the differential cross section [Br 72b, c]

$$\frac{d\sigma}{d\Omega}(\Theta) = P_{if}(\Theta)P_A(\Theta)\left(\frac{d\sigma}{d\Omega}(\Theta)\right)_{cl},$$ (4.4)

where $(d\sigma/d\Omega)_{cl}$ is the classical differential cross section for elastic scattering by the combined Coulomb and nuclear potential [see (3.53)]. The quantity $P_{if}(\Theta)$ is the transfer probability given by

$$[P_{if}(\Theta)]^{1/2} = \frac{1}{i\hbar}\int_{-\infty}^{+\infty} dt \, \langle f | V_{if} | i \rangle \exp\left[\frac{i}{\hbar}(E_f - E_i)t\right],$$ (4.5)

where V_{if} is the interaction responsible for the transfer, and E_i, E_f are the centre-of-mass energies in the initial and final channel. Finally the "attenuation factor" $P_A(\Theta)$ describes the probability that the system will escape absorption into other inelastic channels, which are not considered explicitly. It can be written as

$$P_A(\Theta) = \exp\left\{-\frac{2}{\hbar}\int_{-\infty}^{+\infty} W[r(t)]dt\right\},$$ (4.6)

where W is identified with the imaginary part of the optical model potential, which fits the elastic scattering in channels i and f.

The smooth angular distributions with maxima near the grazing angle for $E > E_C$, and at $\Theta = \pi$ for $E < E_C$, are readily explained using (4.4). First we note that most of the cross section arises from parts of the trajectory which are close to the classical turning point, or distance of closest approach

$D\,(E,\,\Theta)$. If the scattering is dominated by the Coulomb interaction, then D decreases monotonically with increasing scattering angle Θ at constant energy E. The transfer probability P_{if} increases exponentially with decreasing D, and hence leads to a strong increase of the cross section with increasing scattering angle, as long as $\Theta < \Theta_{gr}$. At larger angles, $\Theta > \Theta_{gr}\,(D < R_C)$, the elastic channel is depleted by absorptive processes and therefore $P_A(\Theta)$ drops sharply with a further increase in Θ. Consequently also the transfer cross section decreases again.

The decisive influence of the distance of closest approach D on the transfer cross section can be demonstrated by plotting instead of $d\sigma/d\Omega$ as a function of Θ, the quantity

$$\frac{d\sigma}{dD} = \frac{d\Omega}{dD}\frac{d\sigma}{d\Omega} \approx -\frac{8\pi k}{n}\left(\sin\frac{\Theta}{2}\right)^3\frac{d\sigma}{d\Omega} \tag{4.7}$$

as a function of $D \approx k(1 + \csc(\Theta/2))$, assuming Rutherford trajectories. In this way, kinematic effects of energy and scattering angle are largely removed, and data obtained at different bombarding energies can—after suitable normalization—be combined in a single plot, which peaks sharply at a characteristic interaction distance [Mc 60].

In quantitative work, the extreme semiclassical approach leading to (4.4) must be refined. In particular, elastic scattering of strongly absorbed particles is not well described by the classical cross section, and quantal interference effects are important. A first improvement, in this respect, is obtained by writing

$$\left(\frac{d\sigma}{d\Omega}\right)_{if} = P_{if}\left[\left(\frac{d\sigma}{d\Omega}\right)_{ii}\left(\frac{d\sigma}{d\Omega}\right)_{ff}\right]^{1/2} \tag{4.8}$$

where $(d\sigma/d\Omega)_{ii}$ and $(d\sigma/d\Omega)_{ff}$ are elastic scattering cross sections in the initial and final channel, respectively, as derived either by measurement or by quantal calculation with the optical model. A more accurate representation of the cross section is the partial-wave expansion [Wi 73]

$$\frac{d\sigma}{d\Omega} = |f_{tr}(\Theta)|^2 \tag{4.9}$$

$$f_{tr}(\Theta) = \frac{-i}{2\sqrt{k_i k_f}}\sum_{l=0}^{\infty}{}^{L}P_{if}(l)A_l(i)A_l(f)]^{1/2}\exp\{i[\alpha_l(i) + \alpha_l(f)]\}$$
$$(2l + 1)P_l(\cos\Theta). \tag{4.10}$$

Here the transfer probability P_{if} is still calculated according to (4.5) for a well-defined trajectory; it is, however, no longer associated with a definite scattering angle, but rather with the corresponding partial wave l. The quantities A_l and $\alpha_l = \sigma_l + \delta_l$ are real parameters, which describe the elastic scattering in channels i and f, and were introduced in Chap. 1 [see (1.25) and (1.28)].

Clearly, the semiclassical trajectory concept has now been relaxed to a considerable extent; in fact, (4.10) is formally equivalent to a completely quantal representation of the reaction amplitude in the limit of zero angular momentum and mass transfer and negligible Q value.

The semiclassical theory has been developed in great detail for sub-Coulomb transfer by Alder and Trautmann [Tr 70, Al 71a,b; Al 72] and more generally by Broglia and Winther [Br 72b,c; Wi 73]. Effects due to a finite transfer of mass, charge, and energy and due to mismatch of the initial and final orbits are included approximately by describing the relative motion of the colliding fragments in terms of wave-packets centred on averaged classical trajectories. Since the formalism is rather involved, and only few applications have been made to actual experimental cases, we shall not go into detail here, but refer the reader to the original literature.

The conditions under which either structured or smooth angular distributions are observed in transfer reactions between heavy ions have been the object of considerable discussion in the literature. A very elegant and quite general formulation of the problem was given in 1964 by Strutinsky [St 64] and independently by Frahn and Venter [Fr 64]. Following these authors we can write for the reaction amplitude in the quasi-elastic limit (i.e. for vanishing transfer of angular momentum, mass, and energy)

$$f_{tr}(\Theta) = \frac{1}{2ik} \sum_{l=0}^{\infty} (2l + 1) f(l - l_0) \exp (2i\alpha_l) P_l(\cos \Theta). \tag{4.11}$$

This expression is identical with (4.10), except for some obvious simplifications. The factor $f(l - l_0)$ now describes the effective angular momentum spectrum, which contributes to the transfer reaction. This will usually be concentrated in a relatively narrow region of width Δl around a central value l_0 near the grazing angular momentum. Angular momenta much below l_0 will be eliminated by "absorption" into complicated reaction channels, whereas angular momenta much above l_0 will be associated with comparatively large distances of closest approach and hence small transfer probabilities. The resulting localization of the reaction amplitude in angular momentum space determines the shape of the angular distribution.

For simplicity we assume $1 \ll \Delta l \ll l_0$, and expand the real scattering phase shift α_l to first order around $l = l_0$:

$$\alpha_l \approx \alpha_0 + \frac{1}{2} \bar{\Theta}_0 (l - l_0), \tag{4.12}$$

where

$$\bar{\Theta}_0 = 2\left(\frac{d\alpha_l}{dl}\right)_{l=l_0} = \bar{\Theta}_{l=l_0} \tag{4.13}$$

may be denoted as the quantal deflection function for angular momentum $l = l_0$. In the JWKB limit for α_l, the quantity $\bar{\Theta}_0$ becomes equal to the classical deflection angle for angular momentum $(l_0 + 1/2)\hbar$ [see (3.58)]; therefore its value should be near to the grazing angle. Replacing further in (4.11) the summation over l by integration we obtain

$$f_{tr}(\Theta) \approx \frac{\exp(2i\alpha_0)}{2i\,k} \int_{-\infty}^{+\infty} (2l + 1)\, f(l - l_0)\, \exp[i\bar{\Theta}_0(l - l_0)]$$
$$P_l(\cos \Theta)\, d(l - l_0). \tag{4.14}$$

For large values of l_0 and $\Theta \gg l^{-1} \approx l_0^{-1}$, the Legendre polynomials $P_l(\cos \Theta)$ can be replaced by their asymptotic representations

$$P_l(\cos\Theta) \approx \frac{2\cos\left[\left(l + \frac{1}{2}\right)\Theta - \frac{\pi}{4}\right]}{(2\pi l\,\sin\Theta)^{1/2}}$$
$$= \frac{\exp\left\{i\left[\left(l + \frac{1}{2}\right)\Theta - \frac{\pi}{4}\right]\right\} + \exp\left\{-i\left[\left(l + \frac{1}{2}\right)\Theta - \frac{\pi}{4}\right]\right\}}{(2\pi l\,\sin\Theta)^{1/2}}. \tag{4.15}$$

Inserting (4.15) into (4.14) leads to the following approximate expression for the reaction amplitude:

$$f_{tr}(\Theta) \approx -\frac{\exp(2i\alpha_0)}{k}\left(\frac{l_0}{2\pi\,\sin\Theta}\right)^{1/2}\left\{\exp\left[i\left(l_0 + \frac{1}{2}\right)\Theta + \frac{i\pi}{4}\right]g(\bar{\Theta}_0 + \Theta)\right.$$
$$\left. -\exp\left[-i\left(l_0 + \frac{1}{2}\right)\Theta - \frac{i\pi}{4}\right]g(\bar{\Theta}_0 - \Theta)\right\}, \tag{4.16}$$

where the function g is the Fourier transform of f:

$$g(t) = \int_{-\infty}^{+\infty} f(x)\,\exp(itx)\, dx. \tag{4.17}$$

If $f(l - l_0)$ is symmetric in $l - l_0$, then g is real, and the cross section can be written

$$\frac{d\sigma}{d\Omega} = |f_{tr}(\Theta)|^2 = \frac{l_0}{2\pi k^2\,\sin\Theta}\left\{[g(\bar{\Theta}_0 + \Theta)]^2 + [g(\bar{\Theta}_0 - \Theta)]^2\right.$$
$$\left. + 2g(\bar{\Theta}_0 + \Theta)g(\bar{\Theta}_0 - \Theta)\sin\left[2\left(l_0 + \frac{1}{2}\right)\Theta\right]\right\}. \tag{4.18}$$

A convenient ansatz for $f(l - l_0)$ is a Gaussian, which leads to

$$f(l - l_0) = f_0\,\exp\left\{-\left(\frac{l - l_0}{\Delta l}\right)^2\right\} \tag{4.19}$$

$$g(t) = \sqrt{\pi}\,f_0\,\Delta l\,\exp\left\{-\frac{(\Delta l)^2 t^2}{4}\right\} \tag{4.20}$$

$$\frac{d\sigma}{d\Omega} = \frac{f_0^2(\Delta l)^2 l_0}{2k^2 \sin\Theta} \left\{ \exp\left[-\frac{(\Delta l)^2}{2}(\bar\Theta_0 + \Theta)^2 \right] + \exp\left[-\frac{(\Delta l)^2}{2}(\bar\Theta_0 - \Theta)^2 \right] \right.$$

$$\left. + 2\exp[-(\Delta l)^2(\bar\Theta_0^2 + \Theta^2)] \sin\left[2\left(l_0 + \frac{1}{2}\right)\Theta \right] \right\}. \tag{4.21}$$

In spite of the various approximations made in their derivation, (4.18) and (4.21) contain all the essential qualitative features of angular distributions for quasi-elastic transfer among heavy ions. We begin our discussion by considering the first two terms on the right-hand side, which represent smooth, nonoscillatory components. From a semiclassical point of view, these two terms can be associated with trajectories passing around different sides of the interaction region: the second term yields the well-known bell-shaped peak near the grazing angle, which can be explained in terms of "direct" trajectories, whereas the first term rises towards zero degrees, and may be thought of as arising from deflection to negative angles around the "dark side" of the interaction region [note, however, that (4.18) and (4.21) are not valid close to zero degrees due to the substitution (4.15)].

Although simple explanations of angular distributions in terms of classical trajectories may appeal to our intuition, they nevertheless have to be viewed with great caution. This point has been discussed by a number of authors [St 64, Si 72, Fr 74, Au 75] and may be illustrated by considering the angular width $\Delta\Theta$ of the grazing peak. Defining $\Delta\Theta$ in analogy to Δl [see (4.19)] we obtain from (4.21) the relationship

$$(\Delta\Theta)^2 = \frac{2}{(\Delta l)^2}, \tag{4.22}$$

which shows that the localization of the scattered particles in angular space is inversely proportional to that in angular momentum space. This is a direct consequence of the quantum-mechanical complementarity between angle and angular momentum. Hence a well-defined grazing peak requires an appreciable spread in angular momenta, in apparent contradiction to the classical picture of a well-defined impact parameter. The radial extent of the wave packets close to the interaction region has been studied by Austern [Au 75] and was found appreciable, at least of the order of 1 fm under typical conditions.

It should be noted that the validity of (4.22) is restricted to relatively small values of Δl, since in the derivation of (4.21)—especially in making the expansion (4.12)—we have assumed that only a relatively narrow range of l values around l_0 contributes to the reaction. The changes which occur for larger values of Δl have been discussed by Siemens and Becchetti and by Strutinsky [Si 72, St 73; see also Sc 74]. Equation (4.22) can then be generalized by writing

$$(\Delta\Theta)^2 = \frac{2}{(\Delta l)^2} + \left(\frac{d\bar\Theta}{dl}\right)_{l_0}^2 \frac{(\Delta l)^2}{2}, \tag{4.23}$$

where the first term on the right hand side—as in (4.22)—is the angular spread due to quantal diffraction, whereas the second term is the classical angular spread associated with the range of participating angular momenta. The quantity $(\Delta \Theta)^2$ increases both for small $(\Delta l)^2$ and for large $(\Delta l)^2$, and goes through a minimum value

$$(\Delta \Theta)^2_{\min} = 2 \left| \left(\frac{d\bar{\Theta}}{dl} \right)_{l_0} \right| \approx \frac{4}{n^2} \sin^2 \frac{\bar{\Theta}_0}{2} \tag{4.24}$$

at an intermediate value of Δl given by

$$(\Delta l)^2_{\min} = 2 \left| \left(\frac{d\bar{\Theta}}{dl} \right)_{l_0}^{-1} \right| \approx n^2 \csc^2 \frac{\bar{\Theta}_0}{2}. \tag{4.25}$$

The second parts of (4.24) and (4.25) hold, if the scattering is dominated by the Coulomb interaction, and the nuclear attraction can be ignored. For $(\Delta l)^2 < (\Delta l)^2_{\min}$, the width of the grazing peak is mainly due to quantal diffraction, whereas for $(\Delta l)^2 > (\Delta l)^2_{\min}$ it is mainly due to classical dynamics. Note that $(\Delta l)^2$—and hence the predominance of quantal or classical scattering—can no longer be inferred unambiguously from the width of the grazing peak [Si 72].

We now turn to a brief discussion of the third, oscillatory term in the cross section, as given by (4.18) and (4.21). It is obviously a diffraction phenomenon arising from the interference of waves travelling around opposite sides of the interaction region. The oscillations are concentrated at small angles $[\Theta \lesssim (\Delta l)^{-1}]$ and have a period $\pi/(l_0 + \frac{1}{2})$. The strength of the oscillations is governed by the factor $\exp[- (\Delta l)^2 \bar{\Theta}_0^2]$ and is therefore expected to be significant for $(\Delta l)\bar{\Theta}_0 \lesssim 1$, and absent for $(\Delta l)\bar{\Theta}_0 \gg 1$.

It follows from the preceding discussion that the character of an angular distribution is determined by the following parameter:

$$\bar{\Theta}_0(\Delta l) = \bar{\Theta}_0 l_0 \frac{\Delta l}{l_0} \approx \bar{\Theta}_0 \cot \frac{\bar{\Theta}_0}{2} n \frac{\Delta l}{l_0} \approx 2n \frac{\Delta R}{R}. \tag{4.26}$$

With $\Delta R \approx 0.5$ fm and $R \approx 10$ fm one expects very roughly that the transition from oscillatory to nonoscillatory behaviour should occur somewhere in the region $n = 10$–20, in agreement with experimental observation.

Although in early experiments, performed with tandem accelerators and light projectiles on light targets ($A \lesssim 20$), pronounced oscillatory structure had been observed and successfully analysed in terms of diffraction models or DWBA [see, for example, Bo 66, Oe 69, Sc 72], it was generally felt that only smooth angular distributions could be expected with heavier projectiles and targets. This was a consequence of the fact that the beam qualities available at that time with projectiles of higher energy or heavier mass and the associated detection equipment would not permit resolution of the narrow fine structure. Therefore, the theoretical pioneering work by Strutinsky [St 64] and Frahn and Venter [Fr 64] on angular distribution structures went largely unnoticed. This situation changed in the early seventies, when by a combination of advances in beam energy, beam quality, and detection systems it became possible to study angular distributions of reactions induced by light projectiles on medium-mass targets ($A \gtrsim 40$) in much more detail than before. Experiments then performed at Brookhaven, Argonne, Copenhagen, Heidelberg and in other laboratories [see references quoted in Ka 74 b, Ga 74] revealed the existence even in these heavier systems of sharp oscillatory structure, superimposed on a more or less distinct grazing peak. As examples we show in Figs. 4.5 and 4.6 angular distributions measured by the Brookhaven

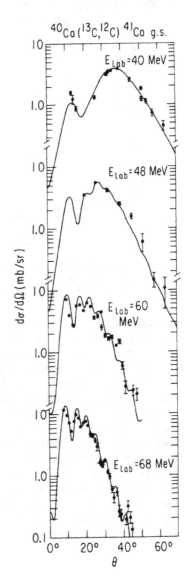

Fig. 4.5. Angular distributions of the neutron-transfer reaction $^{40}Ca(^{13}C, \,^{12}C)^{41}Ca$, populating the ground states of the final nuclei. The centre-of-mass bombarding energies are $E_{CM} = 30.2$ MeV ($n = 10.7$), 36.2 MeV ($n = 9.8$), 45.3 MeV ($n = 8.8$) and 51.5 MeV ($n = 8.2$). (From Bo 73)

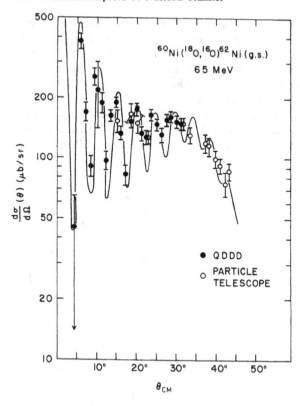

Fig. 4.6. Angular distribution of the two-neutron transfer reaction ^{60}Ni(^{18}O, ^{16}O)^{62}Ni, populating the ground states of the final nuclei. The centre-of-mass bombarding energy is $E_{CM} = 50$ MeV ($n = 18.5$). (From Le 74)

group for the one-neutron-transfer reaction ^{40}Ca(^{13}C, ^{12}C)^{41}Ca [Bo 73] and the two-neutron-transfer reaction ^{60}Ni(^{18}O, ^{16}O)^{62}Ni [Le 74]. These and similar results have been analysed successfully either by DWBA methods [see Sect. 5.2] or by a parametrized partial-wave analysis as proposed originally by Strutinsky [St 64] and Frahn and Venter [Fr 64]. The latter approach has been extended to include finite angular momentum transfers by Strutinsky [St 73] and in more detail by Kahana, Bond and Chasman [Ka 74a, Ka 74b]. Both the experimental results and the theoretical analyses indicate that the overall phase pattern of the oscillations is mainly determined by the odd–even character of the angular momentum transfer l; however, shape and position of the first maximum appear to be sensitive to the actual value of l, similar to the situation in light-projectile-induced transfer. Whether this sensitivity can be used in practice to assign l values is an open question, especially in view of the difficulties associated with taking precise data at extreme forward angles.

4.3 Selectivity Due to Kinematic Conditions

In Sect. 4.2 we discussed a characteristic feature of quasi-elastic transfer reactions between complex nuclei, namely the pronounced bell-shaped peak frequently observed in angular distributions at the "grazing angle". This peak has been explained semiclassically as a consequence of two kinematic conditions: the localization of the transfer reaction in a well-defined radial region, and the continuous motion of the two cores across the interaction region along classical trajectories. These two conditions appear to be effective not only in quasi-elastic, but also in highly inelastic transfer reactions, regardless of the detailed microscopic mechanism. Moreover, for a given incident channel, they impose restrictions both on the angles and on the kinetic energies of specified outgoing fragment pairs and hence on the reaction Q values. As a result, certain regions of excitation are selectively populated in the final nuclei. For quasi-elastic transitions to comparatively low-lying final states, this kinematic selectivity will be superimposed on the spectroscopic selectivity due to nuclear structure effects. At higher excitation, on the other hand, the spectroscopic strength will usually be more or less uniformly distributed and, in addition, multistep processes are more likely to be important. Under these circumstances, the energy spectra of outgoing products are dominated by kinematic effects.

The remainder of this section will be devoted to a detailed discussion of the kinematic conditions leading to an "optimum Q value", or "Q-value window" in heavy-ion-induced transfer, and the relevant experimental evidence. We begin by presenting some examples of experimental data. Figures 4.7 and 4.8 refer to a well-known case, the reaction $^{54}Fe(^{16}O, {}^{12}C)^{58}Ni$, at incident energies of 46 MeV [from Bo 72] and 60 MeV [from Ch 73], respectively. When quasi-continuous "bump-shaped" ^{12}C energy spectra, as shown in Fig. 4.8, were first observed in $(^{16}O, {}^{12}C)$ reactions, it was thought that they might arise from projectile break-up. This was subsequently shown not to be the case, however. At the lower energy of 46 MeV, and a lab angle of 42.5°, final states around 7 MeV excitation (corresponding to a Q value of about -8 MeV) are most strongly excited. At the higher bombarding energy of 60 MeV, on the other hand, the Q window has moved up in excitation, and is now centred at about 13 MeV, or a Q value of about -14 MeV.

In Fig. 4.9 we show energy spectra of one-proton stripping reactions on ^{54}Fe, taken with ^{16}O and ^{14}N projectiles at 80 MeV incident energy and a laboratory angle of 20° [Po 73]. The single-particle orbits to which the proton is transferred are indicated in the figure. Whereas the $(^{16}O, {}^{15}N)$ reaction ($Q_0 = -7.1$ MeV) preferentially excites the ground state of the residual nucleus ^{55}Co, the $(^{14}N, {}^{13}C)$ reaction ($Q_0 = -2.5$ MeV) populates more strongly final states between 2 and 9 MeV excitation. In both cases the centre of the Q window is near $Q = -7.5$ MeV.

We can summarize these and a large body of other results by noting that "stripping" reactions, involving a light donor and a heavy acceptor nucleus,

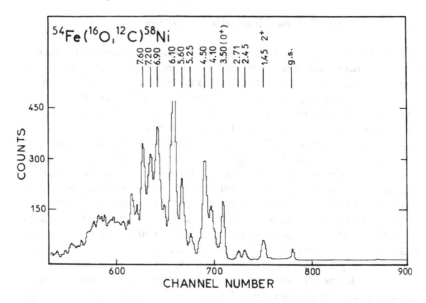

Fig. 4.7. Energy spectrum of ^{12}C ions from the reaction ^{54}Fe(^{16}O, ^{12}C)^{58}Ni at 46 MeV bombarding energy and an angle of 42.5° in the laboratory system. Excitation energies in the final nucleus ^{58}Ni are indicated. (From Bo 72)

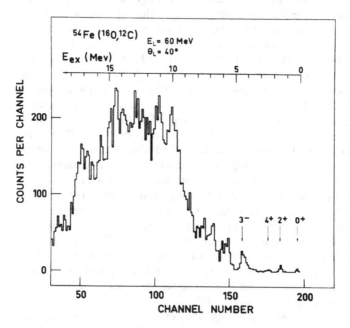

Fig. 4.8. Same as Fig. 4.7, but at a laboratory bombarding energy of 60 MeV and a laboratory angle of 40°. (From Ch 73)

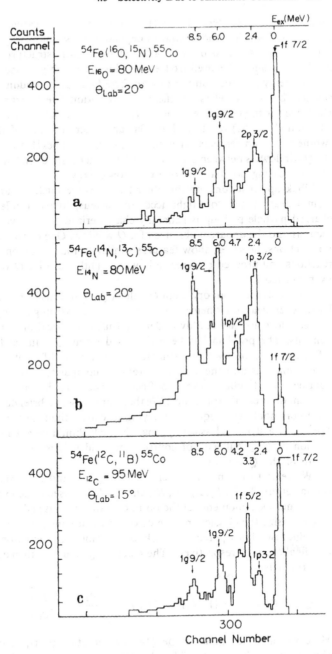

Fig. 4.9. Energy spectra of products from one-proton stripping reactions, produced by bombarding a ^{54}Fe target with ^{16}O, ^{14}N and ^{12}C projectiles at the laboratory energies indicated. (From Po 73)

exhibit a striking tendency to select a certain band of negative Q values. The position and width of the Q window depend in a systematic way on kinematic variables, but seem to be little affected by nuclear structure. For a given incident channel, the preferred energy loss is roughly proportional to incident energy and the number of transferred charges; in addition, there is a smaller, but nevertheless significant contribution due to transferred mass. Evidence for the latter effect comes, for example, from comparisons of (^{16}O, ^{12}C) and (^{16}O, ^{14}C) reactions [Oe 73a] and from a study of the (^{18}O, ^{16}O) two-neutron transfer reaction on the even nickel isotopes [Kö 73]. The width of the Q windows is between 5 and 10 MeV, with a possible tendency to increase with increasing bombarding or excitation energy.

"Pick-up" reactions, on the other hand, where nucleons are transferred from a heavy donor to a light acceptor nucleus, follow a different pattern: charged particle pick-up is generally characterized by relatively small cross sections and the apparent absence of a Q window. One or two neutron pick-up reactions, however, show features similar to the corresponding stripping reactions and proceed preferentially with slightly negative Q values [see, for example, Kö 73]

The existence of an optimum Q value (Q_{opt}) in transfer reactions, and its basic dependence on kinematic variables like bombarding energy, transferred charge and mass, are readily understood using classical or semiclassical arguments. The problem has been discussed extensively in the literature (for references, see below); unfortunately, the use of different notations and approximations and the lack of systematic comparisons with experimental data have led to considerable confusion concerning the range of validity of different theoretical expressions. We therefore present a brief derivation of the most commonly used equations, trying to clarify their mutual relationship. We follow a strictly classical approach, which should be adequate for a discussion of Q_{opt}, but will, of course, not yield information concerning the width of the Q window.

We refer to the notation defined in Fig. 4.1 and 5.3, except that now capital letters R_i, R_f; R_1, R_2; K_i, K_f; and L_i, L_f are used to denote relative distances and local momenta at the point of transfer. This is done—as before—to emphasize the classical nature of our present considerations. We further introduce two-body potentials V_i, V_f to describe the interaction in the initial and final channel, respectively. The reaction Q value can then quite generally be written as

$$Q_{if} = E_f - E_i = \{V_f(R_f) - V_i(R_i)\} + \left[\frac{\hbar^2 K_f^2}{2\mu_f} - \frac{\hbar^2 K_i^2}{2\mu_i}\right]. \qquad (4.27)$$

Here we have decomposed the Q-value into two parts, representing the changes in two-body potential and kinetic energy, respectively, produced by rearrangement of the transferred particle x. In order to identify the right-hand side of equation (4.27) with an optimum Q value, Q_{opt}, we have to make a suitable choice of optimum values for R_i, R_f, K_i, and K_f. These must satisfy the conditions of a continuous trajectory for each constituent fragment, and of maximum transfer probability. We are thus led to assume matching of the initial and final trajectories at their respective classical turning points, or—if no turning point exists in the incident channel ($L_i < L_{gr}$)—at the effective potential-barrier where the radial velocity has a minimum. With this interpretation of R_i and R_f, and substituting angular for linear momenta, we can rewrite (4.27) as

$$Q_{opt} = \{V_f(R_f) - V_i(R_i)\} + \left\{\frac{\hbar^2 L_f^2}{2\mu_f R_f^2} - \frac{\hbar^2 L_i^2}{2\mu_i R_i^2}\right\} - \left[\frac{\hbar^2(K_i^2 R_i^2 - L_i^2)}{2\mu_i R_i^2}\right]. \quad (4.28)$$

Here we have further decomposed the kinetic energy term into an angular-momentum-dependent term and a radial term. We note that the latter is always negative and appears only for low incident partial waves, which are predominantly absorbed into compound nucleus formation. For these partial waves there is an unavoidable mismatch of the incident and outgoing radial momenta, which must be compensated either by the momentum of the transferred particle, or by some independent dissipative mechanism. We shall return to this interesting point below.

As a next step, we have to relate R_i to R_f, and L_i to L_f. This involves assumptions on the position and momentum of the transferred particle at the moment of transfer. In order to keep our discussion as general as possible we introduce the following model-dependent parameters (to be specified later):

$$\gamma_V = \frac{V_f(R_f)}{V_i(R_i)} ; \quad \gamma_L = \frac{L_f^2}{\mu_f R_f^2} \frac{\mu_i R_i^2}{L_i^2} . \quad (4.29)$$

It is further convenient to define variables ξ, χ for each channel by writing

$$\xi = 1 - \frac{\hbar^2 L^2}{2\mu R^2 E} ; \quad \chi = \frac{V(R)}{E} . \quad (4.30)$$

In the limiting case of a pure Coulomb potential, these variables are related to the corresponding scattering angle Θ by

$$\xi = \chi = \frac{2\sin\frac{\Theta}{2}}{1 + \sin\frac{\Theta}{2}} . \quad (4.31)$$

In practice, this simple relationship will be only approximately valid, due to the influence of nuclear forces on the trajectories. A rigorous calculation of Q_{opt} in terms of scattering angles—within our classical model—will therefore involve the corresponding classical deflection function $\Theta(L)$ [see (3.54)–(3.58)].

Using (4.28), (4.29), and (4.30) we can write for Q_{opt}, without loss of generality,

$$Q_{opt} = -E_i[(1 - \gamma_V)\chi_i + (1 - \gamma_L)(1 - \xi_i) + (\xi_i - \chi_i)], \qquad (4.32)$$

where again the three contributions associated with changes in potential energy, orbital, and radial kinetic energy are evident. An alternative formulation, which is frequently found in the literature, makes use of an "effective" (optimum) Q value, obtained by subtraction of the potential energy term from Q_{opt}:

$$\begin{aligned} Q_{eff} &= Q_{opt} - [V_f(R_f) - V_i(R_i)] \\ &= -E_i[(1 - \gamma_L)(1 - \xi_i) + (\xi_i - \chi_i)] \end{aligned} \qquad (4.33)$$

The different model-dependent expressions for Q_{opt} (or Q_{eff}) are now readily obtained from (4.32) and (4.33) by inserting appropriate values for the parameters γ_V and γ_L. Considering first the potential energy and neglecting the nuclear potential, we can write

$$\gamma_V = \frac{Z_1(Z_2 + z)}{(Z_1 + z)Z_2} \frac{R_i}{R_f}. \qquad (4.34)$$

Assuming further a collinear configuration of the transferred particle and the two cores, we can calculate the difference between R_f and R_i [see (5.33) and (5.34)]:

$$\frac{R_f - R_i}{R_f} \approx \frac{m_x}{R_1 + R_2}\left(\frac{R_1}{M_1} - \frac{R_2}{M_2}\right), \qquad (4.35)$$

where R_1 and R_2 are effective radii of the initial and final bound states of particle x.

At bombarding energies below the Coulomb barrier, the cross section is concentrated at back angles ($\Theta \approx \pi$). In this particular situation we have $\chi_i = \xi_i = 1$, and only the potential energy term contributes to Q_{opt} ($Q_{eff} = 0$). Equations (4.32), (4.34), and (4.35) then yield

$$\begin{aligned} Q_{opt} &= -E_i(1 - \gamma_V) \\ &= -E_i\left[\frac{z(Z_2 - Z_1)}{(Z_1 + z)Z_2}\right. \\ &\quad \left. + \frac{m_x}{R_1 + R_2}\left(\frac{R_1}{M_1} - \frac{R_2}{M_2}\right)\frac{Z_1(Z_2 + z)}{(Z_1 + z)Z_2}\right]. \end{aligned} \qquad (4.36)$$

Here we have separated effects due to charge and mass transfer; the former arise from a change in Coulomb potential at fixed distance, whereas the latter are a consequence of "recoil". Both contributions to Q_{opt} are seen to be negative, if charge or mass are transferred from a light core to a heavy core, and positive in the opposite case. The absolute value of Q_{opt} should increase with increasing amount of transferred charge or mass, and with increasing asymmetry of the cores; it should be large for light projectiles incident on heavy targets, and should vanish for identical cores. These predictions, which were first discussed in detail by Buttle and Goldfarb [Bu 71], are well supported by the available experimental evidence on transfer reactions below and near the Coulomb barrier. They also explain the absence of observable Q windows in transfer reactions proceeding from a heavy to a light core as due to a lack of final states corresponding to positive Q values.

At energies above the Coulomb barrier and scattering angles $\Theta < \pi$, the kinetic-energy terms come into play. Somewhat different considerations apply to the angular regions forward and backward of the grazing angle, and we consider first the former region. Here we have no mismatch of radial momenta and—neglecting again the nuclear potential—both χ_i and ξ_i are given by (4.31). From (4.32) we obtain

$$Q_{opt} = - E_i[(1 - \gamma_V)\xi_i + (1 - \gamma_L)(1 - \xi_i)]$$
$$= - E_i[(1 - \gamma_L) + (\gamma_L - \gamma_V)\xi_i] \tag{4.37}$$

In order to compare this equation with experiment, it must be reformulated in terms of the experimental angle of observation Θ. We define

$$\xi = \frac{1}{2}(\xi_i + \xi_f) \approx \frac{2 \sin \frac{\Theta}{2}}{1 + \sin \frac{\Theta}{2}}, \tag{4.38}$$

and eliminate ξ_f using (4.29) and (4.30). Retaining only the first-order term in Q_{if}/E_i, we obtain

$$\xi_i \approx \xi\left\{1 + \frac{1}{2}\left[(1 - \gamma_V) + \frac{Q_{if}}{E_i}\right]\right\} \approx \xi\left\{1 + \frac{1}{2}(\gamma_L - \gamma_V)(1 - \xi)\right\}. \tag{4.39}$$

Inserting (4.39) into (4.37) yields

$$Q_{opt} = - E_i[(1 - \gamma_L) + (\gamma_L - \gamma_V)\xi + \frac{1}{2}(\gamma_L - \gamma_V)^2\xi(1 - \xi)]. \tag{4.40}$$

Similarly the "effective Q value", as defined in (4.33), can be written as

$$Q_{eff} = - E_i(1 - \gamma_L)[(1 - \xi) - \frac{1}{2}(\gamma_L - \gamma_V)\xi(1 - \xi)]. \tag{4.41}$$

It is instructive to express Q_{eff} and the angular momentum transfer $\hbar(L_i - L_f)$ in terms of the relative velocity of the fragments at the turning point. The resulting equations are

$$Q_{eff} = -(1 - \gamma_L)\frac{\mu_i v_i^2}{2}, \tag{4.42}$$

$$\hbar(L_i - L_f) \approx \frac{R_i + R_f}{2v}\left(\frac{R_2 - R_1}{R_2 + R_1}\frac{m_x v^2}{2} - Q_{eff}\right), \tag{4.43}$$

where μ_i and v_i refer to the incident channel, whereas v denotes an average relative velocity for the incident and outgoing channels. In (4.43) terms of higher order in the ratio of transferred to total mass have been neglected.

In (4.37) and (4.39–41), the parameter γ_V is again given by (4.34) and (4.35). The additional parameter γ_L depends on assumptions concerning the matching of angular momenta in the incident and outgoing channels. More precisely, by specifying γ_L, one determines how the momentum of the projectile is shared between the core C_1 and the transferred particle x at the instant of transfer. Various relevant models have been proposed in the literature; the basic assumptions and resulting expressions for the quantity $(1 - \gamma_L)$ are collected in Table 4.2.

Table 4.2.   Comparison of different optimum Q-value models

Designation	Model assumption	Resulting expression for $(1 - \gamma_L)$	
a	$k_i - k_f = \dfrac{m_x}{M_1 + m_x} k_i$	$\dfrac{Mm_x}{(M_1 + m_x)(M_2 + m_x)} \approx$	$\dfrac{M_1 + M_2}{M_1 M_2} m_x$
b	$k_i - k_f = 0$	$\dfrac{(M_1 - M_2)m_x}{M_1(M_2 + m_x)} \approx$	$\dfrac{M_1 - M_2}{M_1 M_2} m_x$
c	$L_i = L_f$	$\dfrac{\mu_f R_f^2 - \mu_i R_i^2}{\mu_f R_f^2} \approx$	$\dfrac{M_1 + M_2}{M_1 M_2} \dfrac{R_1 - R_2}{R_1 + R_2} m_x$
d	$\xi_i = \xi_f(\Theta_i = \Theta_f)$	$= (1 - \gamma_V)$ [see (4.34) (4.35)]	
e	$Q_{eff} = 0$	$= 0$	

Model a, proposed by Siemens et al. [Si 71], assumes that the particle x is at rest with respect to the donor nucleus while being transferred. This yields the following simple result for the most probable effective Q value

$$Q_{eff} = -\frac{1}{2} m_x v^2. \tag{4.44}$$

In contrast to all the other models in Table 4.2, the model of Siemens et al. is not symmetric under exchange of the incident and outgoing channels. This has the remarkable consequence that transfer always results in a loss of kinetic energy and angular momentum from orbital to intrinsic motion, regardless of the direction in which particle x is transferred. In particular, such a loss is also predicted for transfer between identical cores, and for successive transfers in either direction, which do not lead to a net rearrangement of either mass or charge. The model, therefore, implies that nucleon transfer and nucleon exchange may constitute an important mechanism for dissipation of energy and angular momentum in inelastic collisions. An extension of the model, which explicitly includes nonzero intrinsic angular momenta of the transferred particle in the donor nucleus, has been given by Brink [Br 72a] and will be discussed in more detail below.

Models b, c, d, and e in Table 4.2 may be termed "reversible" in the sense that they are symmetric with respect to the incident and outgoing channels; they predict optimum Q values, which change sign upon reversal of the direction of transfer, and vanish for identical cores. It should be noted that these models yield distinctly different results, as will be shown below. Model b, for example, predicts sizeable positive values of Q_{eff} for $M_2 > M_1$, in striking contrast to model a, while model c usually predicts very small values of Q_{eff}, practically indistinguishable from model e. Model c has been applied successfully by Schiffer et al. to $(^{16}O, {}^{12}C)$ reactions on medium-heavy targets near the Coulomb barrier [Sc 73a, An 73a]. Toepffer has performed a "quasi-classical" analysis of the Q-window problem, starting from the DWBA amplitude and using the JWKB approximation [To 72]. He deduced model d $(\Theta_i = \Theta_f)$, and obtained a good fit to ^{15}N-induced multinucleon transfer data, taken well above the Coulomb barrier at forward angles. In this model we have $\gamma_L = \gamma_V$ and Q_{opt} is given by (4.36) independently of scattering angle, as long as pure Coulomb trajectories are assumed. In the absence of a rigorous theory of the Q-window effect, and in view of the possibility that nuclear structure may modify the purely kinematic aspects, we shall follow an empirical approach, judging the various models by their overlap with experimental observation. A comparison of the different predictions with experimental data for two selected cases is given in Figs. 4.10 and 4.11 below.

First, however, we have to discuss the angular region backward of the grazing angle, where the incident classical trajectory leads to absorption into a compound nucleus. Re-emergence of the system after transfer then requires conversion of intrinsic radial momentum to relative radial momentum (or vice versa). We refer to this situation, which will in general result in reduced transfer cross sections, as "radial mismatch". Minimum mismatch, and hence maximum transfer probability, will be achieved, if the transfer occurs near the point of minimum incident radial momentum, and the outgoing radial momentum is approximately zero. The latter condition suggests the following expression for Q_{opt}:

$$Q_{opt} = V_f(R_f)\zeta_f^{-1} - E_i \approx \frac{Z_1(Z_2 + z)}{2R_f}\left(1 + \csc\frac{\Theta}{2}\right) - E_i, \qquad (4.45)$$

where the parameter R_f is mainly determined by the absorptive properties of the incident and outgoing channels, and depends only weakly on angular momentum or energy. Von Oertzen has shown that (4.45)—with empirically adjusted but plausible values of R_f—accounts well for a considerable body of results for ^{12}C- and ^{16}O-induced transfer reactions, including in particular the dependence of Q_{opt} on bombarding energy and the variation of the angle of maximum yield with outgoing energy [Oe 73a, Oe 73b]. Interesting evidence in this context also comes from a study of Larsen et al. of one-nucleon transfer to low-lying final states, induced by 77 to 116 MeV ^{12}C bombardment of a ^{208}Pb target [La 72]. These authors found that the variation of the angle of maximum yield with incident and outgoing energy was characteristically different for neutron pick-up from ^{208}Pb and proton stripping onto ^{208}Pb. In the former case, the data were consistent with classical trajectories corresponding to a constant interaction distance in the outgoing channel, whereas in the latter case, trajectories with a constant interaction distance in the incident channel had to be assumed. This result may be understood in our simple classical picture, if recoil is taken into account: the difference of channel radii $R_f - R_i$ is negative for the pick-up reaction, and positive for the stripping reaction [see (4.35)]. Absorption into the compound nucleus will affect primarily the channel with the smaller radius, and hence the effective cut-off distance is determined by the outgoing channel in pick-up from a heavy target, and by the incident channel in stripping onto a heavy target.

In Figs. 4.10 and 4.11 we compare simple classical predictions for Q_{opt} with experimental results in two cases where complete angular distributions are available. The data have been obtained by Rehm et al. [Re 73] for the reaction ^{90}Zr(^{16}O, ^{12}C)^{94}Mo at 52.5 MeV (close to the barrier) and by Wilczynski et al. [Wi 75] for the reaction ^{58}Ni(^{16}O, ^{12}C)^{62}Zn at 60 MeV (above the barrier); they are not shown in detail, but their trend is indicated by the full points in Figs. 4.10 and 4.11. Curves marked a–e have been calculated using (4.40) and (4.38), with γ_V given by (4.34) and (4.35), and γ_L according to Table 4.2. Curves marked f, on the other hand, have been obtained with the help of (4.45), assuming $R_f = r_0(A_1^{1/3} + A_2^{1/3})$ with $r_0 = 1.5$ fm.

It should be remembered that models a–e are appropriate for angles smaller than the grazing angle (all angles below the Coulomb barrier), whereas model f is applicable to angles larger than the grazing angle. We further note that deflection by the nuclear potential is neglected in all cases.

Several points are immediately obvious from a comparison of the theoretical predictions in Figs. 4.10 and 4.11. Models a–e become identical at back angles, as they only differ with respect to angular momentum matching. Concerning the forward-angle behaviour, three classes of models may be distinguished: models a and d predict a vanishing or small variation of Q_{opt} with angle; models e and c, on the other hand, predict a significant decrease; and model b a still sharper decrease of $-Q_{opt}$ towards small angles. Within each class, there is little hope of differentiating experimentally, as long as (^{16}O, ^{12}C) reactions on medium or heavy targets are considered. Changes in charge-to-mass ratio of the transferred particle, or in mass asymmetry of the cores, will change this pattern, however. In particular, for identical or nearly identical cores, all models except model a (and f) predict $Q_{opt} \approx 0$ for all angles and energies, whereas model a predicts $Q_{opt} = 0$ for $\Theta = \pi$, but negative values of Q_{opt}, increasing in magnitude towards smaller angles, for $\Theta < \pi$ and energies above the barrier.

Comparing now the predictions of the model with experiment, we find that back-angle data at energies below or near the Coulomb barrier are generally well reproduced by all models. Under these conditions, and for cases similar to those shown in Figs. 4.10 and 4.11, recoil effects account for about 25% of $-Q_{opt}$; they must be included in the model predictions, as has been pointed out by several authors [see, for example, Sc 73a and Re 73]. Back-

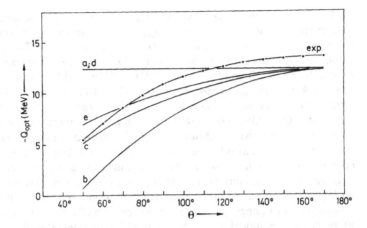

Fig. 4.10. Comparison of experimental results for Q_{opt} in the reaction $^{90}Zr(^{16}O, ^{12}C)^{94}Mo$ at 52.5 MeV bombarding energy (from Re 73) with predictions of different classical models. The letters on the theoretical curves correspond to Table 4.2; for further explanation see text.

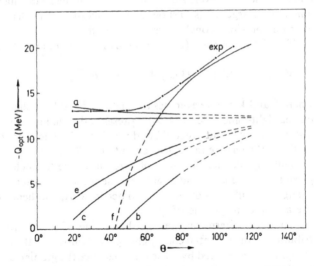

Fig. 4.11. Same as Fig. 4.10, but for the reaction $^{58}Ni(^{16}O, ^{12}C)^{62}Zn$ at 60 MeV bombarding energy (experimental data from Wi 75)

angle data at energies above the barrier, on the other hand, are in qualitative agreement with (4.45) assuming a constant interaction radius. This observation lends strong support to the idea of a radial cut-off due to absorption, as suggested by von Oertzen [Oe 73a, Oe 73b] and is consistent with our earlier discussion of "radial mismatch".

The forward-angle data taken at energies below or near the barrier (Fig. 4.10) exhibit a definite decrease of $-Q_{opt}$ towards smaller angles, and appear to be best reproduced with models c($\Delta L = 0$) or e($Q_{eff} = 0$). The same conclusion was reached earlier by Schiffer et al. [Sc 73a, An 73a] for transfer reactions induced by ^{16}O bombardment of ^{172}Yb, ^{124}Sn, and ^{88}Sr. In contrast the results taken at energies above the barrier (Fig. 4.11) require model a or d, whereas the other models fail dramatically by predicting much too small values of $-Q_{opt}$. Although there is clearly a need for more systematic data, the following tentative conclusions may be drawn on the basis of the examples shown here, as well as other results. At low energies, transfer seems to proceed preferentially in a manner that minimizes the amount of energy and angular momentum converted from relative to internal motion (models c,e). At high energies, on the other hand, transfer seems to be dominated by the requirement of momentum conservation for each of the constituents (model a). In either case, model b is inconsistent with experimental evidence.

It remains to discuss briefly the influence on Q_{opt} of the deflection by nuclear forces, which has been neglected so far. Unfortunately this influence cannot be described by simple analytical expressions, both due to our incomplete knowledge of the nuclear potential and due to mathematical complications. For these reasons, the problem has been largely disregarded in the literature [see, however, Re 73 and Ca 73]. In the following we present semiquantitative arguments to show that an inclusion of nuclear forces should not alter our previous conclusions substantially.

The change in total (Coulomb plus nuclear) potential may be written as

$$V_f(R_f) - V_i(R_i) \approx \Delta V(\bar{R}) + \Delta R \cdot (\text{grad } \bar{V})_R, \qquad (4.46)$$

where $\bar{R}$ and $\bar{V}$ denote averages for the initial and final channel. The first term on the right-hand side of (4.46) represents the change in potential at a fixed distance (i.e. without "recoil") and is affected by nuclear forces in two ways: firstly by the change in nuclear potential at $\bar{R}$ (which may be neglected) and secondly by the fact that $\bar{R}$ itself is reduced due to nuclear attraction for a given incident energy and angular momentum. This latter effect increases the change in Coulomb potential which occurs at $\bar{R}$, and hence tends to increase the absolute magnitude of Q_{opt}.

The second term on the right-hand side of (4.46) gives the recoil contribution to the potential energy change. In contrast to the first term, this term is significantly reduced by nuclear forces since the gradients of the nuclear and Coulomb potentials are generally comparable and of opposite sign at typical interaction distances. As a result, the nuclear effects are partially cancelled, as far as the potential energy is concerned; similar arguments (not reproduced here) can also be given for the change in kinetic energy.

We summarize the discussion of this point by noting that the nuclear potential enters in a rather complex way. It appears that the main features of Q_{opt}—at least in the nonabsorptive region—can be understood without explicitly taking into account the nuclear potential. Caution should be exercised, however, in drawing detailed quantitative conclusions, unless the nuclear effects are included.

So far we have not specified the intrinsic angular momentum of the transferred particle in the donor or acceptor nucleus. This means that absence of any selectivity with respect to intrinsic angular momentum for other than kinematic reasons was implied. In practice, however, the observation of isolated levels with well-defined angular momentum, or a high density of such levels in particular regions of excitation, may impose additional constraints. The question how kinematic matching affects the transfer probability between states of specified intrinsic angular momentum has been discussed by Brink, using a classical description for the rotational velocity of particle x in either nucleus and assuming that the distance of the cores at the instant of transfer is equal to $R_1 + R_2$ [Br 72a]. In the model of Brink, kinematic continuity is achieved by requiring

$$\Delta k_x = \frac{m_x v}{\hbar} - \frac{\lambda_1}{R_1} - \frac{\lambda_2}{R_2} \approx 0, \tag{4.47}$$

$$\Delta\Lambda = \frac{1}{2} \frac{m_x v}{\hbar} (R_1 - R_2) + \frac{Q_{\text{eff}}}{\hbar v} (R_1 + R_2) + \lambda_2 - \lambda_1 \approx 0, \tag{4.48}$$

where Λ, λ_1, and λ_2 are components perpendicular to the reaction plane of the total angular momentum (Λ) and the intrinsic angular momenta of particle x with respect to core 1 (λ_1) and 2 (λ_2), respectively.

These conditions are a consequence of the conservation laws for the momentum of the transferred particle on one hand, and the total angular momentum on the other hand. A third condition

$$l_1 + \lambda_1 = \text{even}; \quad l_2 + \lambda_2 = \text{even}, \tag{4.49}$$

where l_1, l_2 are the intrinsic orbital angular momenta, arises from the collinear approximation for the transfer configuration, together with symmetry properties of the spherical harmonics describing the intrinsic orbital motion.

We note that the conditions (4.47) and (4.48) cannot be satisfied simultaneously for arbitrary combinations λ_1, λ_2 at a given bombarding energy. For a specified value of λ_1, however, they define optimum values of Q_{eff} and λ_2. In particular, for $\lambda_1 = 0$, we obtain the result

$$Q_{\text{eff}} = -\frac{1}{2} m_x v^2; \quad \hbar\lambda_2 = m_x v R_2, \tag{4.50}$$

which shows that in this case the model of Brink is equivalent to that of Siemens et al. [compare (4.44) and (4.43)]. The assumption $\lambda_1 \approx 0$ is probably not unreasonable in general, provided a sufficient density of levels with angular momenta satisfying (4.50) is available at the appropriate excitation energy in the acceptor nucleus.

Equations(4.47–49) have been used to predict or explain certain spectroscopic selectivities observed in high-energy transfer reactions. Examples are provided by one-nucleon stripping reactions at bombarding energies of the order of 10 MeV per nucleon, where high-spin final states are most strongly populated, especially for large negative Q values [Br 72a, Sc 73b, An 73b, An 74]. Since the latter require large values of $\lambda_2 - \lambda_1$ according to (4.48), they favour a sign change of the orbital angular momentum projection λ. Thus, if no spin flip occurs, transitions

$$j_1 = \left(l_1 + \frac{1}{2}\right) \to j_2 = \left(l_2 - \frac{1}{2}\right) \text{ or } j_1 = \left(l_1 - \frac{1}{2}\right) \to j_2 = \left(l_2 + \frac{1}{2}\right)$$

are preferentially induced.

The width of the Q window, i.e. the change in cross section due to kinematic mismatch, has also been discussed in the framework of Brink's model [Br 72a, Sc 73b], on the basis of other semiclassical approaches [Br 72 b, c] and using the DWBA [see, for example, Oe 74]. The results are generally consistent with experimental observation, where comparisons have been made.

References Chapter 4

(For AR 73, HE 69, HE 74, NA 74 see Appendix D)

Al 71a Alder, K.; Trautmann, D.: Ann. Phys. (N.Y.) *66*, 884 (1971)
Al 71b Alder, K.; Trautmann, D.: Nucl. Phys. *A178*, 60 (1971)
Al 72   Alder, K. et al.: Nucl. Phys. *A191*, 399 (1972)
An 73a Anantaraman, N.; Katori, K.; Schiffer, S.P.: AR 73 Vol. II p. 413
An 73b Anyas-Weiss, N. et al.: Phys. Lett. *45B*, 231 (1973)
An 74   Anyas-Weiss N. et, al.: Phys. Rep *.12*, 201 (1974)
As 73a Ascuitto, R.J.; Glendenning, N.K.: Phys. Lett. *45B*, 85 (1973)
As 73b Ascuitto, R.J.; Glendenning, N.K.: Phys. Lett. *47B*, 332 (1973)
As 74   Ascuitto, R.J.; Vaagen, J.S.: NA 74 Vol. 2 p. 257
Au 64   Austern, N.; Drisko, R.; Satchler, G.: Phys. Rev. *133*, B 3 (1964)
Au 75   Austern, N.: Phys. Rev. *C12*, 128 (1975)
Bo 66   Bock, R.; Grosse-Schulte, M.; von Oertzen, W.: Phys. Lett. *22*, 456 (1966)
Bo 72   Bonche, P. et al.: Phys. Rev. *C6*, 577 (1972)
Bo 73   Bond, P.L. et al.: Phys. Lett. *47B*, 231 (1973)
Br 72a Brink, D.M.: Phys. Lett. *40B*, 37 (1972)
Br 72b Broglia, R.A.; Winther, A.: Nucl. Phys. *A182*, 112 (1972)
Br 72c Broglia, R.A.; Winther, A.: Phys. Rep. *4*, 153 (1972)
Bu 66   Buttle, P.J.A.; Goldfarb, L.J.B.: Nucl. Phys. *78*, 409 (1966)
Bu 68   Buttle, P.J.A.; Goldfarb, L.J.B.: Nucl. Phys. *A115*, 461 (1968)
Bu 71   Buttle, D.J.A.; Goldfarb, L.S.B.: Nucl. Phys. *A176*, 299 (1971)
Ca 73   Cassagnou, Y. et al.: AR 73 p. 485
Ch 73   Christensen, P.R. et al.: Nucl. Phys. *A207*, 33 (1973)
Fr 64   Frahn, W.E.; Venter, R.H.: Nucl. Phys. *59*, 651 (1964)
Fr 74   Frahn, W.E.: HE 74 p. 102
Ga 74   Garrett, J.D.: HE 74 p. 59
Ge 74   Gelbke, C.K. et al.: Nucl. Phys. *A219*, 253 (1974)
Go 74   Gobbi, A.: NA 74 Vol. 2 p. 211
Ka 74a Kahana, S.; Bond, P.D.; Chasman, C.: Phys. Lett. *50B*, 199 (1974)
Ka 74b Kahana, S.: NA 74 Vol. 2 p. 189
Ka 75   Kalinsky, D. et al.: Nucl. Phys. *A250*, 364 (1975)
Kö 73   Körner, H.J.: AR 73 Vol. I p. 9
La 72   Larsen, J.S. et al.: Phys. Lett. *42B*, 205 (1972)
Le 74   Le Vine, M.J. et al.: Phys. Rev. *C10*, 1602 (1974)
Mc 60   McIntyre, J.A.; Watts, T.L.; Jobes, F.C.: Phys. Rev. *119*, 1331 (1960)

Oe 69   von Oertzen, W. et al.: HE 69 p. 156

Oe 73a   von Oertzen, W.; Bohlen, H.G.; Gebauer, B.: Nucl. Phys. *A207*, 91 (1973)

Oe 73b   von Oertzen, W.: AR 73 Vol. II p. 675

Oe 74   von Oertzen, W.: In *Nuclear Spectroscopy and Reactions*, Part B, ed. by S. Cerny, p. 279. New York, London: Academic Press 1974

Po 73   Pougheon, F.; Roussel, P.: AR 73 Vol. II p. 637

Re 73   Rehm, K.E. et al.: Phys. Lett. *46B*, 353 (1973)

Re 75   Reisdorf, W.N.; Lau, P.H.; Vandenbosch, R.: Nucl. Phys. *A253*, 490 (1975)

Sa 73   Satchler, G.R.: AR 73 Vol. I p. 145

Sc 72   Schlotthauer-Voos, U. et al.: Nucl. Phys. *A180*, 385 (1972)

Sc 73a   Schiffer, J.P. et al.: Phys. Lett. *44B*, 47 (1973)

Sc 73b   Scott, D.K.: AR 73 Vol. I p. 97

Sc 74   Scott, D.K.: HE 74 p. 165

Si 71   Siemens, P.J. et al.; Phys. Lett. *36B*, 24 (1971)

Si 72   Siemens, P.J.; Becchetti, F.D.: Phys. Lett. *42B*, 389 (1972)

St 64   Strutinsky, V.M.: Sov. Phys. JETP *19*, 1401 (1964)

St 73   Strutinsky, V.M.: Phys. Lett. *44B*, 245 (1973)

To 72   Toepffer, Ch.: Z. Phys. *253*, 78 (1972)

Tr 70   Trautmann, D.; Alder, K.: Helv. Phys. Acta *43*, 363 (1970)

Wi 73   Winther, A.: AR 73 Vol. I p.1

Wi 75   Wilczynski, J. et al.: Nucl. Phys. *A244*, 147 (1975)

5. Quasi-Elastic Transfer Reactions

5.1 Formal Description I: Spectroscopic Aspects

We consider first the simplest possible situation, where the transferred particle x has no internal structure, and occupies definite single-particle states with orbital and total angular momenta l_1, j_1 and l_2, j_2 relative to the cores C_1 and C_2, respectively. The differential cross section can then be written

$$\frac{d\sigma}{d\Omega} = S_A(l_1, j_1)S_B(l_2, j_2) \sum_{l,m} |f_{j_1 j_2 lm}(\Theta)|^2. \tag{5.1}$$

Here S_A and S_B are "spectroscopic factors", which are defined as overlaps of the physical states A and B with pure single particle states $\{C_1 + x\}_{l_1 j_1}$ and $\{C_2 + x\}_{l_2 j_2}$, respectively. The transfer amplitudes f contain the reaction dynamics and depend on bombarding energy, Q value, and emission angle. The quantum number l denotes the angular momentum transfer in the reaction, which is given by

$$l = (I_{C_1} + I_B) - (I_{C_2} + I_A) = j_2 - j_1 = l_2 - l_1, \tag{5.2}$$

provided, as we shall assume in the following, the interaction is independent of spin and hence leaves the spin orientation of the particles C_1, C_2 and x unaffected. Angular momentum conservation requires that any change in total intrinsic angular momentum must be accompanied by a corresponding change in the angular momentum of relative motion. This leads to the relationship

$$l = l_i - l_f, \tag{5.3}$$

where l_i and l_f are the orbital angular momenta of the initial and final channel, respectively.

In a completely analogous manner we can define the "parity transfer" in the reaction

$$\Pi = \Pi_{C_1}\Pi_{C_2}\Pi_A\Pi_B = (-)^{l_1+l_2}. \tag{5.4}$$

Again, parity conservation requires

$$\Pi = (-)^{l_i + l_f}, \tag{5.5}$$

and hence, combining (5.4) and (5.5),

$$l_1 + l_2 + l_i + l_f = \text{even}. \tag{5.6}$$

Equation (5.2) yields the following selection rules for the quantum number l of angular momentum transfer:

$$|j_2 - j_1| \leq l \leq j_2 + j_1, \tag{5.7}$$

$$|l_2 - l_1| \leq l \leq l_2 + l_1. \tag{5.8}$$

We note that parity conservation does not, in general, impose any further restriction on l. Although Π has a definite value, this only fixes the odd–even character of $l_i + l_f$, but not that of l. Therefore both even and odd values of l are allowed simultaneously, as long as they are compatible with (5.7) and (5.8).

An additional selection rule arises, however, if the transfer is assumed to take place while the particle x it located on the line connecting the cores C_1 and C_2. For such a collinear configuration the vectors r_i and r_f are both parallel to the vector r_{12} (see Fig. 4.1), and consequently the angular momenta $l_i = r_i \times k_i$, $l_f = r_f \times k_f$, and $l = l_i - l_f$ all have zero projection along a definite direction in space (here parallel to r_{12}). It follows then from general properties of angular momentum coupling coefficients, that (5.3) can only be satisfied if

$$l_i + l_f - l = \text{even}. \tag{5.9}$$

This condition, combined with (5.6), yields the selection rule

$$l_1 + l_2 + l = \text{even}, \quad \Pi = (-)^l, \tag{5.10}$$

which requires that, for a given transition, l can assume only even or only odd values, depending on the parity change Π. Transitions with l values allowed by (5.10) are often referred to in the literature as "normal parity transitions", as opposed to "non-normal parity transitions". The latter are ascribed to "recoil effects", a phenomenon which will be discussed in Sect. 5.2.

Having summarized the angular momentum selection rules for one-nucleon (or cluster) transfer, we return to (5.1) for the differential cross section. An important property of the right-hand side of (5.1) is that it factorizes into a product of spectroscopic factors $S_A S_B$, and a term which reflects the kinematic aspects of the reaction. The latter depends on the quantum numbers and binding energies of the initial (A) and final (B) bound states of the particle x and the angular momentum transfer l; as stated before, it determines the dependence of the cross section on incident energy, emission angle, and

Q value. The separation of the cross section into a spectroscopic and kinematic term—well known from stripping and pick-up reactions with light projectiles—greatly simplifies the extraction of "spectroscopic information" on the single-particle nature of the states involved from experimental data. In order to obtain quantitative results, however, it is necessary to compute the kinematic part of the cross section, using a more or less approximate reaction theory such as the DWBA (see Sect. 3.4 and 5.2). It turns out that this task is much more difficult in heavy-ion transfer than with light projectiles ($A < 4$), since several approximations break down in the former case which are known to yield satisfactory results in the latter. Nevertheless, considerable progress has been achieved towards a reliable quantitative analysis of transfer cross sections.

Whereas the absolute magnitude of the differential cross section is affected both by the spectroscopic factors S_A, S_B and by the kinematic terms, its angular dependence is determined entirely by the latter. A question of considerable importance is, then, to what extent spectroscopic information can be deduced from angular distributions via the selection rules for the angular momentum transfer l (5.7, 8, 10). In single-nucleon transfer reactions induced by light projectiles ($A < 4$) at energies above the Coulomb barrier, the angular distributions in general exhibit a pronounced structure, which depends characteristically on l. Since further $l_1 = 0$, it is usually possible to assign l_2, at least in cases where only one orbit l_2, j_2 is involved. This straightforward method of assigning l values has contributed in large measure to the experimental verification of the nuclear shell model.

In transfer reactions induced by more complex projectiles, however, the situation is not as simple. With comparatively light target nuclei and projectiles ($A \lesssim 40$) at energies above the Coulomb barrier, pronounced oscillations of the differential cross section as a function of angle are frequently observed. As an example we show in Fig. 5.1 angular distributions measured by von Oertzen et al. for the proton transfer reaction $^{12}C(^{19}F, {}^{20}Ne)^{11}B$ (Oe 69a). Theoretical analyses show that the phase pattern of these oscillations depends on the angular momentum transfer l, although not in such a distinctive manner as with light projectiles. Since, further, usually $l_1 \neq 0$, several values of l may be allowed by the selection rules, and may appreciably contribute to the cross section (especially if "recoil effects" are important). In such cases the structure tends to be washed out and it may be difficult or impossible to deduce l values from an angular distribution. In addition, the oscillating structure may be difficult to resolve experimentally, due to its short period and relatively small amplitude.

At bombarding energies near or below the Coulomb barrier, and generally with systems involving at least one heavy ($A > 40$) fragment, smooth, structureless angular distributions have usually been observed. The distributions are "bell shaped", with a maximum at $\Theta = \pi$ for sub-Coulomb energies, and near the grazing angle for higher energies. Again we show examples in Fig. 5.2. Clearly, these angular distributions do not carry spectroscopic information; their interpretation in semiclassical terms is discussed in Sect. 4.2.

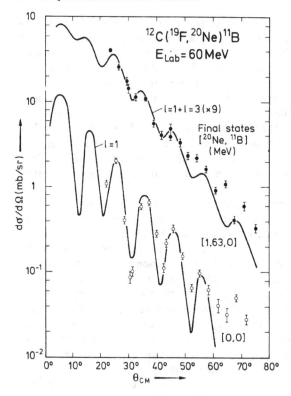

Fig. 5.1. Angular distributions of the proton-transfer reaction $^{12}\text{C}(^{19}\text{F}, ^{20}\text{Ne})^{11}\text{B}$, populating the ground state of ^{11}B and the ground and first excited states of ^{20}Ne. The centre-of-mass bombarding energy is $E_{\text{CM}} = 23.2$ MeV ($n = 4.8$). The solid lines are DWBA calculations. (From Oe 69a)

If the transferred "particle" x consists of two or more nucleons, its internal structure (as well as that of the cores C_1 and C_2) can in general no longer be disregarded. The cross section can then be written, somewhat schematically, as

$$\frac{d\sigma}{d\Omega} = \sum_{l,m} \left| \sum_{\alpha\beta\gamma\lambda} S_A^{1/2}(\alpha, \gamma, \lambda_1) \, S_B^{1/2}(\beta, \gamma, \lambda_2) f_{\alpha\beta\gamma\lambda lm} \right|^2. \tag{5.11}$$

Here α, β, γ denote sets of quantum numbers which characterize the intrinsic states of the cores C_1, C_2 and the transferred subgroup x, respectively, while $\lambda_1 = (n_1, l_1, j_1)$ and $\lambda_2 = (n_2, l_2, j_2)$ refer to the relative motion of the centre of gravity of x with respect to C_1 and C_2 [$\lambda = (\lambda_1, \lambda_2)$]. The "spectroscopic amplitudes" $S_A^{1/2}$, $S_B^{1/2}$ are coefficients in appropriate cluster expansions of $A = C_1 + x$ and $B = C_2 + x$, and contain the nuclear structure dependence of the cross section. They are weighted by the "kinematic amplitudes" $f_{\alpha\beta\gamma\lambda lm}$, which can be interpreted as transition amplitudes between specified initial and final configurations, under specified dynamical conditions, and contain the dependence of the cross section on incident energy, Q value, and angle of emission.

The cross section now involves a coherent summation over internal con-

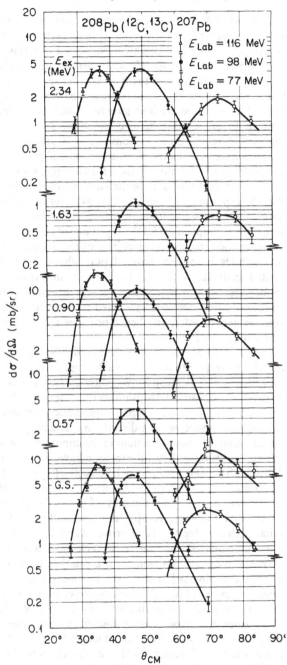

Fig. 5.2. Angular distributions of the neutron-transfer reaction $^{208}\text{Pb}(^{12}\text{C},\ ^{13}\text{C})^{207}\text{Pb}$, populating the ground state of ^{13}C and states at different excitation energies E_{ex} in ^{207}Pb. The centre-of-mass bombarding energies are $E_{CM} = 73$ MeV ($n = 30.5$), 93 MeV ($n = 27.0$), and 110 MeV ($n = 24.8$). (From La 72)

figurations of A, B, and x, and no longer factorizes into a spectroscopic and a kinematic factor for each l value, as in (5.1) (the latter equation is, however, recovered, if only one combination $\alpha,\beta,\gamma,\lambda$ contributes). This considerably complicates the analysis of experimental data, which can no longer be reduced to normalizing a theoretical cross section—calculated from a reaction model essentially without reference to the structure problem—to the data. Instead a complete nuclear structure calculation must be performed in general, before a meaningful comparison between theoretical and experimental differential cross sections can be made.

Procedures to calculate the (quite complex) ingredients of (5.11) have been discussed by several authors [see, for example, Gl 65, Ro 68, Ro 69, To 69, Bo 72a, Ku 73, An 74] and we shall not go into detail here. We note, however, that the expansion of A and B in terms of internal degrees of freedom is not unique. The use of functions describing the intrinsic and relative motion of the subgroup x—implied in (5.11)—is just one of several possible approaches, and in fact introduces certain complications to be discussed below. Other successful methods have been worked out, where the constituent nucleons are treated individually [Ba 72, Bo 73a]. In the present context we prefer the former approach, as we feel that the separation between relative and intrinsic motion of x is more directly related to the simpler case of one-nucleon (or inert cluster) transfer, and hence provides a better intuitive insight into the reaction mechanism.

The transformation from a shell-model description of the transferred nucleons with respect to A and B to a description involving relative and intrinsic coordinates of x is most conveniently carried out with harmonic-oscillator wave functions, using the Moshinsky transformation brackets [Mo 59]. A problem with this procedure is that the harmonic-oscillator functions for the relative motion cannot be used directly in calculating the kinematic amplitudes f due to their incorrect behaviour in the important tail region near and beyond the nuclear surface. Several methods have been devised to overcome this difficulty. The simplest approach consists of joining the harmonic-oscillator functions smoothly on to Hankel-function tails, with logarithmic derivatives appropriate to the binding energy of x, or of replacing the entire functions with radial functions computed numerically for Woods–Saxon wells of suitably chosen depths. An improved method is to start with Woods–Saxon wave functions for the transferred nucleons and to expand these in terms of harmonic-oscillator functions, which can then be transformed to relative and intrinsic coordinates [Dr 66]. Even this more elaborate procedure, however, cannot be expected to yield a reliable description of nucleonic correlations at the nuclear surface, unless the nuclear-structure problem and the calculation of the kinematic amplitudes are treated in a consistent manner. A method of calculating the relative s part of a two-nucleon wave function which does not utilize oscillator functions or the Moshinsky transformation has been given by Bayman and Kallio [Ba 67].

The possibility of probing two- and four-nucleon correlations in nuclei is perhaps the most interesting aspect of multi-nucleon transfer. If such correlations are strong in the initial and final nuclei, one expects qualitatively that terms corresponding to relative s states of the transferred nucleons will contribute significantly and constructively to the coherent sum in (5.11), yielding a large cross section. Weak correlations, on the other hand, will result in small contributions due to relative s states with more or less randomly distributed signs, and hence in a small cross section. These arguments imply that interesting qualitative information on correlations should be obtainable from multi-nucleon transfer studies even without elaborate calculations. Detailed analyses of three- and four-nucleon transfers among light nuclei ($A < 20$) by Bock and Yoshida [Bo 72a] have indeed shown that amplitudes corresponding to transfer in a relative $0s$ state (with no node in the relative radial wave function) usually dominate the stronger transitions, where they interfere constructively. The relative enhancement of the $0s$ amplitudes in these and other cases can be traced back to two reasons. Firstly, configuration mixing due to the residual interaction produces correlations and hence constructive interference of different initial and final configurations associated with a given $0s$ transfer. Secondly, as the $0s$ state is the lowest intrinsic state of the transferred cluster, it also has the smallest binding energy in the initial and final nuclei and therefore a relatively large amplitude in the tail region, where the cross section is determined. In other words, the correlations are especially pronounced at the nuclear surface, where they can be probed with transfer reactions.

Another point to be made here is that the cross section is sensitive to correlations both in the donor nucleus (denoted A) and in the acceptor nucleus (denoted B). This point is of little consequence in light-projectile-induced transfer reactions, where the nucleons in the lighter fragment always occupy relative s states. With heavier projectiles this is no longer true, and therefore different selectivities may be observed, if a given type of transfer, leading to a given final nucleus, is studied using different donor (or acceptor) nuclei. In addition, different selectivities may also be produced by different kinematic properties (masses, Q values) of different fragment pairs (see Sect. 4.3). These considerations lead us to expect a great variety of complex, but potentially interesting phenomena in heavy-ion-induced multi-nucleon transfer.

5.2 Formal Description II: Transfer Amplitudes

The quantitative analysis of quasi-elastic transfer reactions is one of the most challenging topics in nuclear-reaction theory, and an enormous amount of relevant work has been published. One starts by reducing the underlying many-body problem to a three-body problem (see Fig. 4.1) which, however, is still mathematically too complicated to be treated exactly. Additional approximations are necessitated by our incomplete knowledge of the interaction involved, and thus one ends up with reaction models whose accuracy and limits of applicability cannot be derived from first principles, but have to be inferred indirectly by systematic comparison with experimental data.

By far the most successful of these "models" is the "distorted-wave Born approximation" (DWBA), which was introduced in Sect. 3.4 in the context of inelastic scattering. Other approaches include the "molecular state approximation" [Br 59, Br 63, Br 64a, Oe 70, Oe 73c], also referred to as the method of linear combination of nuclear orbitals (LCNO, see Sect. 2.4), various semi-classical approximations (see Sect. 4.2), and the so-called diffraction models [Da 65a, Fr 69]. Some of these models have been shown to be equivalent to the DWBA under certain simplifying conditions, but in general their applicability is more limited than that of the DWBA (assuming one-step transitions). Therefore we shall concentrate throughout this section on the DWBA; at the end, however, we shall briefly comment on the relationship of the DWBA to more approximate models, as well as on its extension to multistep processes.

The DWBA cross section can be written

$$\frac{d\sigma}{d\Omega} = \frac{\mu_i \mu_f}{(2\pi\hbar^2)^2} \frac{k_f}{k_i} \frac{1}{(2I_A + 1)(2I_{C_2} + 1)} \sum_{m_i, m_f} |T_{m_i m_f}|^2, \tag{5.12}$$

where the sum on the right-hand side runs over the intrinsic spin projections of the initial and final fragments ($m_i = m_A, m_{C_2}$; $m_f = m_B, m_{C_1}$). Dropping for simplicity the subscripts m_i, m_f, the transfer amplitudes T are given by the six-dimensional integral

$$T = \int \chi_f^{(-)*}(k_f, r_f) G(r_i, r_f) \chi_i^{(+)}(k_i, r_i) \, dr_i dr_f. \tag{5.13}$$

Here $\chi_i^{(+)}$ and $\chi_f^{(-)}$ are the ingoing and outgoing "distorted waves", which describe elastic scattering in the initial and final channel, respectively, and are generated from appropriate optical model potentials. The "transfer function" $G(r_i, r_f)$ is a matrix element involving the internal coordinates of the initial and final fragments, as far as they are independent of r_i and r_f, and the interaction V_{if} responsible for the transition. Assuming inert constituents C_1, C_2, and x, we have

$$G(r_i, r_f) = \psi_B^*(r_{2x}) V_{if} \psi_A(r_{1x}), \tag{5.14}$$

where ψ_A and ψ_B are the bound state wave functions for particle x bound to the cores C_1 and C_2, respectively. Note that the vectors r_{1x} and r_{2x} can be expressed in terms of r_i and r_f (or vice versa) by means of (4.2) and (4.3).

Next we have to specify the interaction V_{if} which enters the calculation of the transfer function [see (5.14)]. The usual procedure is to start from either of the alternative expressions:

$$V_{if}^{(po)} = V_{BC_1} - U_f \approx V_{C_1x}(r_1) + V_{C_1C_2}(r_{12}) - U_f(r_f), \tag{5.15}$$

$$V_{if}^{(pr)} = V_{AC_2} - U_i \approx V_{C_2x}(r_2) + V_{C_1C_2}(r_{12}) - U_i(r_1). \tag{5.16}$$

Here V denotes the "true" interaction between the particles given as subscripts (here and in the following approximated by an effective two-body interaction), whereas U_f and U_i are the optical potentials used in generating the distorted waves in the final and initial channel, respectively. Equation (5.15) is called the "post representation", and (5.16) the "prior representation" of the interaction. In principle, both representations should yield identical results when inserted into the DWBA formula for the transition amplitude. In practical calculations, however, this equivalence may be destroyed by approximations, which are specific for each representation. For example, it is customary to neglect in (5.15) and (5.16) the difference between the nuclear part of the core–core interaction and the nuclear part of the relevant optical model potential. This leads to the following approximate relationships:

$$V_{if}^{(po)} \approx V_{C_1x}^{(N)}(r_{1x}) + \Delta V_{C}^{(po)} \text{ with } \Delta V_{C}^{(po)} = V_{C_1x}^{(C)} + V_{C_1C_2}^{(C)} - V_{C_1B}^{(C)}, \quad (5.17)$$

$$V_{if}^{(pr)} \approx V_{C_2x}^{(N)}(r_{2x}) + \Delta V_{C}^{(pr)} \text{ with } \Delta V_{C}^{(pr)} = V_{C_2x}^{(C)} + V_{C_1C_2}^{(C)} - V_{C_2A}^{(C)}, \quad (5.18)$$

where the superscripts (N) and (C) denote nuclear and Coulomb potentials, respectively. Moreover the Coulomb terms ΔV_C are often disregarded (recent analyses have shown, however, that these may have a significant influence, especially if a charged particle is transferred [Go 74, De 74, De 75]). A closer examination of (5.15–18) shows that the approximations associated with the post representation should be most appropriate for transfer from a light to a heavy core (light donor, heavy acceptor); conversely, transfers from a heavy to a light core (heavy donor, light acceptor) should be better described in the (approximate) prior representation.

The calculation of the transition amplitude T according to (5.13) must be done numerically on a high-speed computer. In order to make this calculation feasible it has been necessary until recently to reduce the six-dimensional integral on the right-hand side of (5.13) to a three-dimensional integral by suitable approximations in the integrand. Although computer codes for an exact evaluation of the six-dimensional integral are now available, we first discuss some approximate methods, in order to give a better insight into the physical problems involved.

With light donor or acceptor nuclei ($A \leq 4$), the "zero-range approximation" can be employed, where one substitutes for the product $V_{if}\psi_A(r_{1x})$ (if the post representation of V_{if} is used) or $V_{if}\psi_B(r_{2x})$ (if the prior representation is used) a delta function of the appropriate vector. Thus the transferred particle is effectively constrained to coincide with the smaller of the cores at the instant of transfer. The remaining vectors r_i, r_f, and r_{2x}(post) or r_{1x} (prior) are all related by constant factors and are parallel to r_{12} [see Fig. 4.1 and (4.2) and (4.3)]. In the post treatment we have now

$$r_{1x} = 0, \ r_{2x} = r_{12}; \ r_i = r_{12}, \ r_f = \frac{M_2}{M_2 + m_x} r_{12}, \quad (5.19)$$

and in the prior treatment,

$$r_{1x} = -r_{12}, \; r_{2x} = 0; \; r_i = \frac{M_1}{M_1 + m_x} r_{12}, \; r_f = r_{12}. \tag{5.20}$$

As a result, the transition amplitude can be calculated as a three-dimensional integral over r_{12}.

The zero-range approximation is, however, in general not applicable for transfer between heavier cores, especially not if the bound-state wave function of the transferred particle with respect to the smaller core has non-zero orbital angular momentum and hence vanishes at the origin. A method which in such cases still allows the transition amplitude to be reduced to a three--dimensional integral is the "no-recoil approximation", which was introduced by Buttle and Goldfarb [Bu 66; a similar approach has been followed also by Dar and collaborators in their diffraction model, see Da 65a]. The method consists of two essential steps. Firstly one uses the approximation (5.19) or (5.20) for the arguments r_i and r_f of the distorted waves (thus neglecting "recoil effects", see below), but *not* for the arguments r_{1x} and r_{2x} of the transfer function (thus retaining the finite range of the interaction V_{12}). Introducing a "reduced transfer function" or "form factor" $F(r_{12})$, which in the post and prior treatment, respectively, is given by

$$F^{(po)}(r_{12}) = \int \psi_B^*(r_{12} + r_{1x}) V_{C1x}^{(N)}(r_{1x}) \psi_A(r_{1x}) \, d(r_{1x}), \tag{5.21}$$

$$F^{(pr)}(r_{12}) = \int \psi_B^*(r_{2x}) V_{C2x}^{(N)}(r_{2x}) \psi_A(r_{2x} - r_{12}) \, d(r_{2x}), \tag{5.22}$$

the transition amplitude T can be written as

$$T \approx \int \chi_f^{(-)*}(k_f \; r_f) F(r_{12}) \chi_i^{(+)}(k_i, \; r_i) dr_{12}, \tag{5.23}$$

with F, r_i, and r_f given by (5.19–22). Note that T in principle is still a six--dimensional integral, since F implies three-dimensional integration. At this point, however, a second approximation is made by substituting for the bound-state wave functions ψ_A and (or) ψ_B simple functions, which allow the integrals in (5.21) and (5.22) to be evaluated to a large extent analytically. Since the outer tail of the wave functions ψ_A and ψ_B is most important for the transfer reaction, one may use spherical Hankel functions, which correctly describe the asymptotic behaviour of a neutral particle x with appropriate binding energy E_A or E_B, respectively [Bu 66]:

$$\psi_A(r_{1x}) = S_A^{1/2}(l_1, j_1) \phi_A(r_{1x}) Y_{l_1 m_1}(\hat{r}_{1x})$$
$$\approx S_A^{1/2}(l_1, j_1) N_A h_{l_1}^{(1)}(i\kappa_1 r_{1x}) Y_{l_1 m_1}(\hat{r}_{1x}) \tag{5.24}$$

$$\psi_B(r_{2x}) = S_B^{1/2}(l_2, j_2) \phi_B(r_{2x}) Y_{l_2 m_2}(\hat{r}_{2x})$$
$$\approx S_B^{1/2}(l_2, j_2) N_B h_{l_2}^{(1)}(i\kappa_2 r_{2x}) Y_{l_2 m_2}(\hat{r}_{2x}), \tag{5.25}$$

$$\kappa_{1,2} = \left(\frac{2\mu_{A,B}E_{A,B}}{\hbar^2}\right)^{1/2}. \tag{5.26}$$

In the method of Buttle and Goldfarb, the Hankel-function approximation is only used for one bound-state function, namely that associated with the larger core (i.e. ψ_B in the post and ψ_A in the prior treatment). In order to simplify our discussion, and to exhibit the symmetry properties of the reaction more clearly, we use it here for both bound states (see also Oe 74). Making use of an addition theorem for spherical Hankel functions, the integrals (5.21) and (5.22) can now be reduced considerably, and after some manipulation, the cross section for $j_x = \frac{1}{2}$ (one-nucleon transfer) assumes the form

$$\frac{d\sigma}{d\Omega} = \frac{\mu_i\mu_f}{(2\pi\hbar^2)^2}\frac{k_f}{k_i}\frac{2I_B + 1}{2I_{C2} + 1}$$
$$\times \sum_{j_1 j_2 l} N_A^2 S_A(l_1, j_1) N_B^2 S_B(l_2, j_2) C(j_1, j_2, l)|I|^2 \sum_m |T_{lm}(\Theta)|^2. \tag{5.27}$$

Here $C(j_1, j_2, l)$ is a geometrical coefficient, I is a one-dimensional radial integral involving the initial and final bound-state wave functions, and the amplitudes $T_{lm}(\Theta)$ are three-dimensional integrals involving the initial and final distorted waves. The latter are formally identical with the corresponding integrals in the zero-range approximation, and can therefore be computed with standard zero-range codes.

Equations (5.24–26) are not directly applicable if the transferred particle x carries an electrical charge, as in the case of proton transfer. However, Buttle and Goldfarb have shown that the tail of the bound-state wave function for a charged particle can be approximated quite accurately by a Hankel function with appropriately adjusted binding energy [Bu 68]. Including, in addition, the Coulomb interaction ΔV_C [see (5.17) and (5.18)] in the radial integral I, they were able to analyse proton transfer using essentially the same formalism as for neutron transfer.

We note from (5.27) that the spectroscopic factors S_A, S_B enter the cross section only in combination with the asymptotic normalization factors N_A^2, N_B^2 for the appropriate wave functions. This reflects the fact that the physical property probed in heavy-ion transfer is the amplitude of the bound-state wave function near the nuclear surface, which is proportional to the product $NS^{1/2}$ [see (5.24), (5.25)]. A separate determination of N and S, however, invariably involves some model of nuclear structure.

An alternative notation, which has often been used in the past to characterize the spectroscopic strength of a transfer reaction, involves the concept of a "reduced width" [Ma 60]. The term is indicative of the relationship which exists between the decay probability of an unbound state into a given particle channel, and the cross section for production of a bound state by the corresponding transfer reaction. To some degree of approximation, both quantities can be expressed as a product of an "external" or "kinematic" factor, which only depends on properties of the two-body interaction at large distances outside the nuclear surface, and an "internal" or "structure" factor, which reflects properties of the nuclear many-body problem. The latter factor is denoted as reduced width, and is written again as a product of two terms: a spectroscopic factor Θ^2_{lj} (identical with S_{lj}) describing the fractional single--particle strength of the state in question, and the "single-particle reduced width" Θ^2_0, which is related to the intensity of the single-particle wave function ϕ_{lj} at the cut-off radius R by

$$\Theta^2_0 = \tfrac{1}{3} R^3 |\phi_{lj}(R)|^2 \approx \tfrac{1}{3} R^3 N^2 |h^{(1)}_l(i\kappa R)|^2. \tag{5.28}$$

Here we have used the approximation given by (5.24) and (5.25) for ϕ_{lj}, in order to exhibit the relationship between Θ^2_0 and N^2.

Equation (5.28) shows that the single-particle reduced width Θ^2_0 depends in principle on the choice of a cut-off radius R, which is assumed to separate the internal and external regions. In practice both Θ^2_0 and N^2 are usually treated as empirical parameters to be determined from experimental transfer cross sections involving known single-particle states.

We now turn to a discussion of "recoil effects", which have so far been neglected. In general, recoil effects are understood to arise from the difference between r_i and r_f, which results from a finite mass transfer between the two fragments. This difference affects the phase shift $\Delta\alpha$ of the distorted wave in the final channel relative to the initial channel [see, for example, Sa 73]

$$\Delta\alpha \approx \mathbf{K}_i \mathbf{r}_i - \mathbf{K}_f \mathbf{r}_f$$

$$= (\mathbf{K}_i - \mathbf{K}_f) \cdot \tfrac{1}{2} (\mathbf{r}_i + \mathbf{r}_f) + \tfrac{1}{2} (\mathbf{K}_i + \mathbf{K}_f) \cdot (\mathbf{r}_i - \mathbf{r}_f), \tag{5.29}$$

and hence the transfer amplitudes T (5.13). Here $\mathbf{K}_i$ and $\mathbf{K}_f$ are local momenta at the instant of transfer (see Fig. 5.3). The first term on the right-hand side of (5.29) is associated with the momentum transfer, and the second term represents the recoil effect. Contrary to naive expectation, the recoil con-

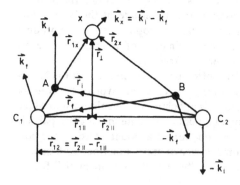

Fig. 5.3.   Vector diagram of transfer reaction $A + C_2 \rightarrow C_1 + B$, showing instantaneous distance and momentum vectors relevant to a discussion of recoil effects

tribution to $\Delta\alpha$ is usually more important for heavy-ion-induced transfer than for transfer induced by light ions ($A \leq 4$), in spite of the fact that the ratio of the transferred mass m_x to the reduced mass of the colliding system is usually smaller in the former case. This can be understood by noting that the average local momentum $\frac{1}{2}(K_i + K_f)$ is roughly proportional to the reduced mass at comparable energies, and therefore much larger for heavy than for light projectiles. This effect is not fully compensated by a correspondingly larger value of $|r_i - r_f|$ in the light systems since there—for reasons related to the validity of the zero-range approximation—the transfer occurs preferentially when the particle x is close to the light core. With increasing bombarding energy the average momentum $\frac{1}{2}(K_i + K_f)$ increases, and therefore the recoil effect becomes more important.

Another interesting aspect of recoil is its effect on angular momentum transfer or—more precisely—on the relationship between linear and angular momentum transfer, and hence on the angular distribution for a given transition. In formal analogy to (5.29) we can express the angular momentum transfer l as

$$l = (r_i \times K_i) - (r_f \times K_f)$$

$$= \frac{1}{2}(r_i + r_f) \times (K_i - K_f) + (r_i - r_f) \times \frac{1}{2}(K_i + K_f). \qquad (5.30)$$

Here again we have on the right-hand side the usual term proportional to linear momentum transfer $(K_i - K_f)$, and in addition the recoil term proportional to $(r_i - r_f)$. The magnitude of this latter term [as well as that of the corresponding term in (5.29)] is roughly given by the share of the transferred particle in the total relative linear momentum, multiplied by a nuclear radius, and is easily verified to be of order unity under typical conditions. Therefore recoil effects cannot in general be neglected.

A question of some importance is to what extent recoil effects are actually included in various approximate treatments of the DWBA amplitude. In the post version of the no-recoil approximation, one uses the basis vectors r_{12} and r_{1x} (see Figs. 4.1 and 5.3) and subsequently neglects r_{1x}, whereas in the prior version one starts off with r_{12} and r_{2x}, and then neglects r_{2x}. Using (4.2) and (4.3) for r_i and r_f, one obtains

$$r_i - r_f \approx + \frac{m_x}{M_2 + m_x} r_{12} \qquad \text{(post)}, \qquad (5.31)$$

$$r_i - r_f \approx - \frac{m_x}{M_1 + m_x} r_{12} \qquad \text{(prior)}. \qquad (5.32)$$

It is obvious that these two approaches imply quite different treatments of the recoil effect, and for cores of comparable mass become equally inadequate. A considerable improvement can be achieved by decomposing the vectors r_{1x} and r_{2x} into components parallel ($r_{1\parallel}$, $r_{2\parallel}$) and perpendicular ($r_\perp$) to r_{12}

as shown in Fig. 5.3. Neglecting now the perpendicular component $r_\perp$, the recoil becomes

$$r_i - r_f \approx \left(\frac{r_{2\|}}{(r_{1\|} + r_{2\|})} \frac{m_x}{(M_2 + m_x)} - \frac{r_{1\|}}{(r_{1\|} + r_{2\|})} \frac{m_x}{(M_1 + m_x)}\right) r_{12} \qquad (5.33)$$

$$\approx \frac{m_x}{(R_1 + R_2)} \left\{\frac{R_2}{(M_2 + m_x)} - \frac{R_1}{(M_1 + m_x)}\right\} r_{12}. \qquad (5.34)$$

Here R_1 and R_2 are effective bound state radii of the particle x when bound to
either core 1 or core 2, respectively. Equations (5.33) and (5.34) represent a
symmetrized version of the no-recoil approximation, which approaches the
post or prior versions in the appropriate limits of core asymmetry.

The physical implication of (5.34) is that the transfer takes place while
particle x is located on the line connecting the two cores, with the ratio of
distances determined by the appropriate bound-state radii. Using an essentially equivalent approach, Buttle and Goldfarb derived "recoil corrections"
for a number of sub-Coulomb transfer reactions [Bu 71]. Thereby large discrepancies which normally exist between the post and prior treatments in the
standard no-recoil approximation could be effectively eliminated. As an
example we show in Fig. 5.4 an analysis of the reaction ^{208}Pb(^{16}O, ^{17}O)^{207}Pb.

A similar method has been developed and applied in analyses of a number
of one- to four-nucleon transfer reactions by Pühlhofer and collaborators

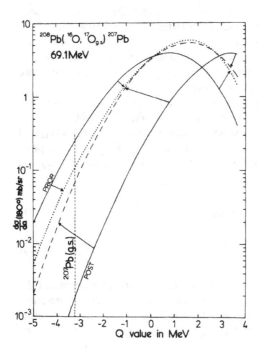

Fig. 5.4. Comparison of post
and prior calculations with
(solid curves) and without
(broken curves) recoil corrections for the reaction
^{208}Pb(^{16}O,^{17}O)^{207}Pb, populating the final nuclei in their
ground states ($l = 3$). The
binding energy of the transferred neutron has been arbitrarily varied, in order to
exhibit the dependence of the
theoretical cross section on
the reaction Q value; the
physical Q value is indicated
by a vertical broken line.
(From Bu 71)

[Pü 70, Sc 75a]. These authors also consider a collinear configuration at the instant of transfer, with a well-defined non-zero distance between the transferred cluster and the lighter core. Their method of analysis is known in the literature as the "fixed-range method".

Figure 5.3 suggests a decomposition of the recoil effect into two components: a *longitudinal* recoil effect, connected with $r_{1\parallel}$ and $r_{2\parallel}$, and a *transverse* recoil effect, connected with $r_\perp$. Considering (5.29) for the recoil phase, one might expect qualitatively that the longitudinal effect is relatively more important at low energies, and the transverse effect at high energies. Moreover, using a symmetrized description as implied by (5.33), the longitudinal effect should be relatively small for nearly symmetric and extremely asymmetric systems. (Note, however, that for nearly symmetric systems the usual no-recoil approximation effectively introduces spurious longitudinal effects of different sign, depending on the treatment chosen.) For reasons given in Sect. 5.1, any violation of the parity selection rule (5.10) must be due to the transverse recoil effect. The collinear approximations discussed so far all neglect the transverse recoil effect, and hence allow only even or only odd values of the angular momentum transfer l, depending on the parity change Π defined by (5.4).

The importance of recoil effects in heavy-ion-induced transfer reactions was first recognized by Dodd and Greider [Do 65, Do 69]. These authors discussed the additional phase shift introduced by recoil [see (5.29)], and how it can be absorbed in an effective transfer function. Using a somewhat simplified "diffraction" model for the ingoing and outgoing distorted waves, and harmonic-oscillator wave functions for the bound states, they were able to derive analytic formulae for the differential cross section. The results show that recoil effects at high energies strongly attenuate the oscillatory structure in angular distributions, mainly as a consequence of relaxing the selection rules on angular momentum transfer. A fall-off of the smooth part of the angular distributions proportional to the inverse third power of linear momentum transfer was predicted, in approximate agreement with then available experimental evidence.

More recently, a number of authors have developed methods of incorporating recoil effects (including the transverse recoil effect) in lowest order in standard DWBA calculations [Na 72, Na 73, Br 74a, Ba 74b, Re 75]. We shall not discuss these approaches here, since computer codes for complete DWBA calculations including recoil are now available (see below). It should be noted, however, that an approximate treatment of recoil effects may considerably reduce the requirements concerning computer size and (or) computing time, while often yielding results equivalent to those of a full calculation.

As a consequence of continuing progress in computer technology and computing techniques, full calculations of the six-dimensional DWBA integrals for heavy-ion-induced transfer reactions became feasible around 1970. The mathematical methods employed may be grouped roughly in three categories [Gl 74].

The first method was formulated in 1964 by Austern et al. [Au 64]. It employs expansions in terms of angular momentum eigenfunctions, which can be integrated analytically, leaving a two-dimensional radial integral to be computed numerically. Computer codes based on this approach have been developed by Kammuri and Yoshida [Ka 69], De Vries [De 73a], and Low and Tamura [Lo 75].

The second method utilizes expansions of the functions under the six-dimensional integral in terms of simple basis functions such as plane-wave functions [Ro 72, Ch 73a], harmonic-oscillator eigenfunctions [Mc 73], or spherical Bessel functions [Gl 74], which can be separated according to two independent vectorial variables. This procedure yields an infinite series of products of two one-dimensional radial integrals.

Finally, the six-dimensional integral may be computed directly. This approach has been followed by Bayman in an analysis of the reaction ^{62}Ni(^{18}O, ^{16}O)^{64}Ni [Ba 74c].

So far, most analyses of experimental results have been performed with the first of these methods. As an example we show in Fig. 5.5 an analysis of the reaction ^{12}C(^{14}N, ^{13}C)^{13}N at $E_{Lab} = 78$ MeV by De Vries [De 73a]. In this case the calculation without recoil predicts a sharply oscillating angular

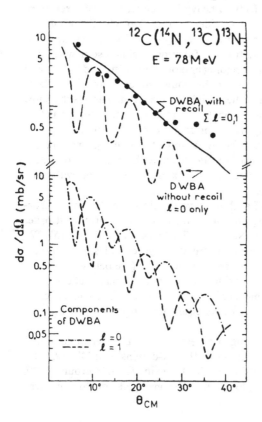

Fig. 5.5. DWBA analysis of the reaction ^{12}C(^{14}N, ^{13}C)^{13}N, populating the final nuclei in their ground states ($E_{lab} = 78$ MeV). In the upper part of the figure the experimental data (full points, from Li 70) are compared with calculations including (full curve) and neglecting (broken curve) recoil effects. In the lower part of the figure the $l = 0$ and $l = 1$ components of the calculation with recoil are shown separately. (From De 73a)

distribution, corresponding to $l = 0$, in contrast to the rather smooth experimental distribution. The calculation with recoil, however, predicts comparable contributions with angular momentum transfer $l = 0$ ("normal parity") and $l = 1$ ("non-normal parity"), which oscillate out of phase. The resulting distribution is much smoother and hence yields a much better fit to the data. It is interesting to note that the $l = 0$ component calculated with recoil is by itself less structured than that calculated without recoil. Further the $l = 0$ component calculated with recoil is identical using either the post or the prior representation of the DWBA matrix element [De 73a]; this confirms the earlier conclusion of Buttle and Goldfarb, that most of the disagreement between post and prior no-recoil calculations for nearly symmetric systems arises from the "no-recoil approximation" [Bu 71]. Another point to be noted here is that the relative strength of the non-normal parity components can be quite different in different reactions, and in fact is sometimes negligible; in general, however, their importance increases with increasing bombarding energy.

Although the DWBA is an extremely successful technique for analyses of transfer reactions, it also has its problems and limitations. One aspect considered unsatisfactory by many workers in the field is its dependence on optical model potentials with their parameter ambiguities. Practical experience shows, however, that comparatively unambiguous transfer analyses are usually possible for kinematically well-matched transitions [see Sect. 4.3] of reasonable spectroscopic strength, provided no particular inelastic channel is strongly coupled to the elastic channel. In such cases potentials deduced from elastic scattering will provide a consistent description of transfer reactions regardless of parameter ambiguities. A useful feature in this context seems to be that the relevant region in radial space is well defined by the onset of strong absorption. In less favourable circumstances, i.e. for poorly matched or spectroscopically weak transitions, arbitrary changes of calculated cross sections with otherwise "reasonable" changes of potential parameters may be encountered. Such observations have occasionally led to the conclusion that transfer data should yield additional information on the optical model potential beyond that obtainable from elastic scattering. It seems much more likely, however, that sensitivities of this kind are indicative of a basic failure of the DWBA with respect to the transition under study.

Another feature of the DWBA which is often considered inconvenient is the necessity to perform lengthy numerical calculations for each parameter combination, and the resulting lack of transparency. An alternative, in this respect, is offered by various semiclassical treatments, which provide analytic expressions for the transfer cross section. Presumably this convenience is achieved at the expense of quantitative precision; however, no systematic comparisons with either recent experimental data or DWBA calculations appear to have been made. Early work in this direction—performed prior to the first successful DWBA analyses by Buttle and Goldfarb [Bu 66, Bu 68]—includes the semiclassical tunnelling theory of Breit and co-workers [Br 56a, b, Br 64a], and the diffraction model of Dar and co-workers [Da 65a, b, Da 66, Va 68].

These models were quite successful in reproducing experimental data available up to about 1965, and in predicting the qualitative features of results to be obtained later.

More recently, the semiclassical methods have been refined and extended by Alder and Trautmann [Tr 70, Al 71a, b] and by Broglia, Winther, and co-workers [Br 72c, d, Br 74b]. Comparisons with experimental data have been made mainly for sub-Coulomb bombarding energies where the agreement is quite satisfactory (see Sect. 5.3). A useful feature of the semiclassical calculations is that they exhibit explicitly the relevant parameters and their qualitative influence on the cross section. Graphs of the functions which enter the cross-section formulae have been published by Alder et al. [Al 72]; these should be helpful in planning experiments.

So far, our discussion of quantitative methods has been restricted to the simplest possible transfer mechanism, namely that of a one-step reaction. In collisions between complex nuclei above the Coulomb barrier, many inelastic channels are usually open and consequently one may expect indirect mechanisms to play a significant part. In fact it seems rather surprising that so many experiments have been successfully analysed in terms of one-step transfer. This may be partly due to the flexibility of the theoretical description in allowing the absorption of indirect admixtures into certain parameters, which are not uniquely defined by independent evidence. More importantly, however, the pronounced selectivity which exists in one-step transfer for reasons of nuclear structure enhances the apparent dominance of these simple transitions. Multi-step transfers, on the other hand, are comparatively non-selective and hence more difficult to identify experimentally.

A quantitative description of transfer reactions proceeding through two or more steps necessarily involves a set of coupled equations, which describe explicitly those channels coupled strongly to the incident or outgoing channel. Such coupled-channel treatments have been developed both in the framework of semiclassical theory [Br 72c, d, Br 74b] and as an extension of the (quantum-mechanical) DWBA, the so-called coupled-channels Born approximation (CCBA; see, for example, As 69, As 74, Ta 74a). The physical background and formal methods are analogous to those encountered in multiple inelastic excitation by Coulomb or nuclear forces (see Chap. 3). If many steps are involved, of course, the reactions cease to be "quasi-elastic", and statistical or even classical models may become useful for their description. This situation will be discussed in Chap. 6.

5.3   Sub-Coulomb Transfer

In 1952 Breit, Hull, and Glückstern published an article entitled "Possibilities of heavy ion bombardment in nuclear studies" [Br 52]. There the authors not only drew attention to various exciting prospects associated with the use of heavy projectiles, such as the possibility of investigating Coulomb-induced nuclear distortion and fission, and producing and studying transuranic elements; they also discussed design characteristics of accelerators which might be capable of producing the necessary heavy-ion beams. One of the proposals put forward was to look for "effects characteristic of the leakage of neutrons and protons out of the two colliding nuclei by wave-mechanical penetration of the region of negative kinetic energy", because "an exploration of these effects should amount to a study of the halo of neutrons and protons surrounding the more compact nuclear interior and might be helpful in determining the number of nuclear particles at the nuclear surface having a given energy." The relevant theoretical considerations given at that time were necessarily of rather qualitative significance; nevertheless there can be no doubt that the 1952 paper of Breit, Hull, and Glückstern marks the beginning of nuclear research with heavy projectiles as we know it today.

The motivation to study transfer reactions at bombarding energies below the Coulomb barrier is based mainly on two arguments, which were recognized already in the early work mentioned above, and have since been thoroughly discussed by many authors. Firstly, a large distance of closest approach eliminates complications arising from the nuclear core–core interaction, including uncertainties due to our incomplete knowledge of that interaction. Secondly, the semiclassical approach should be especially appropriate to heavy systems at relatively low bombarding energies, corresponding to large values of the Sommerfeld parameter n. Both arguments should contribute to a simpler and more accurate formal description of sub-Coulomb as compared to higher-energy transfer, and should be helpful for a better intuitive understanding of the transfer mechanism. The same arguments apply, of course, to Coulomb excitation, and indeed there are many connections and analogies between sub-Coulomb transfer and Coulomb excitation.

In the years following 1952, Breit and collaborators developed systematically the semiclassical theory of neutron tunnelling (in the literature often abbreviated as SCT or SCTT). Initially a strictly classical motion of the colliding nuclei along Coulomb trajectories was considered [Br 56a, b]; later, corrections due to a quantal description of the relative motion [Br 59, Br 64a, b] and the effect of nuclear interactions [Br 67] were included in lowest order. Application was made specifically to the reaction $^{14}N(^{14}N, ^{13}N)^{15}N$, for which experimental data were available from the pioneering work of Reynolds and Zucker, performed with the 63-inch cyclotron at Oak Ridge [Re 56], and from later measurements at the Yale linear accelerator [Be 65] and the Oak Ridge tandem [Hi 65, Ga 66]. Early comparisons between experiment and theory revealed disturbing discrepancies, which were attributed to a probable influence of virtual Coulomb excitation on the transfer mechanism [Br 56a, b, Br 59]. Later it was realized, however, that the discrepancies were largely due to nuclear absorption effects resulting from the use of too high bombarding energies. Indeed, when more accurate data became available at sufficiently low bombarding energies, they were found in satisfactory agreement with the predictions of the semiclassical tunnelling theory of Breit and collaborators [Br 64a, Be 65, Hi 65, Ga 66, Br 67].

Unfortunately, the formulation of the SCT theory in the original papers by Breit and collaborators does not lend itself readily to an analysis of experimental data [other than the reaction $^{14}\mathrm{N}(^{14}\mathrm{N}, {}^{13}\mathrm{N})^{15}\mathrm{N}$] or to a comparison with more recent theoretical work. It has been shown, however, that equivalent results can be deduced by making appropriate approximations in the corresponding DWBA expressions [Br 64b, Bu 66]. In the following we reproduce the most important formulae, starting from (5.27). The extreme semiclassical approach implies completely matched initial and final trajectories ($Q = Q_{\mathrm{opt}}$). Confining our attention to neutron transfer, we therefore set $k_1 = k_f = k$, $n_1 = n_f = n$, $\kappa_1 = \kappa_2 = \kappa$ ($Q = 0$). For simplicity we further assume zero angular momentum transfer ($l = 0$; $j_1 = j_2 = j$), as in $^{14}\mathrm{N}(^{14}\mathrm{N}, {}^{13}\mathrm{N})^{15}\mathrm{N}$. This leaves in the sum on the extreme right-hand side of (5.27) only the term $|T_{00}|^2$, which for large values of the Sommerfeld parameter n assumes the following analytical form [Tr 70, Bu 71]:

$$|T_{00}|^2 = \frac{2\pi^2 n}{\kappa^3 k \sin\left(\dfrac{\Theta}{2}\right)\left[\kappa^2 + 4k^2 \sin^2\left(\dfrac{\Theta}{2}\right)\right]}$$
$$\exp\left\{-4n\left[\arctan\left(\frac{\kappa}{2k}\right) + \arctan\left(\frac{\kappa}{2k \sin\left(\dfrac{\Theta}{2}\right)}\right)\right]\right\}. \qquad (5.35)$$

Evaluation of the remaining factors in (5.27) finally yields for the differential and integrated cross section, respectively,

$$\frac{d\sigma}{d\Omega} = \frac{1}{4\pi} \frac{k_f}{k_i} C_{AB} \kappa^4 |T_{00}|^2, \qquad (5.36)$$

$$\sigma = \int_0^\pi \frac{d\sigma}{d\Omega} 2\pi \sin\Theta \, d\Theta = \frac{\pi^2}{2k_i^2} C_{AB} \exp\left(-8n \arctan \frac{\kappa}{2k}\right). \qquad (5.37)$$

Here k_i and k_f have been re-introduced to take proper account of time-reversal symmetry [$k = (k_i k_f)^{1/2}$], and C_{AB} is a dimensionless factor containing mainly spectroscopic quantities:

$$C_{AB} = \frac{\mu_i \mu_f}{\mu_A \mu_B} \frac{2I_B + 1}{(2I_2 + 1)(2j + 1)} \frac{N_A^2 N_B^2}{\kappa_1^3 \kappa_2^3} S_A S_B. \qquad (5.38)$$

Equations (5.35–37) are equivalent to the quantum-mechanically corrected version of the SCT theory of Breit and collaborators [Br 64a]. We note that in the latter the spectroscopic strength is usually expressed in terms of "reduced widths", which are related to the quantities N_A^2, N_B^2 of (5.38) by (5.28).

The extreme semiclassical limit of the transfer cross section, as described by the SCT theory in its original form [Br 56a, b], is obtained by further simplification of (5.35), which corresponds to letting both n and k go to infinity, while keeping $a = n/k$ constant. The resulting expressions for $|T_{00}|^2$ and σ are

$$|T_{00}|^2 = \frac{\pi^2 n}{2\kappa^3 k^3 \sin^3\left(\dfrac{\Theta}{2}\right)} \exp\left[-2\kappa D(\Theta)\right], \tag{5.39}$$

$$\sigma = \frac{\pi^2}{2k_i^2} C_{AB} \exp\left[-2\kappa D(\pi)\right], \tag{5.40}$$

and the differential cross section is again given by (5.36) together with (5.39). The quantity $D(\Theta)$ denotes the distance of closest approach between the colliding nuclei and is given by [see (1.11)]:

$$D(\Theta) = \frac{n}{k}\left(1 + \csc\frac{\Theta}{2}\right). \tag{5.41}$$

We recognize from (5.39–41) that the dependence of the semiclassical transfer cross section on energy and angle is dominated by the exponential term, which describes a strong increase with decreasing distance of closest approach. This is, of course, just the correlation expected for nucleon tunnelling between two potential wells. In accordance with (4.4), we can define a transfer probability P_{if} by writing

$$P_{if} = \left(\frac{d\sigma}{d\Omega}\right)\bigg/\left(\frac{d\sigma}{d\Omega}\right)_C \tag{5.42}$$

where $(d\sigma/d\Omega)_C$ denotes the differential cross section for Rutherford scattering. Using (5.39), (5.36), and (1.29), we obtain

$$P_{if} = \frac{\pi\kappa}{2nk} C_{AB} \sin\left(\frac{\Theta}{2}\right) \exp\left[-2\kappa D(\Theta)\right]. \tag{5.43}$$

Alternatively, we can express the correlation of the transfer cross section with the distance of closest approach by a transformation of the differential cross section according to (4.7); using again (5.39) and (5.36) this yields

$$\frac{d\sigma}{dD} = \frac{8\pi k}{n}\sin^3\left(\frac{\Theta}{2}\right)\frac{d\sigma}{d\Omega} = \frac{\pi^2\kappa}{k_i^2} C_{AB} \exp\left[-2\kappa D(\Theta)\right]. \tag{5.44}$$

The advantage of plotting either P_{if} or $d\sigma/dD$ instead of $d\sigma/d\Omega$ is that, at a given bombarding energy, most or all of the strong dependence on scattering angle is removed. This has been exploited in many instances to demonstrate the semiclassical nature of transfer reactions and the existence of a radial cut-off due to strong absorption (see, for example, Mc 60, Jo 64, Ch 73d). We note, however, that—contrary to some statements in the literature—the semiclassical theory does not predict either P_{if} or $d\sigma/dD$ to be a universal function of $D(\Theta)$, independent of individual combinations (E_{CM}, Θ) for a given transition. Equations (5.43) and (5.44) show that such a universal dependence on $D(\Theta)$ is expected for the quantities $P_{if}\csc(\Theta/2)$ and $E_{CM}\, d\sigma/dD$.

The application of the DWBA to sub-Coulomb transfer has been discussed in detail by Buttle and Goldfarb [Bu 66, Bu 68, Bu 71], and by Alder and Trautmann [Tr 70, Al 71b]. Comprehensive reviews of relevant work have been presented by Goldfarb at the international conferences held in Heidelberg 1969 [Go 69] and in Nashville 1974 [Go 74; this paper also contains a fairly complete table of references to experimental work on sub-Coulomb transfer published up to 1974]. Among the points which were recognized as important, and have therefore received special attention, are the application of the DWBA to sub-Coulomb proton transfer [Bu 68], effects of including the Coulomb interaction in calculating the transition amplitude [Bu 68, Go 74], effects due to recoil and kinematic mismatch [Bu 71, Go 74], and the sensitivity of extracted spectroscopic factors to parameters of the bound-state potential [Go 68b, Go 74]. Relevant evidence concerning some of these points will be discussed briefly below. In summary we can state that the DWBA approach has been refined considerably over the years, and now seems to be completely adequate for quantitative analyses, as far as the dynamic aspects of

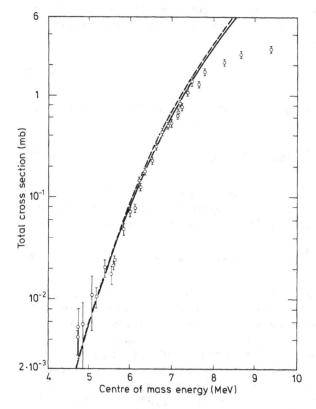

Fig. 5.6. Excitation function for the total cross section of the reaction $^{14}N(^{14}N, \, ^{13}N)^{15}N$. The experimental data of Hiebert et al. (Hi 65) are compared with a DWBA calculation for sub-Coulomb transfer (full curve) and with the corresponding semiclassical approximation (broken curve). (From Bu 66)

sub-Coulomb transfer are concerned. Extended versions of the semiclassical theory have also been worked out [Br 72c,d], but have found practically no application to sub-Coulomb transfer.

In Figs. 5.6 and 5.7, experimental results for the reaction ^{14}N (^{14}N, ^{13}N) ^{15}N are compared with quantal and semiclassical calculations. In this particular case the incident and outgoing channels are kinematically well matched ($Q = +0.28$ MeV; $l = 0$), and therefore the semiclassical calculation and

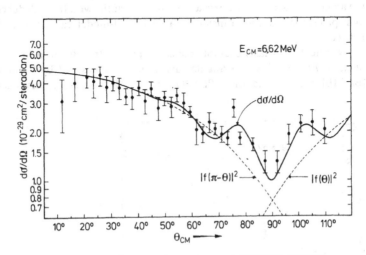

Fig. 5.7. Angular distribution of ^{13}N nuclei from the reaction ^{14}N(^{14}N, ^{13}N)^{15}N, measured at $E_{CM} = 6.62$ MeV (Be 65). The data are compared with the tunneling theory of Breit, Chun, and Wahsweiler (Br 64a, full curve). The broken curves represent incoherent contributions due to projectile stripping and pick-up, respectively. (From Hi 65)

the DWBA treatment yield almost identical results at sub-Coulomb energies. Since the two nuclei in the incident channel are indistinguishable, the differential cross section (Fig. 5.7) must be calculated by coherent superposition of transfer amplitudes corresponding to transfer from the projectile to the target and vice versa. The exchange of projectile and target is equivalent to the substitution $\Theta \to \pi - \Theta$, and the differential cross section is given by [Br 65a]

$$\frac{d\sigma}{d\Omega} = |f_{tr}(\Theta)|^2 + |f_{tr}(\pi - \Theta)|^2 - \frac{1}{3}[f_{tr}(\Theta) f_{tr}^*(\pi - \Theta)$$
$$+ f_{tr}^*(\Theta) f_{tr}(\pi - \Theta)]. \tag{5.45}$$

This equation is identical with the corresponding equation for elastic scattering [(1.32), (2.15)], except for an additional phase factor -1 in the interference term, which arises from the recoupling of intrinsic angular momenta (the general form of this factor for arbitrary angular momentum combinations has been given by Buttle and Goldfarb, Bu 66). Therefore the angular distribution exhibits a conspicuous minimum at $\Theta = \pi/2$, in contrast to elastic scattering.

The integrated cross section, on the other hand, is negligibly affected by the interference term [Br 64a], and hence amounts to twice the cross section for transfer in one direction. Figures 5.6 and 5.7 demonstrate that the theoretical calculations reproduce the angular and energy dependence of the experimental data quite well under sub-Coulomb conditions; discrepancies at higher bombarding energies and extreme forward angles (corresponding to backscattering of the projectile) can be attributed to nuclear absorption. The overall magnitude of the cross section is also correctly predicted, as shown by the agreement between the extracted reduced widths with those deduced from light-projectile-induced transfer data and shell-model calculations [Be 65, Ma 60].

Our next (and last) example concerns proton transfer to a heavy target nucleus. Figure 5.8 shows angular distributions, measured by Barnett et al. [Ba 71b], for the reaction ^{208}Pb (^{16}O, ^{15}N$_{gs}$) ^{209}Bi at a laboratory energy of

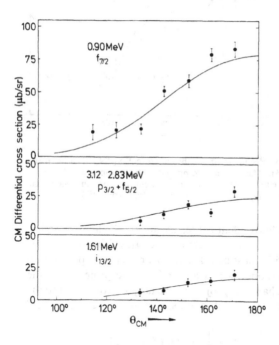

Fig. 5.8. Angular distributions for transitions to different final states in the proton transfer reaction ^{208}Pb(^{16}O, ^{15}N$_{gs}$)^{209}Bi at $E_{Lab} = 69.1$ MeV. Excitation energies and shell model orbits of the final states in ^{209}Bi are indicated; the full curves represent DWBA calculations. (From Ba 71b)

69.1 MeV. Individual states, or groups of states, in the final nucleus ^{209}Bi were resolved and are shown separately in the figure together with DWBA fits. All angular distributions rise smoothly towards 180°, regardless of l value or excitation energy. This behaviour is a general feature of sub-Coulomb transfer (with non-identical target–projectile combinations), and is readily understood as a consequence of the semiclassical relationship between transfer probability and distance of closest approach. Therefore the observed angular distribution shape lends strong support to the assumption of a transfer mechanism by nucleon tunnelling, but does not carry any nuclear structure information. We further note that the relatively small cross sections, of order $10 \, \mu b/sr$, make it difficult to obtain data of good statistical accuracy, although the average Q value for the transitions shown in Fig. 5.8 is not far from the optimum value ($Q_0 = -8.33$ MeV, $Q_{opt} (\pi) \approx -8.25$ MeV).

The reaction ^{208}Pb (^{16}O, ^{15}N) ^{209}Bi is interesting in several respects. Firstly, the doubly magic target nucleus ^{208}Pb should be well represented as an inert core, and the final states in ^{209}Bi are widely spaced and of well-known single-particle character. This makes the reaction an excellent test case for methods of analysing the reaction dynamics and extracting spectroscopic information. In addition, recoil effects and Coulomb contributions to the transfer interaction should be especially important in reactions involving heavy targets [Bu 71, Go 74], and one might therefore hope to learn how to take these effects properly into account.

Using the DWBA with recoil corrections, as proposed by Buttle and Goldfarb [Bu 71], and assuming bound-state well parameters consistent with other evidence, Barnett et al. were able to obtain a good fit to their data (see Fig. 5.8), implying spectroscopic factors close to those derived in other experiments and expected from shell-model arguments. Nevertheless, the quantitative significance of this fit is not quite clear, due to an extreme sensitivity of the extracted spectroscopic strength to the assumed geometrical parameters of the bound-state potential well in ^{209}Bi. Because of this sensitivity it was realized that absolute spectroscopic factors could not be measured with high accuracy; instead the authors suggested "inverting the argument" and deducing bound-state well parameters for ^{209}Bi assuming the relevant spectroscopic factors to be independently known. In a similar spirit, Jones et al. have used their experimental results for the neutron transfer reaction (^{17}O, ^{16}O) on $^{40,44,48}Ca$ to deduce bound-state well parameters and root-mean-square radii of single-neutron wave functions for the calcium isotopes [Jo 74]. As discussed by the authors, one must realize, however, that the quantity directly determined in the experiments is the probability density in the tail of the bound-state wave functions, as expressed by the products N^2S in (5.38) and (5.27). In order to infer matter radii of specified nucleon orbits, not only must the relevant spectroscopic factors be known, but also a simplified model of nuclear structure—usually involving a shell model potential of Saxon–Woods shape—must be invoked. While such a procedure may appear reasonable if one deals with doubly magic cores like ^{48}Ca or ^{208}Pb, it is certainly open to question in more complicated cases. Ultimately, therefore, more sophisticated nuclear structure calculations—including a realistic treatment of the tail region—will be needed to exploit fully the information inherent in sub-Coulomb transfer data. The experiments should then provide a unique possibility to test such calculations.

Another difficulty encountered in the analysis of transfer reactions—not only at sub-Coulomb energies—is that always two bound states enter, one in the donor and one in the acceptor nucleus. For standard reactions, involving, for example, ^{16}O or ^{12}C projectiles, one usually has enough independent information to make fairly reliable estimates of the bound state properties for the lighter partner. Moreover, the cross sections appear to be less sensitive to details of the lighter bound state as compared to the heavier one. Nevertheless, additional uncertainties from this source cannot be completely avoided. A potentially powerful method of dealing with this problem has been discussed by Goldfarb [Go 74]. It consists of performing "triangular experiments", involving three transitions and a total of three bound states. Each of the latter participates in at least two transitions, and therefore the relevant factors N^2S can in principle be determined unambiguously. The practical applicability of this approach seems to be limited, however, and no results have been published to our knowledge.

Summarizing our discussion of sub-Coulomb transfer reactions, we first note that the possibilities of this type of study, as envisaged in some of the early papers, have not been fully realized by the results available so far. In fact, experimental activity in the field has not been expanding in proportion to activities concerned with other aspects of heavy-ion reactions. This may be demonstrated by the fact that by 1974, at the time of the Nashville conference, only about 25 papers on sub-Coulomb transfer experiments had been published in major journals, compared with well over 100 papers on higher-energy transfer. The reasons for this situation are fairly obvious: the comparatively low cross sections in sub-Coulomb transfer, restricted further by the Q--window effect, reduce the number of accessible cases and make the experiments difficult and time consuming. In addition the angular distributions are featureless and do not carry nuclear structure information. Correspondingly more importance must be attached to a precise determination of absolute cross sections, which again is not a trivial matter. On the theoretical side, methods of dealing with the problems of kinematic mismatch and recoil had to be developed; these being now well in hand, the nuclear structure problem must be reconsidered in order to interpret the measured cross sections properly. Facing tedious work on all these issues on one hand, and a rapidly growing energy range of existing heavy-ion accelerators on the other, many experimentalists naturally preferred to look for qualitatively new and exciting phenomena at higher bombarding energies.

Nevertheless we emphasize once more that sub-Coulomb transfer reactions are a unique tool for probing, in an essentially model-independent way, nucleon densities and nucleon–nucleon correlations beyond the nuclear surface. While the difficulties involved both in the experiments and in their theoretical analysis will probably prevent systematic surveys from being performed, a precise and thorough investigation of selected cases should be rewarding. Further, with increasing use of very heavy projectiles, the energy region near the Coulomb barrier and especially the interplay between multiple Coulomb excitation and sub-Coulomb transfer will receive growing attention. For these reasons a revival of interest in the problems of sub--Coulomb transfer may be expected.

5.4 One-Nucleon Transfer Above the Coulomb Barrier

We start this section by noting that the basic mechanism of quasi-elastic nucleon transfer does not change as the bombarding energy is increased above the Coulomb barrier. Consequently major parts of our discussion of sub-Coulomb transfer in the previous section apply also at higher energies. In the following we focus our attention on those aspects where the use of higher bombarding energies is essential in extending the field either qualitatively or quantitatively.

As we go above the barrier, the transfer occurs at smaller internuclear distances close to the strong-absorption distance. At the same time, increasing linear and angular momenta of the colliding nuclei become involved. These physical changes have a number of important consequences for transfer cross sections: above all, the transfer probabilities, and hence the cross-section magnitudes, increase dramatically. This is a great advantage from the experimental point of view which, however, is only available at the expense of theoretical complications. The latter arise from the fact that the nuclear core–core interaction now enters the analysis. This interaction is usually expressed in terms of optical model potentials for the incident and outgoing channels which are inserted into DWBA calculations.

Apart from a general enhancement of cross sections, important changes of kinematic selectivity are observed at higher bombarding energies: transitions corresponding to large transfers of energy (large negative Q values) and (or) angular momentum are increasingly preferred (see Sect. 4.3). This feature enables one to study high-angular-momentum states at high excitation, which may not be accessible either in sub-Coulomb transfer with heavy projectiles or in light-projectile-induced transfer. In addition, the change in selectivity associated with suitable changes of incident fragmentation and energy may be exploited to deduce spectroscopic assignments, such as the orbital and total angular momentum of the final bound state. In the same context, the analysis of oscillatory angular distributions (see Sect. 4.2) may be useful. We shall return to these points in more detail below.

Experimental activity in the field of nucleon transfer reactions expanded rapidly during the period 1965–1975, parallel to the continued development of technical facilities and theoretical methods of analysis. In Table 5.1 we present a—necessarily incomplete—survey of the main centres of activity, together with indications of their principal interests and selected references. Clearly such a schematic survey must do injustice to many excellent individual efforts, both inside and outside the laboratories mentioned in the table. We nevertheless hope that it may provide an instructive outline of the historical development, and serve as a starting point for more detailed studies of the original literature. In the following we comment briefly on the contents of Table 5.1.

Much of the pioneering work on heavy-projectile-induced transfer was done in the mid-sixties at Yale and Heidelberg. Both laboratories concentrated on light nucleus–nucleus systems (mainly 1p-shell nuclei), where nuclear

Table 5.1. Survey of principal laboratories engaged in experimental work on one-nucleon transfer with heavy ions

Laboratory (period)	Beams (Lab. energies)	Targets	Emphasis	References
Yale (1965–67)	^{10,11}B, ^{14}N (10 MeV/u)	^{12,13}C, ^{14,15}N ^{16}O, ^{20}Ne	Reaction mechanism; kinematic and spectroscopic selectivity, charge independence	Sa 65, Bi 67, Po 67
Heidelberg (1966–73)	^{12,13}C, ^{14,15}N ^{16}O, ^{19}F (15–70 MeV)	^{10,11}B, ^{12,13,14}C, ^{16}O	Optical model and DWBA analysis, spectroscopic factors, interference effects (esp. elastic transfer)	Bo 66, Oe 69a,b Sc 72a, Oe 73a (see also Sect. 2.4)
Orsay (1969–71)	^{12}C, ^{14}N (41–113 MeV)	^{11}B, ^{12,13}C	Reaction mechanism; recoil·effect and angular distribution structure	Ja 69, Oe 69c, Oe 70, Li 70, Li 71
Copenhagen (1971–75)	^{12}C, ^{16,18}O (35–66 MeV)	$A = 26 - 96$	Reaction mechanism; kinematic and spectroscopic selectivity; DWBA and semiclassical analysis	Ni 71, Be 73, Ch 73c, Ch 73d, Ba 74a, Ba 75a
Harwell/ Oxford (1972–75)	^{11}B, ^{12}C, ^{14}N (72–174 MeV)	^{11}B, ^{12,13}C, ^{14}N ^{26}Mg, ^{40}Ca, ^{208}Pb	Reaction mechanism; kinematic and spectroscopic selectivity; semiclassical and DWBA analysis	Sc 72c, Sc 73b, An 73c, An 74, Sc 74, Pa 75
Berkeley (1972–75)	^{12}C (78 MeV) ^{16}O(104–216 MeV)	^{54}Fe, ^{62}Ni $A = 99–94$ ^{208}Pb	Spectroscopic factors, angular momentum effects, DWBA analysis	Ko 72, Ko 73b,c Zi 73, Ko 74, Be 74b, Be 75
Argonne (1972–75)	^{13}C, ^{15}N, ^{16,18}O (40–60 MeV)	$A = 40–88$	Spectroscopic factors, angular momentum effects, DWBA analysis	Si 72, Mo 72, An 73a,b Kö 73a, He 75
Brookhaven (1973–75)	^{13}C, ^{14}N, ^{18}O (40–68 MeV) ^{32}S (100 MeV)	^{27}Al, ^{40,48}Ca ^{60}Ni, ^{94,96}Mo	Fine structure of angular distributions, angular momentum effects, optical model and DWBA analysis	Bo 73b, Ch 73b, Sc 73a, Le 74b, Ba 75b, Ga 75
Oak Ridge (1972–75)	^{11}B(72, 75 MeV) ^{12}C(77–116 MeV)	^{208}Pb	Spectroscopic factors, DWBA analysis	La 72, Fo 74, To 76

structure calculations and light-projectile-induced transfer data were available for comparison. Undoubtedly technical considerations—related to low Coulomb barriers and relatively large level spacings—may also have dictated the choice of comparatively light systems. Due to the different energy ranges of the available accelerators, the groups at Yale and Heidelberg focused their interests on somewhat different, but complementary aspects of heavy-ion--induced single-nucleon transfer. The Yale group concentrated mainly on the qualitatively new features associated with the kinematics of high-energy transfer: effects of kinematic matching and recoil and the enhancement of large angular momentum transfers were clearly recognized and discussed within the framework of the then available theoretical models. Pronounced spectroscopic selectivities were also observed, and used to derive information on the relative single-particle strengths of various states and the validity of a charge--independent description of the transfer reactions.

The Heidelberg group, operating at lower energies, performed a series of careful quantitative studies, thereby establishing the applicability of the optical model and the DWBA approach, as developed by Buttle and Goldfarb [Bu 66, Bu 68, Bu 71; see Sect. 5.2], in a consistent analysis of elastic scattering and transfer data [Vo 69, Sc 72a]. Considerable efforts were also devoted at Heidelberg to systematic studies of interference effects, which arise from the coherent superposition of elastic or inelastic scattering and transfer, and of different transfer routes. Part of this work has been discussed in the context of "elastic transfer" in Sect. 2.4.

The work on one-nucleon transfer between light (1p-shell) nuclei was extended to intermediate energies by the Orsay group. Angular distributions, measured at Orsay with carbon and nitrogen beams on boron and carbon targets, provided the first clear evidence for the power and quantitative precision of "exact" DWBA calculations including recoil [De 73a,b; see Fig. 5.5].

Further work on light systems has been performed more recently by the Oxford group, working at Harwell, and at Texas A&M University [Na 75]. Both groups have used variable-energy cyclotrons providing light heavy-ion beams with energies of the order of 10 MeV per nucleon. The Oxford/ Harwell data have been discussed in considerable detail in the framework of a semiclassical approach, with particular emphasis on the kinematic selectivities associated with high-energy transfer [An 74; see also Sc 73b, Sc 74]. The Texas A&M data, on the other hand, were used for a systematic test of "exact" DWBA calculations, including recoil, as developed by Tamura and Low [Ta 73, Ta 74a, Lo 74]. It was concluded that "generally the quality of spectroscopic studies with heavy ion-induced transfer reactions is comparable, if not superior, to that from the corresponding light ion-induced reactions" [Na 75]. We shall return to this point below.

Work on one-nucleon transfer reactions with target nuclei of intermediate mass, in the range $A = 20$–100, was conducted at Copenhangen, Argonne, and Brookhaven with light heavy-ion beams of high quality, but limited energy, from tandem accelerators. The bulk of the data from these laboratories refers to bombarding energies which are-although clearly above the Coulomb barrier—still in the Coulomb-dominated energy region (below about twice the barrier). Main points of interest were initially the applicability of semiclassical and DWBA methods of analysis. The discovery of rapidly

oscillating angular distributions at extreme forward angles then led to considerable activity, directed toward a systematic experimental investigation and
theoretical understanding of these distributions [see Sect. 4.2]. Related to this
activity were efforts to correlate the data with angular momentum transfers
and bound-state angular momenta. These questions were also investigated
in higher-energy work at Berkeley with targets of intermediate mass.

Last, but not least, we have to mention the experiments on one-nucleon
transfer to or from ^{208}Pb, which were performed mainly at the variable-
-energy cyclotrons in Berkeley and Oak Ridge. The unique role played by this
target nucleus due to its doubly magic nature has been discussed in Sect. 5.3;
it makes it especially suitable for quantitative tests of theoretical reaction
models.

For the remainder of this section we shall concentrate on two issues which
have received considerable attention and may ultimately determine the
significance of heavy-projectile-induced one-nucleon transfer as a probe of
nuclear structure. These are firstly the possibility of making spectroscopic
assignments on the basis of relative cross sections or angular distributions,
and secondly the possibility of analysing quantitatively absolute cross sections
by DWBA methods. Obviously both points are closely related. We begin
with a discussion of angular momentum effects.

We consider transfer of a nucleon from a (lighter) projectile to a (heavier)
target nucleus, a situation usually referred to as "stripping" (note, that the
inverse process, "pick-up", can be dealt with by interchanging l_2, j_2 and l_1, j_1).
Normally the angular momentum quantum numbers l_1, j_1 of the initial bound
state are known, and one is interested in assigning quantum numbers l_2, j_2 to
the final bound state.

In conventional light ion reactions we have $A_1 \leq 4$, and hence $l_1 = 0$,
$j_1 = \frac{1}{2}$. The selection rules for the angular momentum transfer l [(5.7) and
(5.8); see also Table 5.2] then require $l = l_2$. Since in light-projectile-
-induced transfer, the shape of the angular distribution is usually characteristic
of l, the quantum number l_2 can be assigned unambiguously. In some cases,
one also finds significantly different angular distributions for total angular
momenta $j_2 = j_> = l_2 + \frac{1}{2}$ and $j_2 = j_< = l_2 - \frac{1}{2}$, as was first shown by Lee
and Schiffer [Le 64]. However, this effect, which arises from the spin–orbit
interaction of the transferred nucleon, is not sufficiently pronounced in general to permit a unique assignment of j_2. A reliable distinction between $j_>$ and
$j_<$ on the basis of heavy-projectile-induced transfer data would therefore be a
new and highly desirable feature.

That such a distinction may indeed be possible becomes clear from a consideration of the selection rules. In Table 5.2 we have listed allowed values of
the angular momentum transfer l for different combinations of (l_1, j_1) and
(l_2, j_2). It can be seen that for a given initial state $(l_1 \neq 0, j_1)$, different l
components will in general contribute to transitions populating final states
with $j_>$ and $j_<$, respectively. Of special practical interest is the case of $p\frac{1}{2}$ transfer, where the dominant l values (i.e. those allowed in the no-recoil approximation) differ by two units. Similarly, for a given final state (l_2, j_2), different

Table 5.2. Allowed values of the angular momentum transfer l for different combinations of the initial (l_1, j_1) and final (l_2, j_2) orbit of the transferred nucleon

l_2, j_2 \ $l_1, j_1 =$	s 1/2	p 1/2	p 3/2
s 1/2	0	1	1
p 1/2	1	0(1)	(1)2
p 3/2	1	(1)2	0(1)2
d 3/2	2	1(2)	1(2)3
d 5/2	2	(2)3	1(2)3(4)
f 5/2	3	2(3)	2(3)4
f 7/2	3	(3)4	2(3)4(5)
g 7/2	4	3(4)	3(4)5
g 9/2	4	(4)5	3(4)5(6)

l values are expected if projectiles with different initial states (l_1,j_1) are used. It follows that any pronounced dependence of the differential cross section on l, either in magnitude or in angular distribution, will be reflected in a dependence on (l_2, j_2) and (l_1, j_1).

Practical applications of j effects have been mainly of an exploratory nature so far. Two different approaches have been advanced, one based on cross-section magnitudes; the other, on angular distribution shapes. We first discuss the former approach.

In a study of the $[^{16}O, \ ^{15}N]$ and $[^{12}C, \ ^{11}B]$ reactions on ^{54}Fe, ^{62}Ni, and ^{208}Pb targets, Kovar et al. at Berkeley noted pronounced j_2 effects [Ko 72]. A general enhancement of $j_>$ final states relative to $j_<$ final states was observed, which, however, was much stronger for the $(^{16}O, \ ^{15}N)$ reaction ($p\frac{1}{2}$ transfer) than for the $[^{12}C, \ ^{11}B]$ reaction ($p\frac{3}{2}$ transfer). This is shown in Fig. 5.9, where experimental cross-section ratios $\sigma(j_>): \sigma(j_<)$ are plotted for both

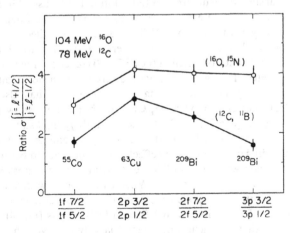

Fig. 5.9 Peak cross-section ratios $\sigma(j_>): \sigma(j_<)$ for proton single-particle states, populated by $(^{16}O, \ ^{15}N)$ and $(^{12}C, \ ^{11}B)$ reactions in different nuclei. The final shell-model orbits are indicated. (From Ko 72)

reactions and a number of different final states. Another way of analysing the data is to consider the cross-section ratio $\sigma(^{16}O, {}^{15}N): \sigma(^{12}C, {}^{11}B)$, for population of the same final state, as demonstrated in Fig. 5.10. The results were explained qualitatively as due to a strong kinematic preference for larger l values, together with the selection rules for $p\frac{1}{2}$ and $p\frac{3}{2}$ transfer. It was concluded that comparative studies of transfer reactions populating the same

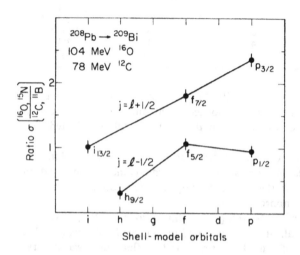

Fig. 5.10. Peak cross-section ratios $\sigma(^{16}O, {}^{15}N): \sigma(^{12}C, {}^{11}B)$ for proton transfer to the same final states in ^{209}Bi. The final shell-model orbits are indicated. (From Ko 72)

final state from different initial orbits could be useful in making spectroscopic assignments. Attempts to reproduce the observed j-dependent effects quantitatively with no-recoil DWBA calculations failed, however; spectroscopic factors deduced for $j_<$ states were consistently too large in the $[^{16}O, {}^{15}N]$ reaction, and too small in the $(^{12}C, {}^{11}B)$ reaction, relative to those deduced for $j_>$ states. The same systematic discrepancy in no-recoil DWBA analyses of $(^{16}O, {}^{15}N)$ data, populating $j_>$ and $j_<$ states, has also been noted by the Copenhagen group [Be 73, Ch 73d], but did not show up in related work at Argonne [Mo 72, Kö 73a], where mainly $j_>$ final states were investigated.

It was realized from the beginning, of course, that a quantitative understanding of the observed j effects had to be reached before j assignments could be made with any confidence. Considerable progress in that direction was achieved when "exact" DWBA calculations with recoil became possible. Becchetti et al. applied such an analysis to Berkeley data on $(^{16}O, {}^{15}N)$ and $(^{12}C, {}^{11}B)$ reactions with ^{54}Fe and ^{62}Ni targets [Be 74b]; they were able to show that the inclusion of recoil largely removed the previously noted discrepancies, and made tentative j assignments to a number of levels in ^{55}Co and ^{63}Cu. Similar calculations were performed, and similar conclusions reached by Tamura and Low [Ta 73, Lo 74] in connection with transfer reactions on ^{208}Pb.

Experiments carried out in Oak Ridge and Berkeley with ^{208}Pb targets at different bombarding energies revealed that the energy dependence of the proton-transfer cross section is also characteristically related to j_1 and j_2 (see Ko 74, Sc 74). In the (^{12}C, ^{11}B) reaction, where the dominant l values are the same for $j_>$ and $j_<$ final states, the corresponding cross sections increase at a similar rate with increasing bombarding energy. In the (^{16}O, ^{15}N) reaction, on the other hand, the ratio $\sigma(j_>) : \sigma(j_<)$ decreases strongly with increasing bombarding energy, and approaches unity for final p and f orbits slightly above 200 MeV. The latter effect has been discussed qualitatively as a consequence of the conditions for kinematic matching (4.48,49) and quantitatively on the basis of semiclassical theory [Sc 74]. These results indicate that not only the dependence on the donor orbit, but also that on bombarding energy may provide a useful basis for j_2 assignments. More work will be needed, however, on the quantitative foundation of such assignments in terms of reaction theory, before their general application can be considered.

The shape of angular distributions can also be used as an indicator of l_2 or j_2 in favourable cases. This method—well known in light-projectile--induced transfer—is based on the existence of oscillatory structure at small angles, which depends characteristically on the angular momentum transfer l. A classical example is provided by $p\frac{1}{2}$-proton transfer to ^{48}Ca, populating the ground $f\frac{7}{2}$ and first excited $p\frac{3}{2}$ states of ^{49}Sc. This process has been studied in different laboratories, using the reactions (^{14}N,^{13}C), (^{15}N,^{14}C) and (^{16}O, ^{15}N) [Sc 73b, Ch 73b, He 75, Ba 75a]. In Fig. 5.11 we show angular distributions of the reaction ^{48}Ca (^{14}N,^{13}C) ^{49}Sc, measured at Brookhaven [Ch73b], together with DWBA fits. The two angular distributions are strikingly different; in particular, the ground-state transition ($l = 4$) exhibits a pronounced maximum at about 10°, whereas the excited-state transition ($l = 2$) exhibits a deep minimum at the same angle. Very similar results have also been obtained for the other reactions, involving proton transfer from an initial $p\frac{1}{2}$ state. In the (^{3}He, d) reaction, on the other hand, which corresponds to an $s\frac{1}{2}$ initial state, the angular momentum transfer is $l = 3$ for the ground state and $l = 1$ for the excited state [Er 66]. Combining these l values with the heavy-ion results, and taking the selection rules into account (see Table 5.2), unique values of j_2 can be assigned. This has been exploited by Henning et al. in a study of the reaction ^{48}Ca (^{15}N, ^{14}C) ^{49}Sc, where j values for some of the higher excited states in ^{49}Sc were determined [He 75]. Another interesting application of the selection rules can be made whenever $l_1 = l_2$. The cases $j_2 = j_1$ and $j_2 \neq j_1$ can then be distinguished by the presence or absence of an $l = 0$ component in the cross section, which normally exhibits a sharp rise towards zero degrees, provided the incident and outgoing channels are kinematically well matched. The feasibility of this approach has been demonstrated by White et al. [Wh 74a,b] for the reaction (^{7}Li, ^{6}He) on ^{62}Ni, where the ground-state transition ($j_2 = \frac{3}{2}^-$) is characterized by a forward maximum, which is absent in the transition to the first excited state ($j_2 = \frac{1}{2}^-$).

It should be noted, however, that the j effects in angular distributions are not always as clear cut as those shown in Fig. 5.11, nor is it always possible to obtain DWBA fits of the same quality. This difficulty seems to be related to the fact that DWBA predictions for angles near zero degrees are often very sensitive to optical model potentials and Q values [Ko 74, Ba 75a, Ba 75b]. Although the positions of maxima and minima appear to be less affected by parameter changes than the magnitude of the cross section at small angles [Ba 75b], more quantitative work is required to establish the conditions under which spectroscopic assignments can be deduced from an analysis of angular distributions in heavy-projectile-induced transfer.

We close this section with some comments on problems of the quantitative analysis of one-nucleon transfer and the extraction of spectroscopic factors. The relationship between the transfer cross section and the intensity of the bound-state wave function at the nuclear surface has been discussed in Sect. 5.2 and 5.3. The resulting sensitivity of deduced spectroscopic factors to the assumed bound-state potential is a general feature of transfer reactions, regardless of projectile mass or bombarding energy. To the extent that somewhat different radial regions may be significant, the precise degree of this sensitivity may, however, vary from case to case.

The methods of DWBA analysis have been refined considerably in recent years. From systematic comparisions between theory and experimental data for a variety of transfer reactions, it has become clear that a full DWBA treatment including recoil effects is an essential prerequisite for the quantita-

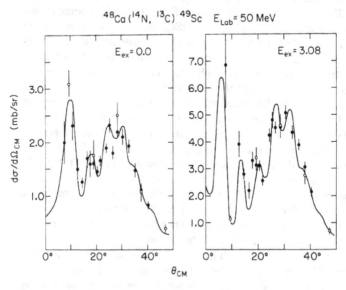

Fig. 5.11. Experimental angular distributions (points with error bars) and DWBA fits (full curves) for the reaction ^{48}Ca(^{14}N, ^{13}C)^{49}Sc at 50 MeV bombarding energy. The ground state in ^{49}Sc (left part) has $j_2, l_2 = \frac{7}{2}^-$ and the 3.08 MeV state (right part) has $j_2, l_2 = \frac{3}{2}^-$. (From Ch 73b)

tive evaluation of heavy-ion data. Careful analyses of this nature have been performed in particular for nuclei in the $1p$ shell by De Vries [De 73a] and by Nair et al. [Na 75], and for nuclei around ^{208}Pb by Tamura and Low [Ta 73, Lo 74], and by Ford et al. [Fo 74]. The conclusion that may be drawn from this work is that the spectroscopic information from heavy-projectile-induced one-nucleon transfer can be of comparable precision to that from light-projectile-induced transfer, provided the most advanced methods of analysis are employed. It appears that spectroscopic factors can be deduced with an uncertainty of about 20—30% in the most favourable cases, i.e. for doubly magic cores and near optimum Q values and l transfers.

Ambiguities in optical model parameters remain an important limiting factor in the quantitative analysis of transfer data taken above the Coulomb barrier. Different parameter sets, although all consistent with elastic scatter-ing data, may produce quite different angular distributions, especially at small angles (see, for example, Fo 74, Ba 75a, Ba 75b). In some instances, it has not been possible to find parameter sets which would reproduce both the elastic scattering and the transfer data, and more or less arbitrary adjustments had to be made in order to obtain acceptable fits to the latter. These difficulties probably reflect basic inadequacies of the DWBA treatment, as applied to the cases in question. In other cases, where acceptable fits could be obtained, this may have been the consequence of some particular, possibly unphysical, choice of parameters. In order to reduce uncertainties of this nature, it would seem generally desirable to select parameters on the basis of systematic trends and physical arguments, rather than for optimum fit to a specific case. Work along these lines has been performed by Baltz et al. at Brookhaven [Ba 75a], who proposed an imaginary potential composed of a strongly absorbing volume part with small diffuseness and a weakly absorbing surface part with larger diffuseness. Such a potential is qualitatively expected as a consequence of strong coupling to the compound nucleus at small separation, and a weaker coupling to quasi-elastic reaction channels at larger separation. The composite potential of Baltz et al. has been used successfully to analyse one- and two-nucleon transfer reactions on targets with $A = 40$—60. Similar concepts have also been developed and applied by Gobbi et al. in analyses of elastic scattering between light nuclei [Go 73] and by Paschopoulos et al. in a careful study of different one-nucleon transfer reactions induced by ^{11}B on ^{26}Mg [Pa 75].

Most of the work on heavy-ion-induced one-nucleon transfer so far has been devoted to the purpose of gaining a qualitative and quantitative under-standing of the reaction mechanism. We have tried to outline in this section the considerable progress which has been achieved towards that goal. In doing so, we have concentrated on simple aspects of the reaction mechanism, and disregarded more complicated ones, like core excitation and two-step pro-cesses. This may be justified by noting that the dominant transitions were usually found to be simple, and that simple features must be understood before complex ones can be discussed. We have also left out applications of transfer reactions to specific problems of nuclear spectroscopy, in accordance with our general practice followed throughout this book. We emphasize, however, that nucleon transfer with heavy projectiles is not only a process of interest in itself, but also a potentially powerful tool for nuclear structure studies.

5.5   Multi-Nucleon Transfer

In reactions between complex nuclei, the simultaneous transfer of several nucleons becomes possible and, indeed, occurs often with a probability comparable to that for one-nucleon transfer. From a spectroscopic point of view, the outstanding importance of these processes lies in the fact that they permit the study of multi-nucleon correlations at the nuclear surface, which may reflect the m-particle n-hole (mp, nh) nature of a given final state or—more specifically—a coherent superposition of (mp, nh) configurations produced by residual interactions. Well-known examples are, in particular, the presence of α-clusters in light nuclei, and of pair correlations and "quartet structures" in medium and heavy nuclei. Multi-nucleon transfer provides a very selective mechanism for the production of such correlated subgroups of nucleons and there are, indeed, clear indications of their existence in the experimental data.

Another interesting application of multi-nucleon transfer is the identification of isobaric analogue states, populated in analogous reaction pairs like (^{6}Li, ^{3}He), (^{6}Li, ^{3}H) [Bi 71, Bi 73, Bi 75], or (^{16}O, ^{14}N), (^{16}O, ^{14}C) [Po 72].

While quasi-elastic multi-nucleon transfer is undoubtedly a potentially powerful tool of nuclear spectroscopy, severe problems still exist in the quantitative analysis of experimental data. These are all related more or less directly to the fact that in general a number of mechanisms of different complexity can contribute to a given transition. Indirect reaction paths may involve sequential transfers of single nucleons or smaller subgroups, or inelastic excitations either preceding or following the transfer. In addition, competing direct paths may exist, corresponding either to components of different internal structure in the wave functions of the transferred nucleon group and the two cores, or to transfer in different directions. The number of such possibilities increases dramatically with the number of transferred nucleons, and consequently it becomes more and more difficult to establish experimental—or even conceptual—criteria for a simple quasi-elastic mechanism. Moreover, even if the mechanism was understood in detail, mathematical difficulties would in many cases prohibit a complete quantitative analysis. We note in passing that such an analysis should, of course, comprise a full finite-range treatment of the transfer including recoil effects, which are expected to be more important in multi-nucleon than in one-nucleon transfer. Under these circumstances, much of the work performed so far had to concentrate first of all on establishing the basic ingredients of the reaction mechanism, and hence the feasibility of studies of this type, and secondly on a qualitative or semiquantitative discussion of the spectroscopic results. At the same time, however, considerable progress has been made in the development of highly sophisticated methods of quantitative analysis; as examples we mention, without further discussion, the DWBA calculations of Yoshida and collaborators [Ka 69, Bo 72a], and of Charlton [Ch 75]; the generator coordinate method of Bonche and Giraud [Bo 72b, Bo 73a], and extensive work on the coupled channels analysis of multistep multi-nucleon transfer by Tamura and Low [Ta 74a], Ascuitto and Glendenning [As 73a, b; As 74], and others [Pi 76a].

We now turn to a brief discussion of systematics observed in multi-nucleon transfer with light target–projectile systems (A_1, $A_2 < 20$), which is based mainly on the work of the Heidelberg [Oe 69a, Hi 70, Sc 72b, Oe 74] and Oxford/Harwell groups [Sc 73b, Sc 74, An 74]. Additional results have been obtained in Yale, Philadelphia, Moscow, and other laboratories. Most of the available data refer to the transfer of two to four nucleons; transfers of larger nucleon numbers, exhibiting symptoms of a direct process, have also been seen, but the reaction mechanism could not be clearly established in these cases. Direct two-nucleon transfer has been observed in a variety of reactions induced by ^{11}B, ^{12}C, ^{14}N, and ^{16}O projectiles, three-nucleon transfer especially in (^6Li, ^3He), (^6Li, ^3H), (^7Li, ^4He), (^{12}C, ^9Be), and (^{19}F, ^{16}O) reactions, and four-nucleon (or α) transfer again in many reactions involving projectiles from ^3He to ^{19}F. The latter process has been exploited extensively for studies of deformed rotational bands in "α-particle nuclei" and will be discussed in somewhat more detail at the end of this section.

The direct nature of these multi-nucleon transfer reactions has been inferred from forward-peaked angular distributions, and from their selectivity, which correlates well with existing nuclear structure calculations and other experimental evidence. The results of the Heidelberg group, obtained at comparatively low bombarding energies ($\leq 5\,MeV/u$), have been analysed by (approximate) DWBA and discussed mainly in terms of nuclear-structure effects. Depending on the structure of the projectile (donor), a strong enhancement of transitions proceeding by transfer of a nucleon group in an s state of internal relative motion could be identified in some cases [Bo 72a; see also Sect. 5.1]; in other cases, however, important contributions corresponding to non-zero internal angular momentum of the transferred nucleon group were indicated. Evidence for such an effect comes, for example, from the strong population of a 2^- state at 8.88 MeV excitation in ^{16}O in the reaction $^{12}C(^{10}B, ^6Li)^{16}O$ [Hi 70]: this state, having "unnatural parity", cannot be reached by direct α transfer and is, in fact, only weakly populated in "good α-transfer reactions" like (^{16}O, ^{12}C) or (^7Li, ^3H). Evidence for a decisive influence of the projectile structure has also been found in the three-nucleon transfer reaction $^{12}C(^{19}F, ^{16}O)^{15}N$, where a $\frac{1}{2}^+$ state in ^{15}N at 5.3 MeV excitation is strongly excited, which is believed to have a close structural relationship to the ground state of ^{19}F [Sc 72b, Oe 74].

The Oxford/Harwell group has studied two- to four-nucleon transfer induced by beams of ^{11}B, ^{12}C, and ^{14}N at bombarding energies of about 10 MeV/u on various light targets. A highly selective population of final states was observed and could be explained as a consequence of kinematic matching conditions [An 74, Bu 74, Sc 74]. The preferred final states in n-nucleon transfer were shown to have a simple shell-model structure, characterized by a group of n nucleons orbiting the core in the lowest unfilled oscillator shell, with minimum internal excitation and angular momenta aligned for maximum resultant. A striking example is provided by Fig. 5.12, which shows a pulse-hight spectrum of ⁹Be particles from the reaction (¹²C, ⁹Be) in a mixed target containing ¹²C, ¹⁶O, and ⁴⁰Ca, taken at a bombarding energy of 114 MeV. There are no more than two strongly excited final states for each target, and in each case the strongest state follows exactly the above prescription: for the ¹²C and ¹⁶O targets we have states corresponding to three nucleons in the $2s1d$ shell coupled to maximum angular momentum, and likewise for the ⁴⁰Ca target with three nucleons in the $2p1f$ shell. It appears therefore, that at these higher bombarding energies (compared with the Heidelberg work) the transfer cross sections are less sensitive to the detailed structure of target and projectile; instead, they appear to be governed by reaction kinematics and the major shells available in the final (acceptor) state.

The Oxford/Harwell data were analysed in the framework of a semi-classical reaction model (based on Brink's kinematic model, Br 72b) and compared in detail with shell-model calculations [An 74]. A very good (qualitative) overall agreement between theory and experiment was achieved, which supports the assumptions underlying the analysis.

Interesting qualitative conclusions concerning the reaction mechanism and the relative importance of direct and sequential reaction paths can also be drawn without complex calculations by merely comparing the orders of magnitude of different transfer cross sections in a systematic way. Anyas--Weiss et al. have pointed out that one-nucleon transfer, (pn)-transfer, (2p1n)--transfer, and (2p2n)-transfer (in light systems at 5–10 MeV/u bombarding energy) all have typical peak cross sections for strong transitions of the order of 1 mb/sr [An 74]. The corresponding values for (2n)-transfer and (2p)--transfer are 20 μb/sr, and for (3n)-transfer, 3 μb/sr. These numbers strongly suggest that a simple one-step mechanism (as implied by the notion of "cluster transfer") is responsible for the former types of transfer reaction. However, there is considerable evidence (to be discussed partly below) that this is also

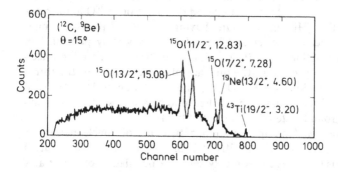

Fig. 5.12. Pulse-height spectrum for the three-nucleon transfer reaction (¹²C, ⁹Be) with a mixed target containing ¹²C, ¹⁶O, and ⁴⁰Ca, at a bombarding energy of 114 MeV ($\Theta = 15°$). Prominent final states with spin–parity assignments are indicated. (From An 74)

true for transfer reactions involving two identical nucleons, at least in favourable cases. The much smaller cross sections for these latter reactions are then probably a consequence of the reduced internal binding energy of the two-nucleon clusters, and the resulting sharper decrease of their centre-of-mass wave function at the nuclear surface. The comparison between transfer cross sections for identical and nonidentical nucleon pairs implies that the latter are predominantly transferred in a relative $S = 1$, $T = 0$ state in the light systems considered here. Further interesting evidence concerning the relative importance of sequential transfer comes from work performed at Argonne by Kovar et al. [Ko 74] on reactions like $^{48}Ca(^{16}O,^{15}C)^{49}Ti$ and $^{48}Ca(^{16}O,^{13}B)^{51}V$. Both of these reactions have forward-peaked angular distributions, with peak cross sections of the order of 20–50 µb/sr in the former case, and about 5 µb/sr in the latter case. Since the nature of the observed products rules out one-step cluster transfer for the former reaction, a sequential two-step mechanism—involving (2p)-transfer in one direction and n-transfer in the other direction—must be assumed. The even smaller cross section for the latter reaction, corresponding to (3p)-transfer, also implies a sequential mechanism. It should be noted, however, that absolute transfer cross sections as well as relative probabilities for one-step and sequential transfers may be strongly affected by kinematic mismatch (and nuclear structure); general conclusions concerning the importance of one or the other mechanism can therefore not be drawn on the basis of simple comparisons of absolute cross sections in isolated cases.

Next we consider the problem of two-nucleon transfer on medium and heavy nuclei. In contrast to the situation in light systems, almost only transfers involving identical nucleon pairs have been studied: Two-neutron transfers by $(^{18}O$, $^{16}O)$ and $(^{12}C$, $^{14}C)$ reactions, and two-proton transfers by $(^{16}O$, $^{14}C)$ and $(^{12}C$, $^{10}Be)$ reactions. The choice of these reactions was probably dictated both by kinematic matching and yield considerations, and by the necessity to resolve individual states in the final nuclei. Since neutrons and protons fill different shells in heavy nuclei $(A \gtrsim 60)$, n–p correlations in low-lying states are presumably weak. On the other hand, strong pair correlations are expected for identical nucleons, and these should produce an observable enhancement of two-nucleon transfer cross sections connecting the ground states of even–even nuclei, which differ by two neutrons or two protons. Theoretical speculations and predictions concerning the possible transfer of a nucleon pair by tunnelling from one heavy nucleus to another at sub-Coulomb bombarding energies [Go 68a, Di 71]—referred to as the "nuclear Josephson effect" in analogy to a well-known effect in solid-state physics—have contributed greatly to the interest in this type of reaction. So far, pair transfer has been studied only near and above the Coulomb barrier and with comparatively light projectiles. Von Oertzen et al. have investigated two-proton transfer to several nuclei with a closed neutron shell $N = 82(^{140}Ce$, $^{142}Nd,^{144}Sm)$ with the reaction $(^{16}O$, $^{14}C)$ at bombarding energies close to the Coulomb barrier [Oe 73b]. Using semiclassical arguments, enhancement factors between 20 and 30 were estimated for the ground-state transitions, when compared either with expectations for uncorrelated (sequential) transfer of two protons, or with experimental (2p)-transfer cross sections in the Fe/Ni region. These estimates are consistent with enhancements observed in light-projectile-induced (2n)-transfer in the Sn region. More recently, Becchetti et al. have studied the two-proton transfer reactions (^{12}C, ^{10}Be) and (^{16}O, ^{14}C) with the doubly magic target nucleus ^{208}Pb well above the Coulomb barrier [Be 74a]. The data were analysed by DWBA and compared with shell model calculations. It was concluded that the strength of the ground state transition corresponds to an enhancement factor of about 8, by comparison with the expected strength for a pure $(g_{\frac{9}{2}})^2$ shell model configuration in ^{210}Po. These results demonstrate, that heavy-projectile-induced two-nucleon (especially two-proton) transfer reactions are indeed sensitive to pair correlations induced by residual interactions in heavy nuclei.

A number of studies concerned mainly with the reaction mechanism in two-nucleon transfer has been conducted with target nuclei in the 2p1f shell [Po 72, Le 73, Le 74a, Le 74b, He 74] as well as with heavier targets [Ch 72, Le 74a, Er 74, As 75, Ya 75, Sc 75b]. The angular distributions observed for moderately heavy targets ($A < 100$), and at bombarding energies between about 1.2 and 2 times the Coulomb barrier, usually exhibit a pronounced oscillatory pattern at forward angles, which is superimposed on a more or less well-developed grazing peak. Some examples, taken from the Saclay work on the reaction ^{64}Ni(^{16}O, ^{14}C)^{66}Zn, are given in Fig. 5.13. The appearance of the small-angle oscillations places restrictions on the absorptive parts of the optical model potentials which must be inserted in DWBA calculations to fit the data [Le 74b, Ba 75b]. As in one-nucleon transfer, the period of the os-

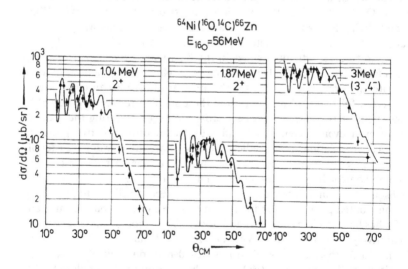

Fig. 5.13. Experimental angular distributions and DWBA fits for different final states in the two-proton transfer reaction ^{64}Ni(^{16}O, ^{14}C) ^{66}Zn, at a bombarding energy of 56 MeV. (From Le 74a)

cillations is determined by the grazing angular momentum, while its phase pattern depends on the angular momentum transfer l.

The single-particle orbits (l,j) which participate in a given two-nucleon transfer reaction may also have a pronounced effect on the cross section [Ko 74]. These orbits determine the relative and absolute weights by which different internal configurations of the transferred pair enter the transition amplitude, and the radial distribution of the transfer "form factor". The influence of these various factors on the cross section depends, in addition, on the kinematic variables. Hence (again as in one-nucleon transfer, but for more complicated reasons), different selectivities may be observed if a given two-nucleon transfer is induced by different projectiles (donors) with different internal structure, or by the same projectile at different bombarding energies. An example is provided by the work of Becchetti et al. on two-proton transfer to ^{208}Pb, where kinematic and structural reasons combine to produce quite different excitation spectra in ^{210}Po, when populated either in the $(^{16}$O, ^{14}C) or in the $(^{12}$C, ^{10}Be) reaction [Be 74a, Ko 74]. This effect demonstrates once more the sensitivity of two-nucleon transfer to nuclear structure; whether it can be used as a general diagnostic tool is an open question, however, in view of its rather complicated nature.

Indirect, two-step mechanisms are much more important in two-nucleon transfer than in one-nucleon transfer. This result is, of course, not unexpected, considering our discussion so far. Clear evidence for two-step contributions has been obtained in a number of recent two-nucleon transfer studies, especially for transitions populating low-lying 2^+ states in transitional or deformed even–even nuclei [Ch 72, Er 74, Le 74a, As 75, Ya 75, Sc 75b]. The most important type of two-step mechanism seems to be inelastic excitation of the lowest 2^+ state in either the target or the residual nucleus, followed (or preceded) by two-nucleon transfer. Other mechanisms, like sequential transfer, cannot be excluded, however. The two-step amplitude may interfere either constructively or destructively with the one-step amplitude, and in general modifies both the magnitude of the cross section and the shape of the angular distribution. These effects are sometimes sufficiently conspicuous to be recognized without an elaborate quantitative analysis. In a number of cases, however, the data were fitted successfully using coupled-channels techniques [Ta 74b, As 74, Er 74, As 75, Ya 75, Sc 75b]. A particularly impressive example, taken from the work of Erb et al. on the reaction ^{186}W$(^{12}$C, ^{14}C) ^{184}W, is shown in Fig. 5.14. The angular distributions of transitions populating the 2^+ and 4^+ members of the ground-state rotational band in ^{184}W exhibit pronounced, but distinctly different interference patterns, which are well fitted by a coupled-channels Born approximation (CCBA) analysis. The interference dip in the 2^+ angular distribution is explained as arising mainly from the destructive interference between the Coulomb and nuclear parts of the (dominating) two-step amplitude. The interference dip in the 4^+ angular distribution, on the other hand, is attributed to destructive interference between the direct and two-step amplitudes, which in this case are of comparable magnitudes. Fits of similar quality have been achieved also in other cases

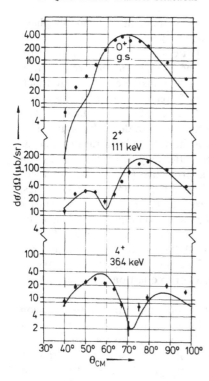

Fig. 5.14. Experimental angular distributions and CCBA fits for the two-neutron transfer reaction ^{186}W(^{12}C, ^{14}C)^{184}W, populating the 0^+, 2^+, and 4^+ members of the ground-state rotational band in ^{184}W. The bombarding energy is 70 MeV. (From Er 74)

[Ko 74, Ta 74b]. Whether or not the remarkable quality of these fits is significant may be open to question; it appears, nevertheless, that the basic mechanisms have been correctly interpreted.

For the remainder of this section we concentrate on four-nucleon transfer reactions. The possible existence of α clusters or other highly correlated four-nucleon substructures in nuclei has been a fascinating problem ever since the early days of nuclear physics. Consequently a large amount of activity has been devoted to four-nucleon transfer, in order to establish the mechanism of various reaction types and to test relevant nuclear-structure models. Much of the earlier work was concerned with questions of the reaction mechanism: whether a given reaction could be described as transferring an α cluster, or just four nucleons, and which of several possible reactions was suited best for α-cluster transfer. Clearly both of these questions over-simplify the situation, since stationary subgroups, identical in all respects with a free α particle, can neither exist in real nuclei nor be transferred in a real nuclear reaction. In a more realistic description of the process, therefore, one has to consider the donor and acceptor nuclei as dynamical systems, characterized by certain probability amplitudes for four-nucleon partitions with definite internal symmetry. A good "cluster transfer" reaction would then be one where the total direct transition probability is dominated by single components of appropriate symmetry both in the initial and in the final state, and where indirect routes are comparatively unimportant.

A large proportion of the earlier experimental work on four-nucleon stripping reactions was performed with ^{6}Li and ^{7}Li beams on light targets, up to about ^{40}Ca. A comprehensive survey of results available by 1969 has been given in a number of laboratory reports, presented at the 1969 Heidelberg Conference [Be 69, Co 69a, Co 69b, Mi 69, Og 69], and in a review article by Bethge [Be 70]. Both the (^{6}Li,d) and the (^{7}Li,t) reactions were found to populate selectively known "α-cluster states", as for example the (4p4h) deformed rotational band in ^{16}O (based on the 6.06 MeV 0^+ level), and the rotational bands in ^{20}Ne based on the ground state ($K = 0^+$) and the 5.79 MeV 1^- level ($K = 0^-$). The (^{7}Li,t) reaction exhibits generally larger cross sections to the strongly excited final states, but less-structured angular distributions. The latter feature is related to the relative P state of (α + t) motion in ^{7}Li, which normally allows more than one value of angular momentum transfer according to the selection rules. Since zero-range DWBA cannot be used under these circumstances, and exact finite-range codes were not available at the time, Pühlhofer et al. developed the "fixed range" method to analyse their data on the reaction ^{12}C(^{7}Li,t)^{16}O [Br 69, Pü 70; see Sect 5.2]. A more recent study of the reactions ^{12}C(^{7}Li, t)^{16}O and ^{16}O(^{7}Li, t)^{20}Ne, which includes a finite-range coupled-channels Born-approximation analysis, has been performed by Cobern et al. [Co 76].

The existence of appreciable indirect admixtures to (^{6}Li,d) and (^{7}Li,t) reactions has been deduced from analyses of angular distributions and excitation functions, and from the population of non-normal parity states in four-nucleon transfer to 0^+ targets. Evidence in favour of a mixed reaction mechanism comes also from studies of angular correlations between outgoing deuterons or tritons (usually observed at zero degrees) and α particles [Ar 72] or γ rays [Ba 71a, Ca 72], emitted in the decay of the final state. Measurements of this type probe the magnetic substate population in the transfer reaction, which is particularly simple for an assumed α cluster transfer mechanism.

The inverse process of Li-induced four-nucleon stripping—four-nucleon pick-up via the reactions (d,^{6}Li) or (^{3}He, ^{7}Be)—has also been studied and interpreted successfully as cluster transfer [Da 64, De 66, Gu 69, Gu 71; Fo 69, De 69, Au 75, Pi 76b]. These reactions yield information on four-nucleon correlations in the ground state of the target nucleus.

The cross sections for Li-induced four-nucleon stripping reactions, and for the corresponding pick-up reactions, exhibit a strong tendency to decrease with increasing mass of the target nucleus. This trend can be traced back to a combination of spectroscopic and kinematic reasons. With increasing target mass, the amplitudes of the relevant four-nucleon configurations at the nuclear surface decrease on the average; at the same time, the kinematic mismatch becomes more severe. The latter circumstance is related to the large core asymmetry in the Li reactions, which results in highly negative optimum Q values for stripping, and highly positive optimum Q values for pick-up. On the other hand, due to the small binding energy of an α cluster in the Li nuclei, the ground-state Q values are normally positive in stripping, and negative in pick-up. Therefore, the advantage gained by the loosely bound nature of the Li nuclei due to a large probability for finding an α cluster at the nuclear surface is strongly counteracted by the kinematic mismatch for transitions to low-lying final states.

The relative merits of different reactions as a spectroscopic tool in α-transfer studies have been the subject of intense discussion in the literature. For some time it was felt that the strong decrease of the cross section for Li-induced α-transfer reactions with increasing target mass would prevent their application to targets beyond about mass forty. Consequently, the interest of most workers in the field shifted to heavier projectiles, where better kinematic matching and hence larger cross sections could be expected. In particular, the (^{16}O, ^{12}C) reaction has been studied extensively at Saclay on target nuclei up to $A = 90$ [Fa 71a, b; Le 73, Bo 73a] and in several other laboratories on targets in the $2s1d$ shell [Ma 72, Br 72a, Er 73, Ch 74, Er 75]. A certain disadvantage of this reaction is the relatively small parentage of the ^{16}O ground state with respect to the partition ^{12}C ground state plus α particle, and the somewhat larger component corresponding to ^{12}C in the first excited state. It was found, however, that the ground-state transition proceeds as a well-behaved direct transfer reaction, and that the population of the excited state in ^{12}C is weak or absent in most cases. The latter effect was explained by finite-range DWBA as predominantly due to kinematic effects [De 73c, d]. The (^{12}C, ^{8}Be) reaction has been studied by several groups as an alternative to (^{16}O, ^{12}C) because of its more favourable spectroscopic factor for the ground-state transition. However, this reaction presents technical problems due to the instability of the ^{8}Be ground state with respect to decay into two α particles. Various schemes have been developed to solve this problem, which involve either rather sophisticated counter arrangements or strongly reduced detection efficiencies [Wo 72, Cr 73, Ho 74]. Thus, although the (^{12}C, ^{8}Be) reaction is undoubtedly a good α-transfer reaction, it will probably not emerge as a first choice for general application.

An impressive come-back of the (^{6}Li, d) reaction has been triggered by recent work of Strohbusch and collaborators, performed at Rochester with incident energies around 30 MeV and target nuclei up to ^{90}Zr [St 74, St 75]. These studies indicate that the disadvantage of a somewhat smaller cross section (of order 10 μb/sr) compared with (^{16}O, ^{12}C) is more than compensated by a number of attractive features: on the technical side, limitations on target thickness and solid angle are less severe than with heavier projectiles, and high-resolution spectra can be obtained with magnetic spectrographs designed for light-particle work. On the physical side, the data give evidence of a clean reaction mechanism, resulting in highly structured angular distributions, which are characteristic of the angular momentum transfer l and are well reproduced by either zero-range or finite-range DWBA. Figure 5.15 shows experimental angular distributions and DWBA fits for a number of transitions in the reaction ^{40}Ca(^{6}Li, d)^{44}Ti. In contrast, relatively featureless angular

Fig. 5.15. Experimental angular distributions and DWBA fits (solid curves: finite range; dashed curves: zero range) for the reaction $^{40}Ca(^6Li,d)^{44}Ti$ at a bombarding energy of 32 MeV. Final state spins and excitation energies are indicated. (From St 75)

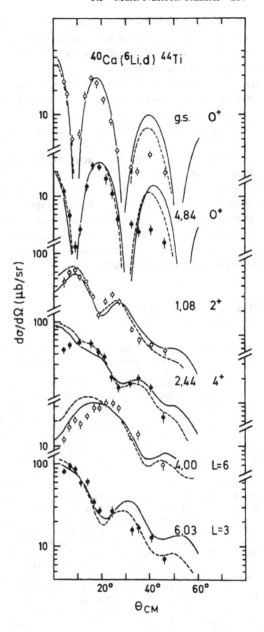

distributions have usually been observed in $(^{16}O, ^{12}C)$ reactions. The question as to which reaction is most suitable for α transfer to medium or heavy nuclei therefore remains open at present.

To summarize our discussion of quasi-elastic two- to four-nucleon transfer we note first that a fairly large body of experimental data has been accumulated for light- and medium-mass nuclei, while problems of low yield and insufficient energy resolution have largely prevented similar studies on heavy nuclei. Extensive information has been obtained on the reaction mechanism which in many cases appears to be dominated by the direct transfer of a correlated nucleon group. Two-step admixtures have, however, been clearly identified in two-nucleon transfer, and are probably even more important (but less conspicuous) in transfers of larger nucleon numbers. In spite of considerable progress, the complex problem of quantitative analysis of multi-nucleon transfer has not yet been adequately solved, and therefore much of the nuclear structure information extracted so far from the data is still of a qualitative nature. Nevertheless, important conclusions were reached on the existence of two-, three-, and four-nucleon correlations in nuclei which could not be discussed in detail here. As examples we mention the identification of deformed rotational bands with pronounced α particle parentage in light nuclei, and of "quartet states" in medium-mass nuclei, which are believed to contain highly symmetric substructures of two protons and to neutrons with pairwise aligned angular momenta.

References Chapter 5

(For AR 73, AS 63, HE 66, HE 69, HE 74, NA 74 see Appendix D)

Al 71a Alder, K.; Trautmann, D.: Ann. Phys. (N.Y.) *66*, 884 (1971)
Al 71b Alder, K.; Trautmann, D.: Nucl. Phys. *A178*, 60 (1971)
Al 72   Alder, K. et al.: Nucl. Phys. *A191*, 399 (1972)
An 73a Anantaraman, N.; Katori, K.; Körner, H.J.: Phys. Lett. *46B*, 67 (1973)
An 73b Anantaraman, N.: Phys. Rev. *C8*, 2245 (1973)
An 73c Anyas-Weiss, N. et al.: Phys. Lett. *45B*, 231 (1973)
An 74   Anyas-Weiss, N. et al.: Phys. Rep. *12*, 201 (1974)
Ar 72   Artemov, K.P. et al.: Sovj. J. Nucl. Phys. *14*, 165 (1972)
As 69   Ascuitto, R.J.; Glendenning, N.K.: Phys. Rev. *181*, 1396 (1969)
As 73a Ascuitto, R.J.; Glendenning, N.K.: Phys. Lett. *45B*, 85 (1973)
As 73b Ascuitto, R.J.; Glendenning, N.K.: Phys. Lett. *47B*, 332 (1973)
As 74   Ascuitto, R.J.; Vaagen, J.S.: NA 74 Vol. 2 p. 257
As 75   Ascuitto, R.J. et al.: Phys. Lett. *55B*, 289 (1975)
Au 64   Austern, N.; Drisko, R.; Satchler, G.: Phys. Rev. *133*, B3 (1964)
Au 75   Audi, G. et al.: Nucl. Phys. *A237*, 300 (1975)
Ba 67   Bayman, B.F.; Kallio, A.: Phys. Rev. *156*, 1121 (1967)
Ba 71a Balamuth; D.P.: Phys. Rev. *C3*, 1565 (1971)
Ba 71b Barnett, A.R. et al.: Nucl. Phys. *A176*, 321 (1971)
Ba 72   Baltz, A.J.; Kahana, S.: Phys. Rev. Lett. *29*, 1267 (1972)
Ba 74a Ball, J.B. et al.: Phys. Lett. *49B*, 348 (1974)
Ba 74b Baltz, A.J.; Kahana, S.: Phys. Rev. *C9*, 2243 (1974)
Ba 74c Bayman, B.F.: Phys. Rev. Lett. *32*, 71 (1974)
Ba 75a Ball, J.B. et al.: Nucl. Phys. *A244*, 341 (1975)
Ba 75b Baltz, A.J. et al.: Phys. Rev. *C12*, 136 (1975)
Be 65   Becker, L.C.; McIntyre, J.A.: Phys. Rev. *138*, B339 (1965)
Be 69   Bethge, K.: He 69 p. 277
Be 70   Bethge, K.: Ann. Rev. Nucl. Sci. *20*, 255 (1970)
Be 73   Becchetti, F.D. et al.: Phys. Lett. *43B*, 279 (1973)
Be 74a Becchetti, F.D. et al.: Phys. Rev. *C9*, 1543 (1974)
Be 74b Becchetti, F.D. et al.: Phys. Rev. *C10*, 1846 (1974)
Be 75   Becchetti, F.D. et al.: Phys. Rev. *C12*, 894 (1975)
Bi 67   Birnbaum, J.; Overley, J.C.; Bromley, D.A.: Phys. Rev. *157*, 787 (1967)

Bi 71 Bingham, H.G. et al.: Phys. Rev. Lett. *26*, 1448 (1971)
Bi 73 Bingham, H.G. et al.: Phys. Rev. *C7*, 57 (1973)
Bi 75 Bingham, H.G. et al.: Phys. Rev. *C11*, 1913 (1975)
Bo 66 Bock, R.; Grosse-Schulte, M.; von Oertzen, W.: Phys. Lett. *22*, 456 (1966)
Bo 72a Bock, R.; Yoshida, H.: Nucl. Phys. *A189*, 177 (1972)
Bo 72b Bonche, P.; Giraud, B.: Phys. Rev. Lett. *28*, 1720 (1972)
Bo 73a Bonche, P.; Giraud, B.: Nucl. Phys. *A199*, 160 (1973)
Bo 73b Bond, P.D. et al.: Phys. Lett. *47B*, 231 (1973)
Br 52 Breit, G.; Hall Jr., M.H.; Gluckstern, R.L.: Phys. Rev. *87*, 74 (1952)
Br 56a Breit, G.; Ebel, M.E.: Phys. Rev. *103*, 679 (1956)
Br 56b Breit, G.; Ebel, M.E.: Phys. Rev. *104*, 1030 (1956)
Br 59 Breit, G.: In *Encyclopedia of Physics*, ed. by S. Flügge, Vol. 51/1, p. 367 ff. Berlin, Heidelberg, New York: Springer 1959
Br 63 Breit, G.: AS 63 p. 97
Br 64a Breit, G.; Chun, K.W.; Wahsweiler, H.G.: Phys. Rev. *133*, B403 (1964)
Br 64b Breit, G.: Phys. Rev. *135*, B1323 (1964)
Br 67 Breit, G.; Polak, J.A.; Torchia, T.A.: Phys. Rev. *161*, 993 (1967)
Br 69 Brommundt, G. et al.: HE 69 p. 293
Br 72a Braun-Munzinger, P. et al.: Phys. Rev. Lett. *29*, 1261 (1972)
Br 72b Brink, D.M.: Phys. Lett. *40B*, 37 (1972)
Br 72c Broglia, R.A.; Winther, A.: Nucl. Phys. *A182*, 112 (1972)
Br 72d Broglia, R.A.; Winther, A.: Phys. Rep. *4*, 153 (1972)
Br 74a Braun-Munzinger, P.; Harney, H.L.: Nucl. Phys. *A223*, 381 (1974)
Br 74b Broglia, R.A. et al.: Phys. Rep. *11*, 3 (1974)
Bu 66 Buttle, P.J.A.; Goldfarb, L.J.B.: Nucl. Phys. *78*, 409 (1966)
Bu 68 Buttle, P.J.A.; Goldfarb, L.J.B.: Nucl. Phys. *A115*, 461 (1968)
Bu 71 Buttle, P.J.A.; Goldfarb, L.S.B.: Nucl. Phys. *A176*, 299 (1971)
Bu 74 Buck, B.: HE 74 p. 152
Ca 72 Carlson, R.R.: Phys. Rev. *C5*, 1467 (1972)
Ch 72 Chasman, C. et al.: Phys. Rev. Lett. *28*, 843 (1972)
Ch 73a Charlton, L.A.: Phys. Rev. *C8*, 146 (1973)
Ch 73b Chasman, C. et al.: Phys. Rev. Lett. *31*, 1074 (1973)
Ch 73c Christensen, P.R. et al.: Nucl. Phys. *A203*, 1 (1973)
Ch 73d Christensen, P.R. et al.: Nucl. Phys. *A207*, 33 (1973)
Ch 74 Charlton, L.A.; Robson, D.: Phys. Rev. Lett. *32*, 946 (1974)
Ch 75 Charlton, L.A.: Nucl. Phys. *A241*, 144 (1975)
Co 69a Comfort, J.R. et al.: HE 69 p. 303
Co 69b Cotton, E.: HE 69 p. 289
Co 76 Cobern, M.E.; Pisano, D.J.; Parker, P.D.: Phys. Rev. *C14*, 491 (1976)
Cr 73 Cramer, J.G. et al.: Nucl. Instrum: Methods *111*, 425 (1973)
Da 64 Daehnick, W.W.; Denes, L.J.: Phys. Rev. *136*, B 1325 (1964)
Da 65a Dar, A.: Phys. Rev. *139*, B1193 (1965)
Da 65b Dar, A.; Koslowski, B.: Phys. Rev. Lett. *15*, 1036 (1965)
Da 66 Dar, A.: HE 66
De 66 Denes, L.J.; Daehnick, W.W.; Drisko, R.M.: Phys. Rev. *148*, 1097 (1966)
De 69 Détraz, C. et al.: HE 69 p. 319
De 73a DeVries, R.M.: Phys. Rev. *C8*, 951 (1973)
De 73b DeVries, R.M.; Kubo, K.I.: Phys. Rev. Lett. *30*, 325 (1973)
De 73c DeVries, R.M.: Phys. Rev. Lett. *30*, 666 (1973)
De 73d DeVries, R.M.: Nucl. Phys. *A212*, 207 (1973)
De 74 DeVries, R.M.; Satchler, G.R.; Cramer, J.G.: Phys. Rev. Lett. *32*, 1377 (1974)
De 75 DeVries, R.M.: Phys. Rev. *C11*, 2105 (1975)
Di 71 Dietrich, K.: Ann. Phys. (N.Y.) *66*, 480 (1971)
Do 65 Dodd, L.R.; Greider, K.R.: Phys. Rev. Lett. *14*, 959 (1965)
Do 69 Dodd, L.R.; Greider, K.R.: Phys. Rev. *180*, 1187 (1969)
Dr 66 Drisko, R.; Rybicki, F.: Phys. Rev. Lett. *16*, 275 (1966)
Er 66 Erskine, J.R.; Marinov, A.; Schiffer, J.P.: Phys. Rev. *142*, 633 (1966)
Er 73 Erskine, J.R.; Henning, W.; Greenwood, L.R.: Phys. Lett. *47B*, 335 (1973)
Er 74 Erb, K.A. et al.: Phys. Rev. Lett. *33*, 1102 (1974)
Er 75 Erskine, J.R. et al.: Phys. Rev. Lett. *34*, 680 (1975)
Fa 71a Farraggi, H. et al.: Ann. Phys. (N.Y.) *66*, 905 (1971)

Fa   71b   Faraggi, H. et al.: Phys. Rev. *C4*, 1375 (1971)
Fo   69   Fortune, H.T.; Zeidman, B.: HE 69 p. 307
Fo   74   Ford, J.L.C. et al.: Phys. Rev. *C10*, 1492 (1974)
Fr   69   Frahn, W.E.; Sharaf, M.A.: Nucl. Phys. *A133*, 593 (1969)
Ga   66   Gaedke, R.M.; Toth, K.S.; Williams, I.R.: Phys. Rev. *141*, 9 (1966)
Ga   75   Garett, J.D. et al.: Phys. Rev. *C12*, 489 (1975)
Gl   65   Glendenning, N.K.: Phys. Rev. *137*, B102 (1965)
Gl   74   Glendenning, N.K.; Nagarajan, M.A.: Nucl. Phys. *A236*, 13 (1974)
Go   68a   Goldanski, V.I.,; Larkin, A.I.,: Sov. Phys. JETP *26*, 617 (1968)
Go   68b   Goldfarb, L.J.B.; Steed, J.W.: Nucl. Phys. *A116*, 321 (1968)
Go   69   Goldfarb, L.J.B.: HE 69 p. 115
Go   73   Gobbi, A. et al.: Phys. Rev. *C7*, 30 (1973)
Go   74   Goldfarb, L.J.B.: NA 74 Vol. 2 p. 283
Gu   69   Gutbrod, H.H.; Yoshida, H.; Bock, R.: HE 69 p. 311
Gu   71   Gutbrod, H.H.; Yoshida, H.; Bock, R.: Nucl. Phys. *A165*, 240 (1971)
He   74   Henning, W. et al.: Phys. Rev. Lett. *32*, 1015 (1974)
He   75   Henning, W. et al.: Phys. Lett. *55B*, 49 (1975)
Hi   65   Hiebert, J.C.; McIntyre, J.A.; Couch, J.G.: Phys. Rev. *138*, B346 (1965)
Hi   70   Hildenbrand, K.D. et al.: Nucl. Phys. *A157*, 297 (1970)
Ho   74   Ho, H. et al.: Nucl. Phys. *A233*, 361 (1974)
Ja   69   Jacmart, J.C. et al.: HE 69 p. 128
Jo   64   Jobes, F.C.; McIntyre, J.A.: Phys. Rev. *133*, B893 (1964)
Jo   74   Jones, G.D. et al.: Nucl. Phys. *A230*, 173 (1974)
Ka   69   Kammuri, T.; Yoshida, H.: Nucl. Phys. *A129*, 625 (1969)
Ko   72   Kovar, D.G. et al.: Phys. Rev. Lett. *29*, 1023 (1972)
Kö   73a   Körner, H.J. et al.: Phys. Rev. *C7*, 107 (1973)
Ko   73b   Kovar, D.G.: AR 73 Vol. I p. 59
Ko   73c   Kovar, D.G. et al.: Phys. Rev. Lett. *30*, 1075 (1973)
Ko   74   Kovar, D.G.: NA 74 Vol.2 p. 235
Ku   73   Kurath, D.: AR 73, Vol.I p. 221
La   72   Larsen, J.S. et al.: Phys. Lett. *42B*, 205 (1972)
Le   64   Lee, L.L.; Schiffer, J.P.: Phys. Rev. Lett. *12*, 108 (1964)
Le   73   Lemaire, M.C.: Phys. Rep. 7, 279 (1973)
Le   74a   Lemaire, M.C. et al.: Phys. Rev. *C10*, 1103 (1974)
Le   74b   LeVine, M.J. et al.: Phys. Rev. *C10*, 1602 (1974)
Li   70   Liu, M. et al.: Nucl. Phys. *A143*, 34 (1970)
Li   71   Liu, M. et al.: Nucl. Phys. *A165*, 118 (1971)
Lo   74   Low, K.S.; Tamura, T.: Phys. Lett. *48B*, 285 (1974)
Lo   75   Low, K.S.; Tamura, T.: Phys. Rev. *C11*, 789 (1975)
Ma   60   Macfarlane, M.H.; French. J.B.: Revs. Mod. Phys. *32*, 567 (1960)
Ma   72   Maher, J.V. et al.: Phys. Rev. Lett., *29*, 291 (1972)
Mc   60   McIntyre, J.A.; Watts, T.L.; Jobes, F.C.: Phys. Rev. *119*, 1331 (1960)
Mc   73   McMahan, C.A.; Tobocman, W.: Nucl. Phys. *A212*, 465 (1973)
Mi   69   Middleton, R.: HE 69 p. 263
Mo   59   Moshinsky, M.: Nucl. Phys. *13*, 104 (1959)
Mo   72   Morrison, G.C. et al.: Phys. Rev. Lett. *28*, 1662 (1972)
Na   72   Nagarajan, M.A.: Nucl. Phys. *A196*, 34 (1972)
Na   73   Nagarajan, M.A.: Nucl. Phys. *A209*, 485 (1973)
Na   75   Nair K.G. et al.: Phys. Rev. *C12*, 1575 (1975)
Ni   71   Nickles, R.J. et al.: Phys. Rev. Lett. *26*, 1267 (1971)
Oe   69a   von Oertzen, W. et al.: HE 69 p. 156
Oe   69b   von Oertzen, W. et al.: Nucl. Phys. *A133*, 101 (1969)
Oe   69c   von Oertzen, W. et al.: Phys. Lett. *28B*, 482 (1969)
Oe   70   von Oertzen, W. et al.: Nucl. Phys. *A143*, 34 (1970)
Oe   73a   von Oertzen, W.: AR 73 Vol. II p. 675
Oe   73b   von Oertzen, W.; Bohlen, H.G.; Gebauer, B.: Nucl. Phys. *A207*, 91 (1973)
Oe   73c   von Oertzen, W.; Nörenberg, W.: Nucl. Phys. *A207*, 113 (1973)
Oe   74   von Oertzen, W.: In *Nuclear Spectroscopy and Reactions*, Part B, ed. by J. Cerny, p. 279. New York, London: Academic Press 1974
Og   69   Oglobin, A.A.: HE 69 p. 231
Pa   75   Paschopoulos, I. et al.: Nucl. Phys. *A252*, 173 (1975)

Pi 76a Pisano, D.J.: Phys. Rev. *C14*, 468 (1976)
Pi 76b Pisano, D.J.; Parker, P.D.: Phys. Rev. *C14*, 475 (1976)
Po 67 Poth, J.F.; Overley, J.C.; Bromley, D.A.: Phys. Rev. *164*, 1295 (1967)
Po 72 Pougheon, F. et al.: Nucl. Phys. *A193*, 305 (1972)
Pü 70 Pühlhofer, F. et al.: Nucl. Phys. *A147*, 258 (1970)
Re 56 Reynolds, H.L.; Zucker, A.: Phys. Rev. *101*, 166 (1956)
Re 75 Reisdorf, W.: Nucl. Phys. *A242*, 406 (1975)
Ro 68 Rotter, I.: Nucl. Phys. *A122*, 567 (1968)
Ro 69 Rotter, I.: Nucl. Phys. *A135*, 378 (1969)
Ro 72 Robson, D.; Koshel, R.D.: Phys. Rev. *C6*, 1125 (1972)
Sa 65 Sachs, M.W.; Chasman, V.; Bromley, D.A.: Phys. Rev. *139*, B 92 (1965)
Sa 73 Satchler G.R.: AR 73 Vol. I p. 145
Sc 72a Schlotthauer-Voos, U.C. et al.: Nucl. Phys. *A180*, 385 (1972)
Sc 72b Schlotthauer-Voos, U.C. et al.: Nucl. Phys. *A186*, 225 (1972)
Sc 72c Scott, D.K. et al.: Phys. Rev. Lett. *28*, 1659 (1972)
Sc 73a Schneider, M.J. et al.: Phys. Rev. Lett. *31*, 320 (1973)
Sc 73b Scott, D.K.: AR 73 Vol.1 p. 97
Sc 74 Scott, D.K.: HE 74 p. 165
Sc 75a Schneider, W.F.W. et al.: Nucl. Phys. *A251*, 331 (1975)
Sc 75b Scott, D.K. et al.: Phys. Rev. Lett. *34*, 895 (1975)
Si 72 Siemssen, R.H. et al.: Phys. Rev. Lett. *28*, 626 (1972)
St 74 Strohbusch, U. et al.: Phys. Rev. *C9*, 965 (1974)
St 75 Strohbusch, U.; Bauer, G.; Fulbright W.W.: Phys. Rev. Lett. *34*, 968 (1975)
Ta 73 Tamura, T.; Low K.S.: Phys. Rev. Lett. *31*, 1356 (1973)
Ta 74a Tamura, T.: Phys. Rep. *14*, 59 (1974)
Ta 74b Tamura, T.; Low, K.S.; Ugadawa, T.: NA 74 Vol.1 p. 70
To 69 Towner, I.S.; Hardy, J.C.: Adv. Phys. *18*, 401 (1969)
To 76 Toth, K.S. et al.: Phys. Rev. *C14*, 1471 (1976)
Tr 70 Trautmann, D.; Alder, K.: Helv. Phys. Acta *43*, 363 (1970)
Va 68 Varma, S.: Nucl. Phys. *A106*, 233 (1968)
Vo 69 Voos, U.C.; von Oertzen, W.; Bock, R.: Nucl. Phys. *A135*, 207 (1969)
Wh 74a White, R.L. et al.: Phys. Rev. Lett. *32*, 892 (1974)
Wh 74b White, R.L.; Kemper K.W.: Phys. Rev. *C10*, 1372 (1974)
Wo 72 Wozniak, G.J. et al.: Phys. Rev. Lett. *28*, 1278 (1972)
Ya 75 Yagi, K. et al.: Phys. Rev. Lett. *34*, 96 (1975)
Zi 73 Zisman, M.S. et al.: Phys. Rev. *C8*, 1866 (1973)

6. Deep-Inelastic Scattering and Transfer

6.1 Introduction

In this chapter we discuss a class of processes which is intermediate between the quasi-elastic collisions considered in previous chapters, and compound nucleus formation, or "complete fusion". We start this introductory section by recalling the general features of different reaction mechanisms, as observed with light projectiles, and then turn to the special role played by intermediate mechanisms in collisions involving two heavy fragments. Subsequently we give a brief survey of the historical development.

In experiments with light projectiles ($A_1 \leq 4$) and medium-weight or heavy target nuclei ($A_2 > 50$), at bombarding energies of the order of 10 MeV/u, one usually observes two quite distinct types of interaction: direct reactions and compound-nucleus reactions. The experimental criteria used in assigning one or the other mechanism are distributions of the reaction products in energy, angle, mass, or charge and excitation functions for specified products. In direct reactions, the distributions are strongly reminiscent of the entrance channel, with preferential emission at forward angles and velocities close to the projectile velocity. In contrast, compound-nucleus reactions are characterized by "evaporation spectra" centered at much lower energies, typically of the order of 1 MeV for neutron emission and near the Coulomb barrier for charged-particle emission, and by angular distributions with forward--backward symmetry. Excitation functions are comparatively smooth for direct reactions; for compound-nucleus reactions, they exhibit narrow resonance structure at excitation energies near the nucleon-emission threshold, which dissolves at higher energies due to increasing overlap of levels. Averaged over resonances, compound-nucleus cross sections for a given final state or species rise from threshold to a maximum and then fall off exponentially with a further increase in bombarding energy.

We note that transitions from a given initial to a given final channel can in general proceed by either mechanism, with relative weights depending both on kinematic and structural parameters. The different observable characteristics make it possible, however, to decompose the total yield into a direct and a compound-nucleus component, provided sufficiently detailed experimental information is available.

The qualitative features of the two mechanisms enumerated above are easily understood by identifying direct reactions as "fast" processes, and compound-nucleus reactions as "slow" processes. In the former, the total interaction time is too short to permit a major rearrangement of the nucleonic structure of the fragments, whereas the latter involve the formation of an intermediate system with sufficiently long lifetime for complete internal equilibrium to be reached. Its decay then becomes independent of the properties of the entrance channel, except for conserved quantities like total mass, energy and angular momentum.

It is possible to obtain crude estimates of the time scale for each mechanism. Appropriate formulae and orders of magnitude are collected in Table 6.1, where we have introduced two universal time constants τ_0 and τ_1. The former of these (τ_0) is relevant to the motion of a nucleon within the nucleus, and he latter (τ_1) to the motion of two nuclei on a Rutherford trajectory. They are defined as follows (with m the nucleon mass, $r_0 = 1.2$ fm):

$$\tau_0 = \frac{mr_0^2}{\hbar} = 2.28 \times 10^{-23}\text{s}, \tag{6.1}$$

Table 6.1. Characteristic times

Quantity	Formula	Order of magnitude
Transit time of nucleon with Fermi momentum through nuclear diameter	$\dfrac{2m_0R}{p_F} = \left(\dfrac{64A}{9\pi}\right)^{1/3}\tau_0$	1×10^{-22} s
Vibrational period	$\dfrac{2\pi}{\omega_{\text{vib}}}$	$5 \times 10^{-21}\text{s}$ [a] $\quad$ $3 \times 10^{-22}\text{s}$ [b]
Compound nucleus lifetime $\quad$ at $E_{ex} = 10$ MeV $\quad$ at $E_{ex} = 100$ MeV	$\dfrac{\hbar}{\Gamma} = \dfrac{\hbar\rho(E,I)}{N(E,I)}$	$10^{-16} - 10^{-19}\text{s}$ [c] $\quad$ $10^{-20} - 10^{-21}\text{s}$ [c]
Rotational period of scattering system	$\pi\left(\dfrac{2\mu_{12}R_{12}^2}{E - V(R_{12})}\right)^{1/2}$ $= \pi\left(\dfrac{A_{12}(A_1^{1/3} + A_2^{1/3})^3}{Z_1Z_2}\right)^{1/2}\left(\dfrac{V(R_{12})}{E - V(R_{12})}\right)^{1/2}\tau_1$	$3 \times 10^{-21}\text{s}$ [d]
Interaction time of scattering system	$2\left(\dfrac{2r_0\mu_{12}R_{12}}{2[E - V(R_{12})] - R_{12}(\text{grad}\,V)_{R_{12}}}\right)^{1/2}$ $= 2\left(\dfrac{A_{12}(A_1^{1/3} + A_2^{1/3})^2}{Z_1Z_2}\right)^{1/2}$ $\left(\dfrac{V(R_{12})}{2[E - V(R_{12})] - R_{12}(\text{grad}\,V)_{R_{12}}}\right)^{1/2}\tau_1$	$10^{-21} - 10^{-22}\text{s}$ [d]

[a] Low-energy quadrupole mode.
[b] "Giant" dipole or quadrupole modes.
[c] The lower limit applies to light ($A \approx 30$) and the upper limit to heavy ($A \geq 200$) compound nuclei.
[d] For $A_1, A_2 \approx 100$, $E \approx 1.5\,V(R_{12})$.

$$\tau_1 = \left(\frac{2mr_0^3}{e^2}\right)^{1/2} = 1.58 \times 10^{-22} \text{s.} \tag{6.2}$$

The transit time of a nucleon with Fermi momentum through the nuclear diameter is a measure of the response time of the intrinsic nucleonic structure to external disturbances, and may therefore serve as a standard of reference. We shall see later that a significant number of nucleons can be exchanged between two nuclei in close contact during this time; this implies that the time required for dissipation of collective (relative) momentum is of the same order of magnitude, if nucleon transfer is the dominant mechanism of relative energy loss.

An alternative mechanism for energy dissipation is provided by the coupling between radial motion and collective vibrations of the fragments, as discussed by Broglia et al. [Br 76b]. We have seen in Sect. 3.2 that the low-energy quadrupole vibrations which give rise to the lowest 2^+ states in even-even nuclei and are usually associated with large-scale deformations of the nuclear surface are too slow to permit appreciable energy transfer during a collision. This is not the case, however, for the "giant quadrupole vibrational mode" whose frequency is about one order of magnitude higher and which is described microscopically as a coherent superposition of particle–hole excitations across two major shells (see, for example, Bo 74a). Typical periods of the latter mode are of the order of 3×10^{-22} s and hence comparable to interaction times in nucleus–nucleus collisions (see below).

Compound nucleus lifetimes can be deduced from the resonance structure of scattering and reaction cross sections at energies near the nucleon-emission threshold. In this region of excitation ($E_{ex} \approx 10$ MeV) they typically span the range from about 10^{-16}s for heavy nuclei ($A \gtrsim 200$) to about 10^{-19}s for light nuclei ($A \approx 30$). A limited amount of information is also available for somewhat higher excitation energies ($E_{ex} \approx 20 - 30$ MeV) and comparatively light compound nuclei from statistical analyses of fluctuating excitation functions [Eb 71]. At still higher excitation energies and for heavy compound nuclei one has to rely on statistical model calculations which relate the lifetime of the compound nucleus to its level density $\rho(E_{ex}, J)$ and the effective number of open decay channels $N(E_{ex}, J)$ (see Chap. 7). The success of this approach in reproducing experimentally determined relative decay widths gives some confidence in its use in extrapolating also absolute decay widths to excitation energies of the order of 100 MeV. Such calculations indicate a dramatic decrease of compound-nucleus lifetimes with increasing excitation energy, especially for heavy nuclei, and yield values around 10^{-20} to 10^{-21}s at $E_{ex} \approx$ 100 MeV.

For collisions between complex nuclei, where two fragments are re-emitted without compound nucleus formation, two characteristic times may be defined: a rotational period, which corresponds to the time required for a (hypothetical) complete revolution of the two touching fragments, and an "interaction time" (or "contact time"), during which the fragments interact strongly by nuclear forces. For a given initial system, both times will in general depend on bombarding energy, relative angular momentum, Q value, and final fragmentation. If the interaction time is much shorter than the rotational period, then we expect asymmetric angular distributions peaked in the vicinity of the "grazing angle". Symmetry with respect to $90°$ will result, however, if the interaction time is long compared with the rotational period. The latter situation, which implies the existence of an "orbiting" quasi-molecular system, would in this respect be indistinguishable from compound-nucleus decay. Intermediate cases, where the two fragments stick together for a time comparable with the rotational period, have been identified experimentally by the observation of products deflected to "negative angles" (see Fig. 6.8).

The formulae given in Table 6.1 refer to trajectories with a classical turning point equal to the half-density distance R_{12}, as appropriate to deep inelastic collisions. Changes in relative kinetic energy, angular momentum, nuclear shape, or fragmentation during the collision are disregarded, and the interaction time is identified with the time spent at distances between R_{12} and $R_{12} + r_0$ ($r_0 = 1.20$ fm). These assumptions are obviously crude, but should yield correct orders of magnitude.

We can conclude from Table 6.1 that typical interaction times in collisions between heavy nuclei are significantly larger than the times needed for nucleonic rearrangements, and are comparable with periods associated with collective excitations. They are not much shorter, on the other hand, than compound-nucleus lifetimes at the rather high excitation energies which are usually involved. Under these circumstances, it will not be possible to make a clear-cut distinction between "fast" and "slow" collisions, and a broad range of intermediate phenomena must be expected. In contrast, in nucleon–nucleus collisions (at comparable incident velocities), "direct" and "compound" reactions proceed on quite different time scales, and are therefore much more easily distinguished.

The fact that the time scales of external and internal motion are comparable in collisions between heavy nuclei has a number of interesting consequences, to be discussed in more detail in the following sections. For example, the energy spectra of reaction products may exhibit the typical characteristics of compound-nucleus decay, with a Maxwellian distribution peaked near the Coulomb energy of two touching fragments. At the same time, however, conspicuous symptoms of fast direct processes may be observed: angular distributions sharply peaked near the grazing angle and mass distributions concentrated near the original fragment masses. These results imply that the system retains its memory throughout the collision with respect to certain degrees of freedom (direction of motion, mass fragmentation) but not with respect to others (radial kinetic energy, charge-to-mass ratio). We conclude that deep-inelastic collisions between nuclei—in spite of many complicating features—offer a unique possibility to study the relative time scales of equilibration for different degrees of freedom.

The history of deep-inelastic collisions starts with the pioneering work of Kaufmann and Wolfgang [Ka 59, Ka 61], performed around 1960 at the Yale university heavy-ion linear accelerator. These authors bombarded medium-weight targets with ^{12}C, ^{14}N, ^{16}O, and ^{19}F beams at energies up to 10 MeV per nucleon, and observed radioactive products (positron emitters) formed by transfer of one or several nucleons to or from the projectile. The experimental techniques were extremely simple by present standards, and yet quite effective, involving carefully designed arrangements of target and catcher foils. Cross sections, excitation functions, and angular and range distributions were deduced by off-line counting of annihilation radiation from the decay of ^{11}C, ^{13}N, ^{15}O, or ^{18}F product nuclei. Figure 6.1 shows angular distributions of different products (plotted as $d\sigma/d\Theta = 2\pi \sin\Theta \, d\sigma/d\Omega$) from the bombardment of a rhodium target with 160 MeV ^{16}O ions, and Fig. 6.2 shows excitation functions for the same system.

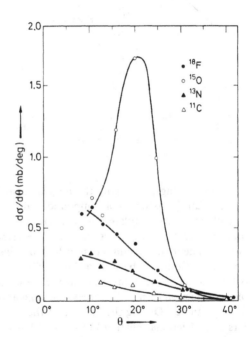

Fig. 6.1. Angular distributions ($d\sigma/d\Theta$) for different product nuclei emitted in the interaction of 160 MeV ^{16}O projectiles with a rhodium target. (From Ka 61)

Two principal results emerged from these studies:

i) Transfers of several nucleons, in particular (pn)-, (p2n)- and (2p3n)-transfers, occur with large cross sections, which amount typically to 10 – 20 % of one-nucleon transfer cross sections. The ratios of these cross sections are approximately equal for different target nuclei, and hence the underlying mechanism appears to be insensitive to Q values or nuclear structure effects.

ii) Angular distributions of multi-nucleon transfer reactions are strongly forward peaked, decreasing exponentially with increasing angle. One-nucleon transfer reactions show in addition a more or less well-developed peak or shoulder at the grazing angle which is superimposed on a smooth forward rise.

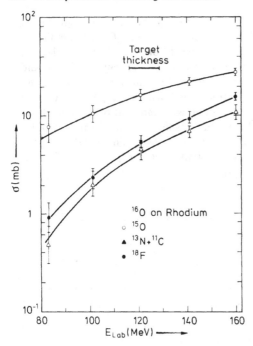

Fig. 6.2. Excitation functions for different product nulcei from the ¹⁶O bombardment of a rhodium target. (From Ka 61)

These findings, especially the angular distributions, were not consistent with the familiar patterns of compound or direct reactions and led Kaufmann and Wolfgang to postulate a new reaction mechanism, denoted as "contact transfer" or "grazing reactions" and described by the "grazing contact model". The model assumes that at high bombarding energies, after penetration of the Coulomb barrier, a dumbbell-shaped system with finite lifetime is formed under the influence of attractive nuclear forces. As a consequence of the deflection by nuclear forces, the angular distribution is shifted towards forward angles. During the lifetime of the composite system, nucleon transfers and "frictional heating" of the overlap volume were thought to occur; the latter was supposed not only to induce energy and angular momentum dissipation, but also to weaken the nuclear cohesion with the result that the fragments eventually separate due to dominating Coulomb and centrifugal forces.

It is remarkable that most of these ideas were later rediscovered and refined on the basis of much more complete experimental evidence, and then led to the development of the so-called friction models and diffusion models of nuclear fusion and deep inelastic collisions. Many concepts that are now familiar, such as the "intermediate complex", deflection to "negative angles", "neck formation and stretching", and the qualitative effects of surface tension and "friction" can be traced back to the work of Kaufmann and Wolfgang.

The early experiments of Kaufmann and Wolfgang were followed by more detailed investigations and extended to include the observation of target residues [Re 62, St 67]. The recoil-catcher technique was further refined and combined with radiochemical techniques. Figure 6.3 shows schematically how measurements of recoil-range distributions, average recoil ranges and excitation functions for target residues allow to discriminate between products of compound nucleus and quasi-elastic reactions. Applying criteria of this nature, Strudler et al. performed a systematic study of products formed in the interaction of ^{12}C and ^{14}N projectiles with ^{115}In at energies up to 10.5 MeV/u [St 67]. They were able to show that inelastic scattering (to isomeric levels) and one-nucleon transfers proceed predominantly by a quasi-elastic, and three- to five-nucleon transfers predominantly by a "grazing contact" mechanism, with two-nucleon transfers displaying intermediate behaviour. Again appreciable cross sections were observed for multi-nucleon transfer.

The grazing contact model was put on a quantitative basis by assuming "partial absorption and quasi-elastic scattering" of the projectile. In this version of the model, there are two components of momentum transfer to the target residue: one due to the transferred nucleons (partial absorption), and one due to Coulomb scattering of the projectile residue (quasi-elastic scattering). The resulting average recoil range of the target residue (projected along the beam axis) as a function of bombarding energy depends on the transferred fraction of the projectile as indicated in Fig. 6.3 b. Semiquantitative agreement with experimental average ranges of multi-nucleon transfer products was obtained [St 67].

Further and complementary evidence on intermediate mechanisms in heavy-ion induced reactions was obtained by several other groups throughout the 1960s. We mention here only a few representative examples. Britt and Quinton, working also at Yale, bombarded ^{197}Au and ^{209}Bi targets with high-energy ($\leq$ 10.5 MeV/u) ^{12}C, ^{14}N, and ^{16}O beams and studied the emission of light charged particles using counter techniques [Br 61]. In particular the cross section for α emission was found to be large, of the order of 30% of the total reaction cross section at 10.5 MeV/u, and to be predominantly

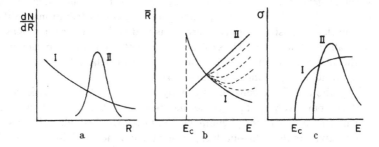

Fig. 6.3. Schematic dependence of different observables on the reaction mechanism. I: Quasi-elastic reaction; II: Compound nucleus reaction; **a** Differential range distribution; **b** Average range; **c** Excitation function

of a direct nature, with strongly forward-peaked angular distributions. The energy spectra of "direct" α particles were found to peak at velocities slightly below the projectile velocities, decreasing with increasing angle of observation. Consequently, the direct α particles were identified as projectile residues and attributed to some kind of break-up process of the projectile, thought to be induced by the interaction with the surface of the target nucleus.

Volkov and collaborators, using light heavy-ion beams (C,N,O) with energies of the order of 8 Mev/u from the 310 cm heavy-ion cyclotron at Dubna, performed systematic studies of transfer reactions on a number of medium and heavy target nuclei [Wi 67, Vo 69, Ar 69a, Ar 69b]. Their work was aimed both at a better understanding of the reaction mechanism and at the production of new isotopes. In one series of experiments, pick-up reactions with ^{16}O ions on ^{27}Al, ^{51}V, and ^{93}Nb targets were investigated, using counter telescope techniques to obtain energy, angular, and element distributions [Ar 69b]. The results reveal the presence of two different mechanisms: quasi-elastic reactions, peaking at the grazing angle and energies near those expected for the ground-state transitions, and inelastic reactions with much lower average Q values and angular distributions decreasing smoothly with increasing angle. Both mechanisms were found to contribute to reactions involving the transfer of a single charge. Only the inelastic component could be identified, however, in the yields of elements further away from the projectile. The cross sections were found to decrease steadily—but not dramatically—with increasing number of transferred charges. Appreciable cross sections for multi-nucleon transfer were also observed in experiments directed specifically at the production of new isotopes. In these, ^{232}Th targets were bombarded with different ion beams, and mass and charge spectra of the reaction products were obtained with a highly selective detection system, combining magnetic analysis with a $\Delta E/E$ counter telescope. In this way, a large number of neutron-rich isotopes of light elements up to Ne could be produced and identified, corresponding to transfers of up to six nucleons [Ar 69b, Ar 71a,b].

Clear evidence for the existence of two mechanisms in one-nucleon transfer was also provided by results of Lefort and collaborators, who bombarded silver targets with 7–8 MeV/u C and N ions from the Orsay variable-energy cyclotron [Le 69, Ga 70]. Energy and angular distributions of different one-nucleon transfer products were measured with counter telescopes down to small angles near 10°. Thus the quite different angular distribution patterns of the high-energy (quasi-elastic) and low energy (inelastic) components could be convincingly demonstrated (Fig. 6.4).

By about 1970, a considerable body of data on deep-inelastic collisions with comparatively light projectiles had been collected, and a certain qualitative understanding of the phenomenon had been reached. Nevertheless, large-scale activity in the field—both experimental and theoretical—only developed after suitable beams of argon, krypton, and even heavier ions became available at Dubna, Orsay, and Berkeley in the early 1970s. This new and explosive development was mainly triggered by the results of two important experi-

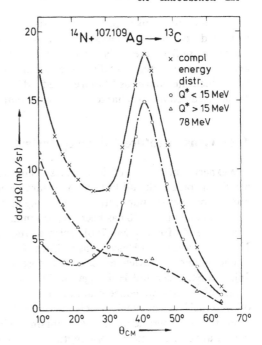

Fig. 6.4. Angular distribution for ^{14}N-induced one-proton pick-up from a silver target at 78 MeV bombarding energy. The "quasi-elastic" $(Q-Q_0 > -15$ MeV, dash-dotted curve) and "inelastic" $(Q-Q_0 < -15$ MeV, dashed curve) contributions are shown separately. (From Ga 70)

ments; firstly, a systematic study of the system ^{40}Ar + ^{232}Th by the Dubna group [Ar 73b] and its interpretation by Wilczynski in terms of friction and scattering to negative angles [Wi 73]; and secondly, the discovery of "quasi-fission" following the bombardment of heavy targets with krypton ions by Lefort and collaborators at Orsay [Le 73]. These studies and other more recent work will be discussed in later sections.

We close this section with some remarks on terminology. Different expressions have been used by different authors to characterize the class of reactions (or subgroups thereof) discussed in this chapter: "grazing reactions" [Ka 61], "deep-inelastic transfer reactions" [Vo 74b], "quasi-fission" [Le 73], "strongly damped collisions" [Wo 74], and others. Each of these expressions emphasizes a specific aspect of the phenomenon involved which may or may not be confined to a specific subgroup of systems or a specific range of bombarding energies.

There is, however, a basic mechanism which—in our present understanding—is common to all of the above reaction types. This involves the approach of the two fragments to within a distance close to the half-density distance, the complete loss of any radial momentum surviving at that point (possibly associated with a partial transfer of angular momentum), the exchange of some nucleons (not necessarily leading to a net transfer of mass or charge), and the eventual separation of the fragments under the influence of Coulomb and centrifugal forces. This mechanism is, of course, deduced by indirect arguments, but seems to provide a consistent explanation of the data. In this picture, differences in observable features like angular, mass, or charge distributions arise from the variation of characteristic time constants with system and bombarding energy, as discussed earlier. In the following we use the term "deep-inelastic collisions" (abbreviated DIC) for the entire class of phenomena discussed in this chapter, in order to emphasize their basic similarity.

6.2 Gross Features of Deep-Inelastic Collisions

The experimental study of deep-inelastic collisions has been a major field of activity at practically all heavy-ion research centres in recent years, and the number of relevant publications is growing at a rapid rate. The status of the field by September 1976 has been well summarized in invited papers by Galin [Ga 76b] and Moretto [Mo 76b], given at the 1976 European Conference on Nuclear Physics with Heavy Ions in Caen. Further information can be found in comprehensive review articles by Lefort [Le 75, Le 76, Le 78], Péter [Pe 76], and Schröder and Huizenga [Sc 77b].

In this section we concentrate on certain simple and outstanding features of deep-inelastic collisions, which emerge from a qualitative inspection of the more recent experimental results, obtained since about 1970. We discuss mainly angular and energy distributions of reaction products, and try to deduce systematic trends as a function of entrance channel and bombarding energy. Subsequently, we introduce some concepts needed for a quantitative discussion of the data.

We begin with a brief discussion of the most important experimental methods and problems. In the simplest case, a single counter (usually a surface-barrier counter) is used to measure energy spectra of uncorrelated products at different angles of emission [Ha 74, Wo 74, Va 76 a, Va 76 b]. In this way, fast surveys can be made without a direct mass or charge identification; it is necessary, however, to ascertain the two-body character of the reaction and the approximate mass range of the detected fragments from independent evidence. The next step is to include a transmission counter and to derive the atomic number Z of the detected particle from the measured energy loss ΔE. Either solid-state or gas-filled counters (ionization chambers) can be used, and element resolution can be achieved up to about $Z = 40 - 50$. The $\Delta E/E$ method is still comparatively simple and yet quite powerful; it has been used to study many systems, especially at Dubna [Vo 69, Gr 70, Ar 73 b] and Berkeley [Mo 75a, Ga 75a, Ba 76a, Mo 76a]. Another possible refinement of the single-counter technique is the determination of the product mass from a combined measurement of energy and time of flight [Ba 75 a, Ga 75b, Co 76, Ta 76]. With flight paths of the order of 2 m or more, the mass resolution is essentially determined by the resolution of the energy measurement, and isotope resolution becomes possible up to about $A = 100$. Since long flight paths require some sacrifice in solid angle compared to the use of "conventional" $\Delta E/E$ counter telescopes (see, however, Ba 75b), most studies so far have been performed with shorter flight paths of the order of 1 m, and correspondingly reduced mass resolution.

In another type of experiment two heavy products are detected in coincidence ("kinematic coincidence"). From the correlated energies or velocities, the mass ratio of the fragments and their total kinetic energy can be deduced by applying simple kinematic arguments, and assuming two-body break-up. In addition, the latter assumption can be checked by requiring full momentum transfer to the two detected fragments. This method is most effective for nearly symmetrical mass splits $[|(A_2 - A_1)/(A_2 + A_1)| \lesssim 0.5]$. While earlier applications usually involved a pair of surface-barrier counters [Le 73, Wo 74], the method has recently been refined by the use of large-area, position-sensitive gas-filled detectors [Sa 77].

It should be realized that in principle all measurements—detecting either one or two fragments—refer to the final, particle-stable products and not to the primary, excited products of a deep-inelastic collision. Any conclusions about mass, charge, or energy of the latter must therefore be based on assumptions (or measurements) concerning their de-excitation by emission of light particles (or gamma rays). Without going into detail we note that the need to correct the measured energies for light-particle emission may introduce significant uncertainties with respect to the properties of the primary fragments and the reaction Q-value.

A question of considerable importance is how best to present experimental data in order to exhibit simple, systematic features. Further one would like to relate observable quantities, like energies, angles, mass and charge transfer, to quantities of theoretical interest, like angular momentum, distance of closest approach, and interaction time. In the following we discuss some approaches to the former problem, and shall return to the latter at the end of this section.

Single-counter data are most frequently presented as one-parameter distributions of yield (for a specified product or group of products) versus angle of emission (integrated over energy) or energy (at a fixed angle). In many cases, measured angles and energies are converted from the laboratory system to the centre-of-mass system; this conversion, however, is subject to some uncertainties, as it involves assumptions on the two-body character of the reaction and the primary masses and kinetic energies. Complications may also arise from the fact that products observed at a given laboratory angle correspond to different centre-of-mass angles depending on Q value and mass transfer. While these problems have no effect on the gross features to be discussed in this section, they must be carefully taken into account in detailed quantitative analyses.

More instructive than one-parameter distributions are contour plots of the yield as a function of two parameters. A well-known example is given by plots of $d^2\sigma/dEd\Theta$ as a function of centre-of-mass angle Θ and total kinetic energy E. This technique of presenting data was first applied to deep-inelastic collisions by Wilczynski in his famous analysis of the Dubna results for the system $^{40}Ar + {}^{232}Th$ [Wi 73], and is often referred to as "Wilczynski plot".

These plots are usually constructed from single-counter data, and are correspondingly affected by uncertainties with respect to conversion from the laboratory to the centre-of-mass system.

Another type of contour plot has been introduced by Lefort and collaborators at Orsay, in order to present mass and energy distributions obtained with the kinematic coincidence technique [Le 73, Pe 75, Ta 75]. Ideally, it would give $d^2\sigma/dEdA$ as a function of the mass A of one fragment and the total kinetic energy E, at a fixed centre-of-mass angle Θ. In this case, the plot would be symmetrical with respect to $A = \frac{1}{2}(A_1 + A_2)$, if A is defined as the primary mass. In practice, however, the fragment with mass A is usually detected at a fixed laboratory angle; therefore, different points in the E–A plane correspond to different centre-of-mass angles, and the symmetry with respect to $A = \frac{1}{2}(A_1 + A_2)$ is lost. In addition, corrections for light-particle emission enter the accurate determination of A and E. Nevertheless, this type of plot has been extremely useful in demonstrating the main features of the mass and energy distribution in deep-inelastic collisions (see Figs. 6.9–11).

In the following, we give some representative examples of experimental data obtained with ^{40}Ar and heavier projectiles. We note that reactions induced by lighter projectiles, like ^{16}O or ^{12}C, exhibit qualitatively similar features at comparable ratios of bombarding energy to interaction barrier (see, for example, Ar 73a, Ey 73, Mo 75a, Co 76, Na 77). However, the relative yield of the deep-inelastic channel tends to increase with increasing reduced mass and charge product of the interacting system.

The next three figures show results obtained by Volkov and collaborators at Dubna for the system ^{40}Ar + ^{232}Th [Ar 73b]. The bombarding energies were 288 and 379 MeV, and a $\Delta E/E$ counter telescope was used to measure angular and energy distributions for individual elements from carbon to calcium. The angular distributions (Fig. 6.5) exhibit a smooth forward rise, with superimposed "bumps" near the entrance channel grazing angle for elements in the vicinity of the projectile. The corresponding energy spectra can be decomposed into a high-energy (quasi-elastic) and a low-energy (deep-inelastic) component, where the former is responsible for the "bumps" and the latter only contributes to the smooth part of the angular distribution. This is particularly evident at the higher bombarding energy and for elements close to the projectile (Fig. 6.6). The potassium data at 388 MeV were used by Wilczynski to construct the original "Wilczynski plot" (Fig. 6.7). The yield contours in the energy-versus-angle plane clearly exhibit two ridges, which appear to meet near zero degrees. This has been explained as a consequence of deflection by nuclear forces from the grazing trajectory towards smaller angles, and eventually across the beam direction to negative angles, for angular momenta below the grazing angular momentum. A schematic representation of this effect, also due to Wilczynski, is shown in Fig. 6.8 [Wi 73].

In reactions between ^{40}Ar and heavy targets ($A_2 \approx 150 - 210$), apart from the light products discussed so far, one also observes a broad distribution of product masses centered at about half the combined target and projectile masses. As an example, we show in Fig. 6.9 a contour plot of the yield

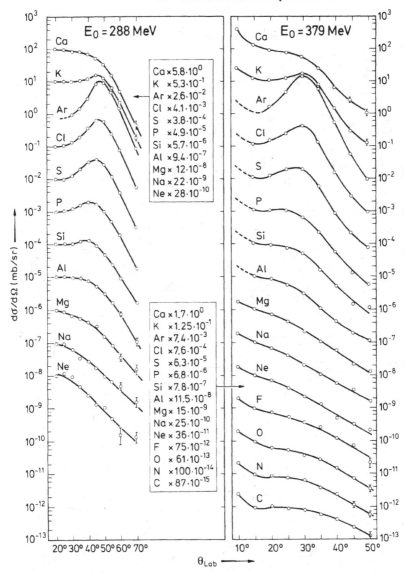

Fig. 6.5. Angular distributions of light elements emitted in the interaction of ^{40}Ar with ^{232}Th at 7.20 and 9.48 MeV/u ($E/B_{int} = 1.45, 1.90$). (From Ar 73b)

versus total kinetic energy and mass number of one fragment (observed at $\vartheta = 59°$) for the system ^{40}Ar + ^{165}Ho at a bombarding energy of 226 MeV [Ta 75]. The products with intermediate masses exhibit all the characteristics which are expected for fission fragments formed in the decay of the appropriate compound nucleus. Hence they have been attributed to reactions involving

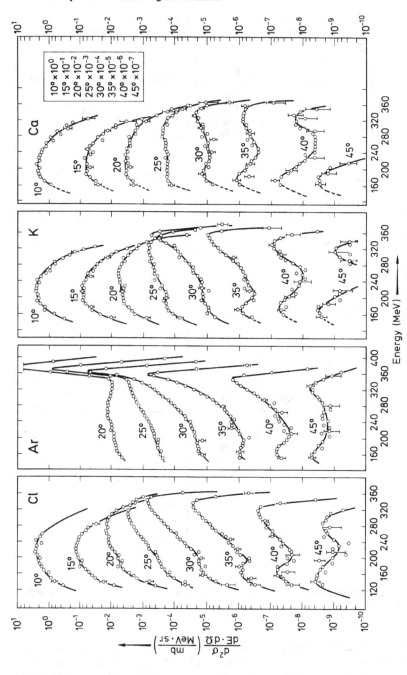

Fig. 6.6.   Energy spectra of different elements emitted in the interaction of ^{40}Ar with ^{232}Th at 9.48 MeV/u. (From Ar 73b)

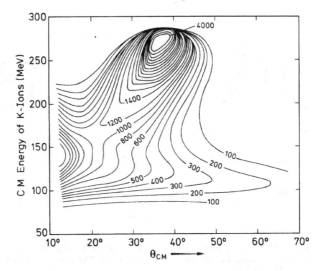

Fig. 6.7. Contour plot of the differential cross section $(d^2\sigma/dEd\Theta)_{CM}$ in μb MeV^{-1} rad^{-1} for the emission of potassium nuclei in the interaction of ^{40}Ar with ^{232}Th at 9.48 MeV/u. (From Wi 73)

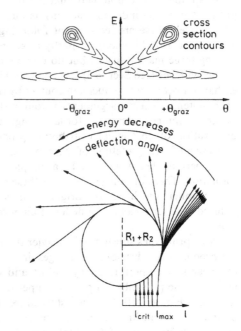

Fig. 6.8. Pictorial representation of the reaction mechanism in deep--inelastic scattering of argon from thorium. (From Wi 73)

complete fusion of target and projectile, followed by fission of the compound nucleus ("fusion-fission"). Results similar to those given in Fig. 6.9 have also

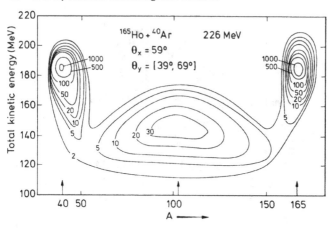

Fig. 6.9. Contour plot of the yield as a function of total kinetic energy and mass of one fragment (detected at 59°) in the interaction of ^{40}Ar with ^{165}Ho at 5.65 MeV/u ($E/B_{int} =$ 1.36). (From Ta 75)

been obtained for the systems ^{40}Ar + ^{209}Bi [Ta 75] and ^{40}Ar + ^{197}Au [Ng 77].

It is important to note here that the interpretation of intermediate mass products in argon-induced reactions as due to fusion-fission is not unique. Alternatively these products (in the following referred to as the "symmetric component") could wholly or partly arise from deep-inelastic collisions involving large mass transfer, but no compound nucleus formation. There is ample evidence (see Sect. 6.3) that appreciable numbers of nucleons can be exchanged between two nuclei in contact by diffusion across the neck. Moreover, potential-energy arguments favour a drift of the two-nucleus system towards symmetry, especially for large angular momenta (see Fig. 6.21). The observation of an angular distribution proportional to $(\sin \Theta)^{-1}$ would be indicative of a long lifetime of the composite system (compared to the rotational period), but would not be a unique signature of compound nucleus formation. Therefore the evolution of these systems from the entrance to the exist channel, and hence the origin of the symmetric component in argon-induced reactions, cannot be deduced unambiguously from the experimental data.

With projectiles substantially heavier than argon and heavy target nuclei, the product-mass distribution changes dramatically. First studies with krypton beams were performed by Lefort and collaborators at Orsay, aimed mainly at the possibility to produce superheavy compound nuclei by fusion reactions. Figure 6.10 shows a historical result: the yield distribution versus product mass and kinetic energy, observed in the bombardment of a ^{209}Bi target with ^{84}Kr ions of 500 MeV [Le 73]. In striking contrast to similar experiments carried out with argon beams, the symmetric component of the product-mass distribution is completely missing. Instead, highly inelastic events are observed with product masses in the vicinity of target and projectile, and kinetic energies somewhat below the entrance channel interaction barrier. Since the latter characteristic is reminiscent of a highly asymmetric type of fission, the term "quasi-fission" was introduced to characterize these events. Clearly these reactions cannot involve compound nucleus formation, a conclusion which is substantiated by the observation of an asymmetric angular distribution, with a pronounced maximum near the grazing angle [Ha 74; see also Fig. 6.12].

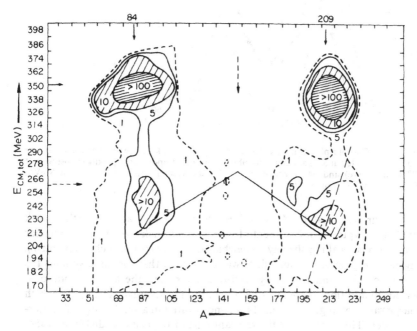

Fig. 6.10. Contour plot of the yield as a function of total kinetic energy and mass of one fragment (detected at 54°) in the interaction of ^{84}Kr with ^{209}Bi at 5.95 MeV/u ($E/B_{int} =$ 1.21). (From Le 73)

The same reaction—^{84}Kr + ^{209}Bi—was also studied at the higher bombarding energy of 600 MeV by Wolf et al. at Berkeley [Wo 74]. The highly asymmetric, inelastic reaction yield ("strongly damped collisions") was found to account for a major fraction of the total reaction cross section. Mass distributions deduced from coincidence measurements for two detection angles of the light fragment are given in Fig. 6.11 and show that the distribution tends to become less asymmetric towards larger deflection angles (corresponding, presumably, to smaller angular momenta and larger interaction times). The average total kinetic energy in the exit channel was found to be almost independent of angle and significantly lower than the interaction barrier, suggesting strong deformation of the fragments prior to scission.

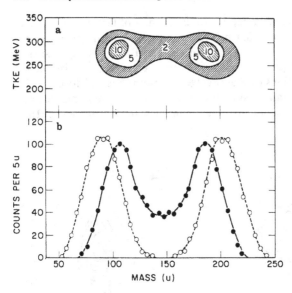

Fig. 6.11. Contour plot of the yield as a function of total kinetic energy and mass of one fragment (detected at 59°) and mass distributions of the deep-inelastic products observed at 59° (solid curve) and 34° (broken curve), for the system $^{84}Kr + {}^{209}Bi$ at 7.14 MeV/u ($E/B_{int} = 1.45$). (From Wo 74)

Similar results as for $^{84}Kr + {}^{209}Bi$ were obtained also in further work involving krypton bombardment of heavy target nuclei. Asymmetric angular distributions, peaking sharply somewhat below the entrance channel grazing angle, were generally observed. It was concluded that for these systems the interaction time must be considerably shorter than the average rotational period. In addition, there must be some kind of "focussing effect," by which a large range of angular momenta is concentrated in a small range of scattering angles. This means that the relevant part of the classical deflection function $\Theta(L)$ must be rather flat.

The influence of the bombarding energy on the angular distribution pattern has also been studied (see Fig. 6.12). The yield of projectile-like products at angles below the grazing angle was found to increase systematically with increasing bombarding energy [Ta 76, Va 76a]. This trend was interpreted as evidence for an evolution towards "orbiting".

In several experiments performed with argon, krypton, and intermediate projectiles, the disappearance of the symmetric component in the mass yield was investigated in more detail. In krypton bombardments of ^{165}Ho targets, the symmetric component was barely observable at 492 MeV [Pe 75], but was strongly present at 600 Mev [Wo 76]; for targets with $A > 180$, it was practically absent at both energies. In two cases, similar composite systems were produced by either argon or krypton bombardment: $^{40}Ar + {}^{209}Bi \rightarrow {}^{249}_{101}Md$, $^{84}Kr + {}^{165}Ho \rightarrow {}^{249}_{103}Lw$ [Pe 73]; and $^{40}Ar + {}^{238}U \rightarrow {}^{278}110$; $^{84}Kr + {}^{186}W \rightarrow {}^{270}110$ [Pe 75]. The symmetric component was consistently present

Fig. 6.12. Angular distributions of krypton-like products from deep-inelastic collisions of ⁸⁴Kr with ²⁰⁸Pb, measured at 5.88 MeV/u $(E/B_{int} = 1.21)$, 6.07 MeV/u $(E/B_{int} = 1.25)$ and 8.55 MeV/u $(E/B_{int} = 1.76)$. (From Va 76a)

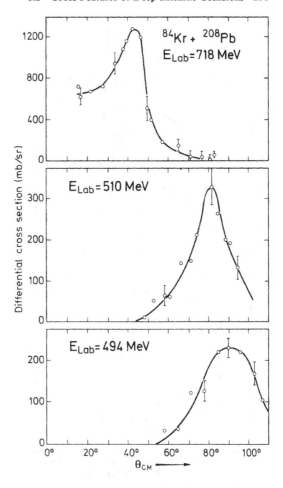

in argon bombardments and absent in krypton bombardments. These observations give strong evidence of an entrance-channel effect. Moreover, they seem to support a compound-nucleus interpretation of the argon results: it appears unlikely that the composite systems formed in argon bombardments should survive large mass transfers to symmetry, while the more symmetric krypton systems—which presumably must be passed as an intermediate stage in the former cases—decay promptly before reaching symmetry.

We conclude this survey of experimental data with some results for xenon-induced reactions. The main features seen in krypton experiments with heavy targets—deep-inelastic scattering as the dominant reaction channel, mass yields sharply peaked near target and projectile, "side-peaked" angular distributions—are even more pronounced with these still-heavier projectiles. As an example of raw data, we show in Fig. 6.13 energy spectra of the reaction ²⁰⁸Pb + ¹³⁶Xe, measured at different angles with a single surface-barrier counter [Va 76b]. At angles below the grazing angle (approximately 30° in

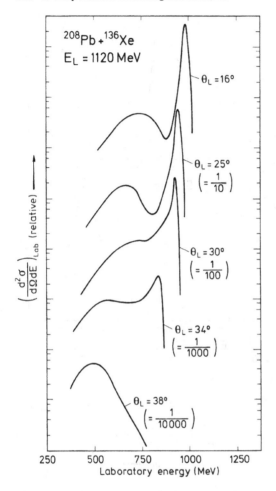

Fig. 6.13. Energy spectra of reaction products observed at different forward angles in the scattering of ^{136}Xe from ^{208}Pb ($\varepsilon = 8.24$ MeV/u; $E/B_{int} = 1.65$). (From Va 76b)

the laboratory system), a distinct group of inelastic events appears, which is well separated from a much stronger elastic peak. Near the grazing angle, the relative intensity of the elastic peak becomes smaller, and a continuous distribution of quasi-elastic events fills the valley between the elastic and inelastic peaks. Finally at somewhat larger angles only the inelastic component remains. Based on reaction kinematics and on separate measurements with counter telescopes, the inelastic peak observed in this particular region of angles is attributed predominantly to projectile-like fragments. As shown in a similar study of the system ^{136}Xe + ^{209}Bi, the total yield of inelastic (i.e. quasi-elastic plus deep-inelastic) projectile-like fragments peaks sharply in the vicinity of the grazing angle and essentially exchausts the total reaction cross section [Sc 76].

We now turn to a discussion of the energy loss in deep-inelastic collisions

—more precisely, of the energy transferred from relative motion to internal excitation, or the (negative) reaction Q values. This energy loss is invariably found to be of the order of the difference between the kinetic energy in the entrance channel and the interaction barrier in the exit channel, a feature which might be considered the signature of deep inelastic collisions [see, for example, Ba 75a, Ga 75a, Mo 75a, Ba 76a, Ga 76b, Mo 76a]. Alternatively, the same result can be expressed by stating that the centre-of-mass kinetic energy of the emerging fragments is roughly given by their mutual potential energy at contact, with little dependence on either bombarding energy or angle of observation.

As an example, we show in Fig. 6.14 average kinetic energies (and widths of energy distributions) for different fragments emitted in the interaction of 288 MeV Ar ions with an ^{197}Au target [Mo 76a]. There is clearly an overall correspondence between experimental and calculated exit channel energies. Systematic comparisons of this kind indicate that the kinetic energies of deep-inelastic products are consistent with the potential energies of touching spheres for systems of intermediate mass, but are significantly (circa 10–25 %) lower for systems involving argon or krypton and heavy targets [Pe 76, Ga 76b].

The interpretation of this result is, however, not as simple as might appear at first sight. Even if complete damping of the initial radial kinetic energy is assumed, angular momentum conservation requires that a certain fraction

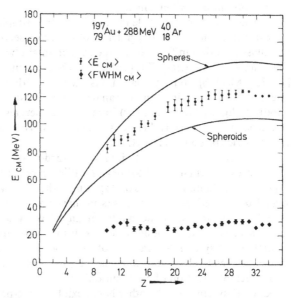

Fig. 6.14. Average kinetic energies and widths of energy distributions for different elements emitted in the interaction of ^{40}Ar with ^{197}Au at 7.20 MeV/u. The measured energies are compared with theoretical values expected from Coulomb repulsion of two touching spheres and of two touching spheroids at equilibrium deformation. (From Mo 76a)

of the initial rotational energy survives the collision and contributes to the kinetic energy in the exit channel. The magnitude of this contribution depends on the coupling between relative and intrinsic angular momenta. In a simple, macroscopic picture the coupling is due to "tangential friction", and the angular momentum loss can be predicted for certain limiting situations ("rolling" and "sticking"; see Sect. 6.4). Such—admittedly crude—estimates and the available experimental evidence on angular momentum transfer in deep--inelastic collisions (to be discussed in Sect. 6.3) indicate that on the average about 50% of the rotational energy in the entrance channel will be dissipated during the early stages of the collision. A further reduction in rotational (as well as potential) energy may result from a strong deformation of the fragments prior to scission.

We conclude that complete radial damping does not automatically imply a close correspondence between exit-channel kinetic energies and potential energies, calculated for some simple configuration like touching spheres. The apparent existence of such a correlation for many systems could in part be due to a cancellation of effects caused by rotation (leading to an increase in kinetic energy) and deformation (leading to a decrease in kinetic energy). For a light system—^{20}Ne + ^{27}Al—comparable contributions of the centrifugal and potential terms to the total kinetic energy in the exit channel have been identified [Eg 76, Na 77]. For heavy systems, on the other hand, the centrifugal term is usually unimportant, and the observed lowering of the kinetic energy below the entrance-channel interaction barrier gives clear evidence for strong deformation in the exit channel.

A question of considerable interest is whether simple systematic patterns are evident in the overall dependence of the cross section on angle, energy, and mass partition in the exit channel, which might be correlated with some characteristic parameter. This problem has been discussed by several authors [Ga 76b, Mo 76b]. In particular, it has been noted that two more or less distinct classes of scattering systems can be distinguished. Class I comprises comparatively light systems and (or) bombarding energies well above the interaction barrier; conversely, heavier systems and (or) bombarding energies close to the barrier belong to class II.

Class I systems are characterized by forward-peaked angular distributions (often in excess of $(\sin \Theta)^{-1}$ for $d\sigma/d\Omega$, falling exponentially with increasing angle), with definite symptoms of "orbiting"; their decay products are usually spread over relatively broad ranges in mass and charge, and usually compound-nucleus formation accounts for an appreciable fraction of the total reaction cross section (besides deep-inelastic scattering and transfer). Apparently these systems are capable of forming—in a certain band of angular momenta—relatively stable nucleus–nucleus configurations which survive an appreciable fraction of a revolution.

Class II systems, on the other hand, exhibit "side-peaked" angular distributions which steeply fall off towards both larger and smaller angles. Their decay products are concentrated near the target and projectile mass, and deep-inelastic collisions account for a major fraction of the total reaction cross section, whereas compound-nucleus formation is either weak or absent. This is the domain of quasi-fission; clearly, here, we have short-lived composite systems in the sense that their lifetimes are short compared to the rotational period for the relevant range of angular momenta.

Very roughly, most of the systems studied so far with projectiles up to and including ^{40}Ar belong to class I, while those formed in collisions with heavier projectiles belong to class II. It might appear, therefore, that the division should be made on the basis of a parameter related primarily to target and projectile mass, for example the charge product $Z_1 Z_2$. A closer examination of the experimental evidence reveals, however, a significant influence of the bombarding energy. With increasing energy, class II symptoms are reduced and class I symptoms are enhanced for a given system (see, for example, Fig. 6.12). Consequently, it was suggested that scattering systems should be classified according to the ratio of bombarding energy to interaction barrier: systems with $E/B_{int} < 1.6$ were assigned to class II, and systems with $E/B_{int} > 1.6$, to class I [Mo 76b]. If this classification is correct, then all systems should exhibit class II properties near the interaction barrier, and class I properties at sufficiently high energy. An alternative classification introduces the "Sommerfeld-like parameter"

$$n' = \frac{Z_1 Z_2 e^2}{\hbar v_B} \approx \frac{L_{gr}}{2(E/B_{int} - 1)} , \qquad (6.3)$$

where v_B is the relative velocity of the fragments at the distance of closest approach. This parameter has been interpreted as proportional to the ratio of the Coulomb force to the friction force; it was suggested that systems with $n' < 200$ belong to class I, and systems with $n' > 250$ belong to class II, while those with $200 < n' < 250$ display intermediate behaviour [Ga 76b].

For a more quantitative discussion, we have to relate the experimental results to quantities like angular momentum and interaction time. The latter are not directly observable, but play an essential role in the theoretical interpretation of the data. The relationship between angular momentum and scattering angle is classically expressed by the "classical deflection function" $\Theta(L)$ which has been introduced in Sect. 3.4 [see (3.53–58)]. In describing deep-inelastic collisions we must take into account that the relative energy and angular momentum are not conserved along the trajectory, and that the effective two-body potential may change in time due to fragment deformation. The deflection function may then be written formally as

$$\Theta(L) = \pi - \int_\infty^{D_L} \left(\frac{L(r)}{K_L(r)} \right)_\downarrow \frac{dr}{r^2} - \int_{D_L}^\infty \left(\frac{L(r)}{K_L(r)} \right)_\uparrow \frac{dr}{r^2} , \qquad (6.4)$$

where $K_L(r)$ and $L(r)$ are local, instantaneous momenta and angular momenta at distance r, and the subscripts $\downarrow$ and $\uparrow$ denote the ingoing and outgoing part of the trajectory, respectively. The functions $L(r)$ and $K_L(r)$ may be computed numerically in the so-called friction models of deep-inelastic collisions, depending on specific assumptions about two-body potentials and dissipative

forces (see Sect. 6.4). We note that mass or charge transfer are usually neglected in applications of (6.4), but can be included in principle; the function
$\Theta(L)$ then describes the deflection averaged over all projectile-like fragments.

The qualitative evolution of the deflection function, as one goes from
class I to class II systems, is illustrated in Fig. 6.15. Curve a refers to a typical
class I system which approaches orbiting as the angular momentum is lowered
from infinity towards a critical value. At still lower angular momenta fusion

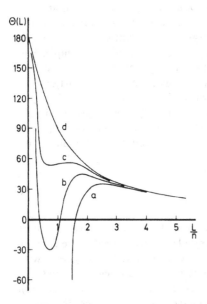

Fig. 6.15. Different types of classical
deflection function $\Theta(L)$(qualitative representation; for explanation of symbols
a, b, c, d see text)

occurs, and hence no fragments re-emerge after the collision. Curve b describes an intermediate situation, where the nuclear attraction is strong enough
for deflection to negative angles in a certain angular momentum window, but
orbiting or fusion are prevented by dominating repulsive forces. Curve c is
appropriate to a typical class II system; here no deflection to negative angles
occurs, and the balance between attractive and repulsive forces produces
focusing of a large range of angular momenta into a narrow range of deflection angles. Finally curve d describes the limiting case of pure Coulomb scattering without loss of energy or angular momentum. In this case, the deflection function is given by

$$\Theta_c(L) = 2 \text{ arc tan} \left(\frac{n}{L}\right). \tag{6.5}$$

In trying to correlate Fig. 6.15 with experimental results one should be aware
that any description of a system in terms of a single, well-defined deflection
function is an idealization. For a given entrance channel with sharp asymptotic angular momentum, the emergent fragments will be characterized by dis-

tributions in deflection angle, correlated with distributions in energy, angular momentum, and mass partition. Therefore one should expect a "deflection distribution" rather than a deflection function.

There appears to be a close connection between energy loss and deflection, and hence also between energy loss and angular momentum. This is best demonstrated by presenting experimental results in the form of a Wilczynski plot. Figure 6.16 shows schematically two such plots, corresponding to cases a and c of Fig. 6.15. The shaded areas in Fig. 6.16 indicate regions in the E–Θ plane, where the differential cross section $d^2\sigma/dEd\Theta$ exceeds a certain value. As shown by Schröder et al. [Sc 77a] and discussed below, plots of this kind can be converted to experimental deflection functions.

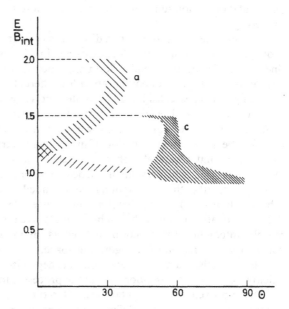

Fig. 6.16. Schematic Wilczynski plots for a system which exhibits orbiting (a) and a system which doesn't (c). The symbols a and c refer to the corresponding deflection functions in Fig. 6.20.

The scattering cross section is related to the deflection function by [see (3.53)]

$$\frac{d\sigma_{cl}}{d\Omega} = \dlambda^2 L \left[\sin\Theta\left(\frac{d\Theta}{dL}\right)\right]^{-1}. \tag{6.6}$$

We rewrite this equation in a form which exhibits explicitly the general connection between cross section and angular momentum [see (3.37)]:

$$\frac{d\sigma}{d\Theta} = \pi\dlambda^2 \frac{dL^2}{d\Theta}. \tag{6.7}$$

Equation (6.7) could be used to deduce the function $L(\Theta)$ from experimental angular distributions if it was single-valued or (equivalently) if the deflection

Θ was a monotonic function of L. This is, however, usually not the case, due to the existence of one or two "rainbow-angles", where $d\Theta/dL = 0$.

Alternatively, we can replace Θ in(6.7) by any other observable for which a correlation with L can be expected. Taking, in particular, the energy loss $\Delta E = -Q$, we obtain

$$\frac{d\sigma}{d\Delta E} = \pi \dot{\lambda}^2 \frac{dL^2}{d\Delta E}. \tag{6.8}$$

Using this relationship and assuming further a monotonic dependence of ΔE on L, Schröder et al. were able to deduce the functions $L(\Delta E)$ and hence $\Theta(L)$ from experimental data for a number of heavy systems [Sc 77a]. Such experimental deflection functions can then be used to test theoretical scattering models.

We conclude this section with a discussion of interaction times. This point is of special significance since—as explained in Sect. 6.1—the study of deep--inelastic collisions in principle offers unique possibilities of deriving information on the time scales of equilibration for different nuclear degrees of freedom. In practice, establishing an absolute time scale turns out to be difficult for conceptual as well as technical reasons. The interaction between the fragments is expected to vary strongly with distance and deformation; it will therefore be a continuous function of time along the trajectory, peaking near the turning point. Observable effects, like energy loss, mass, and angular momentum transfer, will presumably depend on time integrals over effective interactions, rather than interaction times bounded by schematic, more or less arbitrary limits. One may, therefore, introduce effective interaction times, defined in relation to some observable quantity like the deflection angle. Conversely, interaction times may be defined as lifetimes of the system within some particular region of configuration space, as derived from a model calculation. In either case, the interaction time depends—either explicitly or implicitly—on model assumptions about the processes to be studied.

The approach followed so far in the literature is essentially of the former type: the interaction time is defined as the change in deflection angle caused by nuclear interactions, divided by an angular velocity appropriate to the touching fragments. This would be equivalent to an absolute interaction time in a schematic model, where the total deflection angle is composed of Coulomb contributions for the ingoing and outgoing parts of the trajectory, and an angle of rotation at constant separation during the interaction time.

The simplest possible ansatz is that used by Nörenberg in his analysis of the Dubna results for the system ^{40}Ar $+$ ^{232}Th [Nö 74, Nö 76b]:

$$\tau(\Theta) = \frac{\Theta_{gr} - \Theta}{\bar{\omega}} = \frac{\mathscr{I}(\Theta_{gr} - \Theta)}{\hbar \bar{L}}. \tag{6.9}$$

Here the grazing angle Θ_{gr} is adjusted to fit the experimental data, and $\bar{\omega}$ is the average angular velocity of the touching fragments. The quantity $\mathscr{I}$ is an effective moment of inertia of the two-fragment system which depends on assumptions about angular momentum coupling (see Table 6.4), and is in practice usually calculated with the "sticking" assumption ($\mathscr{I} = \mu R_{12}^2 + \mathscr{I}_1 + \mathscr{I}_2$). The average angular momentum in the incident channel, $\bar{L}$, is calculated from experimental cross sections using the relationship

$$\bar{L} = \frac{2}{3} \frac{L_{\text{int}}^3 - L_{\text{fu}}^3}{L_{\text{int}}^2 - L_{\text{fu}}^2}, \tag{6.10}$$

where L_{int}, L_{fu} denote limiting angular momenta appropriate to the total reaction cross section and the fusion cross section, respectively.

This type of analysis can be applied whenever the relevant range of angular momenta is spread over an appreciable range of scattering angles, i.e. to systems of class I. A method which is useful for class II systems was introduced by Huizenga and collaborators (Hu 75, Sc 77a). Here one defines interaction times which depend explicitly on angular momentum, rather than on scattering angle:

$$\tau(L) = \frac{\mathscr{I}(L)[\Theta_{\text{c}}(L) - \Theta(L)]}{\hbar L}. \tag{6.11}$$

The total classical deflection function $\Theta(L)$ is deduced from the experimental energy-loss distribution as discussed above, and the Coulomb deflection function $\Theta_{\text{c}}(L)$ is composed of contributions from the initial (i) and final (f) channel according to

$$\Theta_{\text{c}}(L) = \arctan\frac{n_{\text{i}}}{L_{\text{i}}} + \arctan\frac{n_{\text{f}}}{L_{\text{f}}}. \tag{6.12}$$

The effective moment of inertia $\mathscr{I}(L)$ depends on the angular momentum transfer from relative to intrinsic motion, and hence on angular momentum $L (= L_{\text{i}})$. In the absence of direct evidence on this dependence, however, a constant value $\mathscr{I}$ (corresponding, for example, to the sticking limit) is normally used. We note that this type of analysis effectively relates interaction times to measured energy losses.

A similar, but more elaborate, method of calculating angular-momentum-dependent interaction times was developed by Wolschin and Nörenberg [Wo 78a]. These authors deduce empirical deflection functions and final angular momentum distributions from a simultaneous fit to experimental angular distributions and energy loss spectra. The method should be applicable to systems of either class I or class II.

A comparison of experimental results with theoretical models of deep-inelastic collisions, based on the concepts introduced in this section, will be given in Sect. 6.4.

6.3 Detailed Studies

The number of studies concerned with a detailed investigation of the charge, mass, and angular distribution of reaction products in deep-inelastic collisions is increasing at a rapid rate. It seems neither possible nor desirable to attempt a complete survey of relevant work in this book. Instead we present in this section a selection of examples which may serve to illustrate the experimental methods and the analysis of the results. In addition we discuss the most important conclusions which emerge from the data without application of specific theoretical models. Theoretical approaches will be summarized in Sect. 6.4.

We start with a series of experiments in which thick ^{238}U targets were bombarded with ^{40}Ar, ^{84}Kr, or ^{136}Xe beams at the Berkeley Super-Hilac, and the reaction products were identified with radio-chemical techniques, i.e. fast chemical separations followed by gamma spectroscopy of radioactive nuclides [Kr 74, Kr 76, Ot 76]. Advantages of the radio-chemical method are a unique identification of elements and isotopes for all masses, a large dynamic range in A and Z, low background and, with thick targets, very high sensitivity. The latter is achieved, however, at the expense of information on incident energy, product energy, and angle of emission. In addition, stable products and products with very short or very long half-lives cannot be detected and therefore their yields must be estimated by smooth extrapolation from measured yields. Finally, the primary yields must be inferred by taking into account not only fast particle evaporation from highly excited products (as in counter measurements), but also fast radioactive decays which may precede the observed decay.

The features of the radio-chemical method make it especially useful in two different contexts. These are firstly the identification of rare events with extremely small cross sections ($< 10^{-30}$ cm^2), and secondly survey experiments in which the branching of the total reaction cross section into different basic mechanisms is studied. The Berkeley experiments mentioned above and discussed in the following are of interest mainly in the latter context.

Figure 6.17 shows the results for the system ^{84}Kr $+$ ^{238}U, obtained at a bombarding energy of 605 MeV and based on measured yields of 156 different nuclides [Kr 74]. The total yield distribution in A and Z has been decomposed by the authors in different components, labelled A to G, which are thought to originate from different mechanisms. The narrow spikes E and F, centred at the target and projectile ("rabbit ears"), are associated with quasi-elastic transfer reactions. The same mechanism may lead to sequential fission of the target-like product, if the latter is excited above the fission barrier. This gives rise to a double-humped mass distribution of fission fragments, labelled B. Similarly components C, G, and D are attributed to deep-inelastic collisions between projectile and target, or quasi-fission. The rabbit ears (C,G) are now broader than in the quasi-elastic case, and the fission fragments exhibit a symmetric mass distribution due to the higher excitation energy of the fissioning nucleus. In addition all three components are more proton-rich than the corresponding quasi-elastic component (see Fig. 6.17 b), presumably as a consequence of enhanced neutron evaporation. The existence of the sequential fission components, B and D, should be correlated with a depletion of the corresponding heavy rabbit ears, F and G; in the deep inelastic case, sequen-

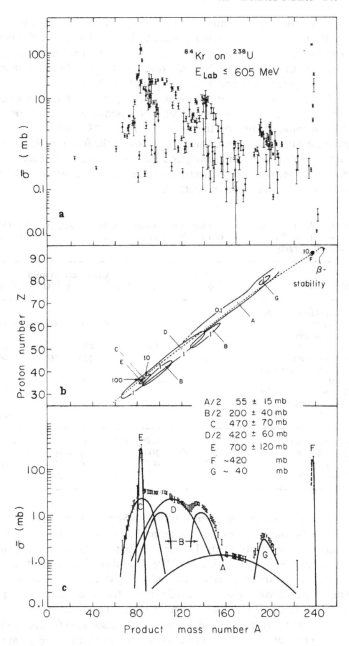

Fig. 6.17. Average thick-target cross sections for different products from 605 MeV ^{84}Kr bombardment of ^{238}U. (a) Yields of individual isotopes as a function of mass number; (b) Contour lines of constant yield in the Z/A plane for classes of products attributed to different reaction mechanisms; (c) Integrated yields as a function of mass number. For an explanation of the symbols A-G see text and Table 6.2. (From Kr 74)

tial fission apparently removes most of the yield near $A = 238$ with the result that the residual rabbit ear, component G, is shifted to $A \approx 195$. Finally, complete fusion of target and projectile followed by fission of the compound nucleus (fusion-fission) is believed responsible for a broad distribution of fission fragments centred at $A \approx 160$ and labelled A.

The cross sections for the different components shown in Fig. 6.17 are given in Table 6.2 together with the corresponding cross sections obtained in companion studies of the systems $^{40}Ar + {}^{238}U$ and $^{136}Xe + {}^{238}U$. It is interesting to observe the dramatic decrease of the fusion-fission cross section with increasing projectile mass, and the corresponding increase in the cross section for deep inelastic scattering, or quasi-fission. These features have been discussed in more detail in Sect. 6.2.

Table 6.2. Cross sections of different components in the interaction of ^{40}Ar ($E_{max} = 288$ MeV; Kr 76), ^{84}Kr ($E_{max} = 605$ MeV; Kr 74) and ^{136}Xe ($E_{max} = 1150$ MeV; Ot 76) with thick ^{238}U targets

Mechanism	Label (Fig. 6.17)	Cross section [mb]		
		$^{40}Ar + {}^{238}U$	$^{84}Kr + {}^{238}U$	$^{136}Xe + {}^{238}U$
Quasi-elastic transfer induced fission	B/2	150 ± 30	200 ± 40	~185
Quasi-fission (light product)	C	~100	470 ± 70	600 ± 125
Quasi-ternary fission	D/2		420 ± 60	400 ± 25
Quasi-elastic transfer (light product)	E	~400	700 ± 120	600
Quasi-elastic transfer (heavy product)	F	~202	~420	~415
Quasi-fission (heavy product) G			~ 40	~140
Fusion-fission	A/2	620 ± 150	55 ± 15	< 20
Quasi-fission	C = D/2 + G	100 ± 50	470 ± 70	600 ± 125
Quasi-elastic transfer	E = B/2 + F	400 ± 120	700 ± 120	~600
Total reaction	A/2 + E + C	1120 ± 300	1225 ± 200	1200 ± 200

Another point of interest in Table 6.2 is the occurence of sequential fission, or "quasi-ternary fission", of the target-like product with appreciable cross sections (components B, D). This process has also been observed in deep inelastic collisions of ^{63}Cu with ^{197}Au [Pe 77] and of ^{136}Xe with ^{197}Au [Mo 76b, Ru 77], and is therefore not restricted to highly fissionable targets like ^{238}U. The fission probability of an excited primary fragment depends on its composition, excitation energy, and angular momentum; once these dependences are quantitatively understood, measured cross sections for sequential fission can in principle provide information on the excitation energy and angular momentum deposited in the primary heavy fragment.

More recently, the uranium beam produced by the Unilac accelerator at Darmstadt was used to study the interaction of ^{238}U with ^{238}U. The reaction products were analysed applying both counter techniques [Hi 77] and radio-chemical methods [Sc 78a]. In the latter study a thick uranium target was bombarded with 7.5 MeV/u uranium ions at intensities up to 2.5×10^{11} particles per second. Following chemical separation the activities generated in the target were counted over a period of several months. One of the main points of interest in this experiment was the possibility of multi-nucleon transfer, leading to exotic transuranium nuclei and perhaps to superheavy elements with $Z > 110$. Figure 6.18 shows the measured cross sections for formation of a number of heavy products with $Z = 94 - 100$ and, for comparison, results of a similar study of the system $^{136}Xe + ^{238}U$ by the same authors. These results demonstrate the extreme sensitivity of the radio-chemical method, which allows cross sections of the order of 10^{-35} cm² to be measured. The strong decrease of the cross sections with increasing atomic number of the product can be explained as due to a combination of small transfer probabilities for large nucleon numbers and small survival probabilities (against fission) for large values of Z. An important restriction

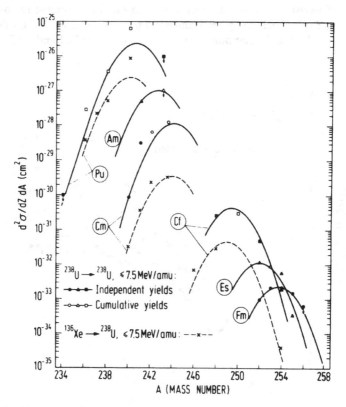

Fig. 6.18. Production cross sections for transuranium elements in interactions of ^{136}Xe (dotted curves) and ^{238}U (full curves) projectiles with thick uranium targets at 7.5 MeV/u. (From Sc 78a)

seems to arise from the fact that the surviving products are those produced at low excitation, whereas transfers of many nucleons are generally correlated with large losses of relative energy, and hence large excitation energies. Although no products with $Z > 100$ could be identified, it was shown that the cross sections for transuranium production are much more favourable for the system $^{238}U + ^{238}U$ than for the system $^{136}Xe + ^{238}U$ [Sc 78a].

We now turn to counter telescope experiments, where energy and angular distributions have been measured for resolved elements up to about $Z = 50$. The method has been discussed briefly in Sect. 6.2 and, like the radio-chemical method, combines high efficiency with the capability to extract detailed information. The results of the Dubna group for the system $^{40}Ar + ^{232}Th$ have already been reported in Sect. 6.2. Further extensive studies with projectiles ranging from ^{14}N to ^{136}Xe and Ag, Tb, and Au targets were performed by Moretto, Thompson and collaborators in Berkeley [Ga 75a; Mo 75a,b; Ba 76a; Mo 76 a,b,c; Ru 77]. As an example we show in Fig. 6.19 Z distributions of products emitted after ^{84}Kr bombardment of a ^{159}Tb target [Mo 76b].

Here and in other cases two main components can be distinguished. The quasi-elastic component which is concentrated in a limited region of angles and Z values near the grazing angle and projectile charge, and the deep-inelastic or "relaxed" component which is forward peaked and exhibits a broad

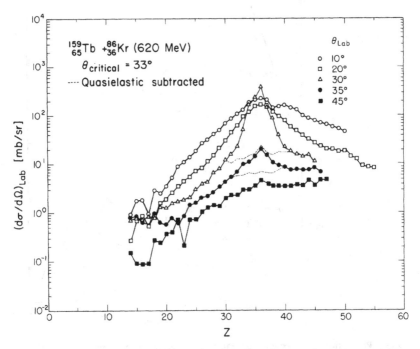

Fig. 6.19. Differential cross section as a function of product atomic number and laboratory angle for the system ^{86}Kr (620 MeV) + ^{159}Tb. (From Mo 76b)

Z distribution extending from well below the projectile charge up to symmetry. From a detailed and systematic analysis of the data—which cannot be reproduced here—it becomes clear that a continuous transition exists between the quasi-elastic and deep-inelastic regimes, and that definite entrance-channel effects persist in the charge and angular distributions of the deep-inelastic component. This general result has been interpreted by Moretto in terms of an evolution of the composite system along the mass-asymmetry mode which proceeds slowly compared to the much faster equilibration of the relative kinetic energy and charge-to-mass ratio of the fragments (see, in particular, Mo 76b). In this picture, changes in mass asymmetry are assumed to occur by nucleon diffusion across the neck. The rate and direction of this diffusive process should then be determined by the entrance-channel asymmetry ("injection asymmetry"), the potential energy of the composite system as a function of mass asymmetry, and the excitation energy or "temperature" of the composite system. Attempts to formulate such a diffusion model quantitatively will be discussed in Sect. 6.4.

An important question that remains open in the studies reported so far concerns the origin of the "symmetric component" in the decay of initially asymmetric systems: does it result from fission of a classical, equilibrated compound nucleus, or from a surface diffusion mechanism, or are both mechanisms operative? This problem has been discussed briefly in Sect. 6.2 in the context of argon-induced reactions with medium and heavy target nuclei. It exists, however, also with lighter projectiles and compound systems in the mass region $A \approx 70 - 130$.

The detailed study of "fission-like events" observed with medium-mass compound systems has been the object of experiments performed at Heidelberg and at College Station, Texas. The Heidelberg group investigated the systems $^{32}S + {}^{48,50}Ti$ and identified the reaction products by mass and charge from combined E, ΔE, and time-of-flight measurements [Ba 75a, Br 76a]. Natowitz and collaborators at Texas A & M University performed systematic experiments with ^{12}C projectiles on a number of medium-weight target nuclei, detecting both fission-like products and evaporation residues with $\Delta E/E$ counter telescopes [Na 76]. Figure 6.20 shows Z distributions obtained in the ^{12}C bombardment of a silver target, where a peak corresponding to symmetric mass division is clearly evident. The results of these studies may be summarized as follows:

i) The average fragment energies follow the systematics expected for compound nucleus fission.

ii) The angular distributions tend towards $(\sin \Theta)^{-1}$ near symmetry, but are more forward peaked near the projectile or target mass. The latter observation indicates a continuous evolution of the reaction mechanism.

iii) The cross sections for symmetric mass divisions increase strongly with bombarding energy.

iv) The angular momenta associated with the symmetric component appear to be those for which the potential (including rotational) energy of the composite system as a function of mass partition favours an evolution towards

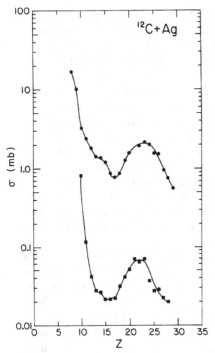

Fig. 6.20. Integrated cross section as a function of product atomic number for the system ^{12}C + Ag at bombarding energies of 107 MeV (squares) and 197 MeV (circles). (From Na 76)

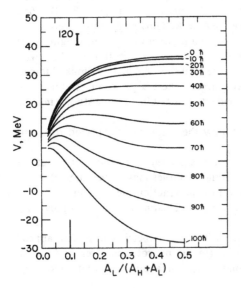

Fig. 6.21. Calculated potential plus centrifugal energies as a function of mass division and relative angular momentum for the composite system ^{120}I. The mass division corresponding to the entrance channel ^{12}C + Ag is indicated by a vertical line. (A_L = mass number of light fragment, A_H = mass number of heavy fragment). (From Na 76)

symmetry (see Fig. 6.21). This implies that injection asymmetries less than a certain "critical asymmetry", at which the effective potential energy goes through a maximum, will lead to symmetric divisions.

Thus the available evidence indicates that the symmetric products observed in the decay of composite systems with intermediate mass originate at least partly from a surface mechanism, i.e. deep-inelastic transfer. The presence of contributions from compound-nucleus fission cannot be excluded, however. A similar conclusion has been reached also for argon-induced reactions with heavy target nuclei from detailed angular distribution studies, involving either counter telescope techniques [Ng 77, Ga 77] or measurements of delayed activities [Ot 78]. It is important to note here that compound-nucleus decay and deep-inelastic transfer proceed on a similar time scale at the high excitation energies encountered in these experiments (compare Table 6.1). Therefore our inability to separate the two mechanisms experimentally may reflect the more fundamental problem that we are no longer dealing with two clearly distinct physical processes.

Next we discuss a series of remarkable experiments performed at Dubna by Volkov and collaborators, where thorium targets were bombarded with different projectiles up to ^{40}Ar, and reaction products were observed at a fixed angle, usually 40° [Ar 69a, Ar 70, Ar 71a,b,c; Ar 73a, Ar 76]. In these experiments, individual isotopes were resolved by a combination of magnetic analysis with a $\Delta E/E$ counter telescope. Large numbers of different isotopes could be identified, and it was found that very neutron-rich isotopes of elements below the projectile were preferentially produced. Table 6.3 gives a survey of the most neutron-rich isotopes which were observed with significant yields. These isotopes are characterized by neutron-number-to-proton-number ratios, N/Z, near those of the composite systems (approximately 1.5) for elements close to the projectile, and significantly higher for lighter elements. On the average, the products were less neutron-rich than the composite systems, but more neutron-rich than the projectiles.

Two important qualitative conclusions emerge from these results. Firstly, there is a fast mechanism of charge equilibration between projectile and target which tends to increase the specific charge of the target-like fragment at the expense of the charge of the projectile-like fragment. Secondly, the systems under consideration exhibit a strong preference for nucleon transfer from the projectile to the target, compared to transfer in the opposite direction.

A quantitative analysis of the yields of different isotopes indicated a close correlation with the corresponding Q value for the ground-state transition, Q_{gg} [Ar 71 a,b]. More specifically, it was found that the differential cross sections for production of different isotopes of a given element could be approximately reproduced with the formula [Vo 74 a,b]

$$\frac{d\sigma}{d\Omega} \sim \exp\left(\frac{Q_{gg} - \delta(n) - \delta(p)}{T}\right), \tag{6.13}$$

where $\delta(n)$ and $\delta(p)$ are pairing energies for neutrons and protons, respec-

Table 6.3. Most neutron-rich isotopes identified in Dubna work on deep-inelastic transfer reactions

Isotope	Initial system	Nucleons transferred to target	References
^{8}He	^{15}N $+$ ^{232}Th	5p + 2n	Ar 71b
^{11}Li		4p	
^{12}Be		3p	
^{15}B		2p − 2n	
^{18}C	^{22}Ne $+$ ^{232}Th	4p	Ar 73a
^{20}N		3p − 1n	
^{22}O		2p − 2n	
^{23}F		1p − 2n	
^{25}Ne		− 3n	
^{27}Na		−1p − 4n	
^{30}Mg	^{40}Ar $+$ ^{232}Th	6p + 4n	Ar 71c
^{33}Al		5p + 2n	
^{36}Si		4p	
^{38}P		3p − 1n	
^{40}S		2p − 2n	
^{42}Cl		1p − 3n	
^{44}Ar		− 4n	

tively. By inclusion of the pairing energies, the Q_{gg} term is corrected for the fact that unpaired excited states are normally populated in deep-inelastic transfer. The temperature parameter T was found to assume values between 1.7 MeV and 2.2 MeV for the cases studied [Vo 74 a]. As an example for the application of (6.13), we show in Fig. 6.22 the empirical "Q_{gg} systematics" for the system ^{22}Ne $+$ ^{232}Th [Vo 74 b]. It can be seen that the data for this

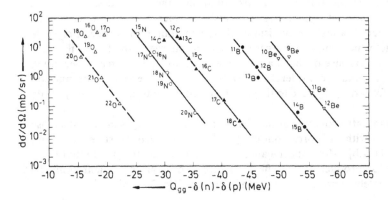

Fig. 6.22. Semilogarithmic plot of differential cross sections for production of different isotopes as a function of ground state Q value ("Q_{gg} systematics"; $\delta(n)$, $\delta(p)$ = pairing corrections). The initial system is ^{22}Ne (174 MeV) $+$ ^{232}Th, and the angle of observation is 12°. (From Vo 74b)

particular energy and angle are well reproduced by the simple relationship (6.13), except for the stable and less neutron-rich isotopes of elements close to the projectile.

The form of (6.13) is suggestive of a statistical interpretation and has, in fact, been shown to follow from the assumption of "partial statistical equilibrium" in the composite system [Bo 71]. This can be understood semiquantitatively by considering the total excitation energy U_{34} available in the composite system formed by touching final fragments 3 and 4:

$$U_{34} = Q_{gg} + \Delta E_{diss} - (V_{34} - V_{12}) - [E_{34}(\text{rot}) - E_{12}(\text{rot})]. \tag{6.14}$$

Here ΔE_{diss} is the energy dissipated in the incident channel and the two terms in brackets on the right-hand side correspond to the changes in relative potential and rotational energy induced by the transfer, respectively. The sum of these latter terms can be identified with the optimum Q value following from trajectory-matching arguments [compare (4.28) and (4.29); note, however, that the most probable Q value now includes the energy loss term ΔE_{diss}]. In a statistical model, the cross section will be proportional to the total number of intrinsic final states contained within the kinematically allowed Q-value window. In lowest order, this leads to the following estimate [Vo 74 a,b]:

$$\sigma_{34} \sim \exp\left(\frac{U_{34} - \delta(\text{n}) - \delta(\text{p})}{T}\right) \sim \exp\left(\frac{Q_{gg} - \Delta V_C - \delta(\text{n}) - \delta(\text{p})}{T}\right), \tag{6.15}$$

where the rotational contribution to U_{34} has been neglected and the potential--energy contribution is approximated by the change in Coulomb potential, ΔV_C. Equation (6.15) generalizes (6.13) to include different final elements and is consistent with the general systematics for production of neutron-rich isotopes below the projectile mass, as observed in the Dubna experiments. Significant discrepancies have also been noted, however, in a number of cases [Ar 73 a, Ja 75].

It is important to realize at this point that the measured isotopic yields are not necessarily representative of the primary production cross sections. Since, under typical conditions, excitation energies of the order of 100 MeV are deposited in the primary fragments, light-particle evaporation will usually occur before a reaction product reaches the counter. Bondorf and Nörenberg have pointed out that—assuming comparable fractions of the total excitation energy in the light and heavy product—the experimentally observed isotopic yields should be considerably modified with respect to the primary ones [Bo 73]. Quantitative corrections for this effect are complicated by the fact that the exact sharing of the excitation energy between the primary fragments is not known. Several authors have given indirect arguments, however, for the relative unimportance of evaporation as far as the neutron-rich light products are concerned [Ja 75, Ar 76].

A further difficulty arises from angular distribution effects. Since the differential production cross section and degree of energy relaxation for different elements or isotopes will in general depend differently on angle (see Figs. 6.5 and 6.6), yields observed at a single laboratory angle may not be strictly comparable.

Interesting evidence on the angular dependence of the isotopic yields has been obtained by Jacmart et al. at Orsay who studied the classical system ^{40}Ar + ^{232}Th (E_{Lab} = 295 MeV) at laboratory angles of 18° and 40° [Ja 75]. Figure 6.23 shows the measured differential cross sections and Fig. 6.24 the corresponding most probable centre-of-mass kinetic energies for separated isotopes of elements from aluminium to scandium. Both the elemental and isotopic yield distributions are considerably narrower at 40° (near the grazing angle) than at 18°, in accordance with the assumption of a predominant quasi--elastic mechanism at the former angle, and of a deep-inelastic mechanism at the latter angle. The kinetic energies also exhibit striking qualitative differences: at 40° they mainly depend on the product mass number and are sharply peaked at the projectile mass; at 18°, on the other hand, the most probable kinetic energies systematically increase with the product atomic number and

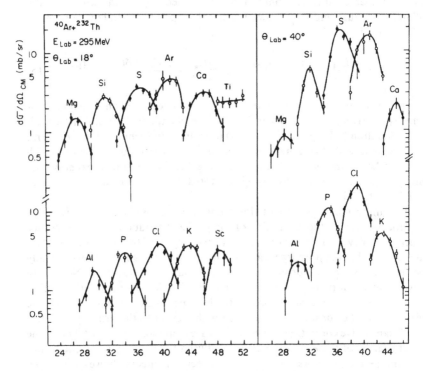

Fig. 6.23. Differential cross sections at laboratory angles of 18° (left) and 40° (right) for the system ^{40}Ar (295 MeV) + ^{232}Th and production of different final isotopes. Elements with even and odd atomic numbers are shown separately to facilitate presentation. (From Ja 75)

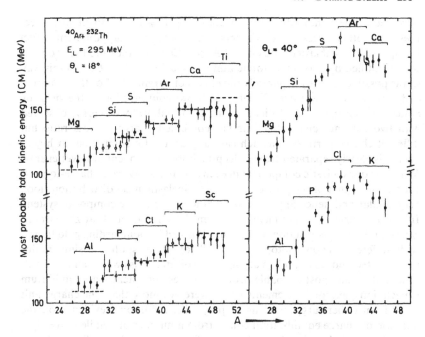

Fig. 6.24. Most probable total kinetic energies of different final isotopes at 18° (left) and 40° (right). The initial system is ^{40}Ar (295 MeV) + ^{232}Th ($E_{CM} = 247$ MeV corrected for energy loss in the target). The horizontal dotted lines represent 80% of the spherical Coulomb barrier calculated with a radius parameter $r_0 = 1.4$ fm. (From Ja 75)

depend little on A for a given Z. Whereas the 18° data are indicative of complete relaxation with respect to relative motion, the 40° data appear to be characteristic of an intermediate state in the dynamical evolution of the inelastic collision, where the energy loss due to dissipative processes is strongly correlated with mass transfer. We note that the trend of the most probable energies with product mass and charge number above those of the projectile, as seen in the 40° data, is at variance with simple predictions of Q_{opt} on the basis of trajectory matching arguments. This indicates the existence of additional restrictions on Q, imposed by the dynamics of energy dissipation.

Clearly studies like those shown in Fig 6.23 and 6.24, involving multi-parameter correlations between angle, energy loss, charge and mass transfer, are a potentially powerful tool for probing the anatomy of deep-inelastic collisions. Similar work has been performed with argon and calcium projectiles and targets ranging from ^{40}Ca to ^{208}Pb [Ga 75b, Ag 77a, Ba 77, Ga 77, Ro 77, Ba 78c]. The analysis and interpretation of the results, however, is complicated by the interdependence of the dynamic variables, and therefore remains a challenge to experimentalists and theorists alike.

We now return to the charge-to-mass ratio of the deep-inelastic products and its dependence on entrance channel. As mentioned before, most probable values N/Z between those of the projectile and of the composite system were

observed in the bombardment of thorium targets with various light projectiles. In another series of experiments, performed at Orsay, different nickel
isotopes were bombarded with either ^{40}Ar or ^{40}Ca projectiles and the products
were identified by mass and atomic number [Ga 75 b,c, Ga 76 a]. In this way
it was possible to study separately the effects of varying N/Z of the projectile
and of the composite system. Figure 6.35 shows contour plots of the yields of
potassium isotopes as a funticon of mass and laboratory energy, observed
with two entrance channels of similar combined charge-to-mass ratio but
different charge partition. In each case a quasi-elastic component at higher
energy is clearly separated from a deep-inelastic component at lower energy.
While the quasi-elastic component in each case has an average charge-to-mass
ratio close to that of the projectile, the deep-inelastic mass distributions look
very similar and hence appear to be characteristic of the composite system.
In another series of experiments, performed by Kratz et al. at Darmstadt,
charge and mass distributions of products formed by bombarding gold targets
with different xenon isotopes (^{129,132,136}Xe) were studied using radio-
chemical techniques [Kr 77]. Varying the projectile mass was found to result
in a shift of the most probable product masses consistent with equilibrium
expectations, except for elements which were not more than one charge unit
away from either target nucleus or projectile. Additional evidence on the
question of charge equilibration comes from a number of detailed investigations of argon-induced multi-nucleon transfer reactions with medium to heavy
target nuclei [Bi 75, Ga 77, Ba 78c, Ca 78]. It should be noted that the results
of all of these experiments are affected—in varying degrees—by light-particle
evaporation from the excited primary fragments, which must be taken into
account in their quantitative interpretation. This is usually done by means of

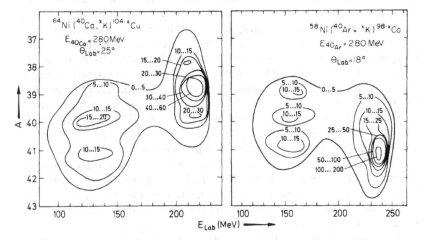

Fig. 6.25. Contour plots of the yield of potassium isotopes versus mass number and
laboratory energy, resulting from two different entrance channels with similar total N/Z
ratio. (From Ga 75b)

statistical evaporation calculations, which involve assumptions on initial excitation energies and angular momenta for the individual fragments, and are thus subject to uncertainties. Nevertheless, the results show rather convincingly that the average charge-to-mass ratio of the deep-inelastic products is close to its equilibrium value for each mass partition, as calculated from potential-energy arguments. Moreover, this seems to hold also for a major fraction of the yield associated with processes intermediate between quasielastic and deep-inelastic scattering. It may be concluded that the charge partition between the fragments reaches statistical equilibrium within characteristic times which are comparable with or shorter than the relaxation times of the relative kinetic energy, and considerably shorter than the times required for appreciable changes in mass asymmetry.

An interesting correlation has been observed between energy loss and charge dispersion of projectile-like products in the inelastic scattering of krypton and xenon from heavy targets [Sc 76, Hu 76]. For these heavy systems it was not possible to resolve individual elements or isotopes, and charge distributions were measured with a resolution of about three units using detector telescopes. As discussed in Sect. 6.2, the projectile-like group of fragments is concentrated in a comparatively narrow range of charge and mass values and kinematically well separated from other products. Figure 6.26 shows measured charge distributions of Xe-like fragments from the interaction of 1130 MeV ^{136}Xe projectiles with a ^{209}Bi target. The widths of the charge distributions are seen to increase strongly with increasing loss of total kinetic energy, while the centroid of the distributions remains close to the charge of the projectile ($Z = 54$). If the energy loss is plotted as a function of the variance σ_z^2 of the charge distribution, then points representing different systems seem to follow approximately a universal curve as shown in Fig. 6.27 [Hu 76]. This remarkable result will be discussed further below and in Sect. 6.4.

We can summarize the results reported so far in this section as follows.
i) Secondary processes, like light-particle evaporation from, or sequential fission of, heavy primary fragments, are often important and must be taken into account in interpreting experimental data. Investigations aimed specifically at these secondary processes can yield important supplementary information on excitation energies or intrinsic angular momenta of the primary fragments (see also the discussion below).
ii) The results give evidence of different time scales involved in the equilibration of different degrees of freedom. Substantial changes in mass asymmetry or macroscopic deformation (eventually leading to either fusion or scission) appear to require times which are comparable with typical rotational periods of the composite systems ($10^{-21} - 10^{-20}$s; see Table 6.1). The relative kinetic energy and charge partition, on the other hand, are found to reach their respective equilibrium values within significantly shorter times.
iii) There are striking correlations between the total kinetic energy loss on one side and the mass or charge transfer in a collision on the other side (see Figs. 6.24, 6.26, 6.27). This suggests that nucleon transfer between the fragments

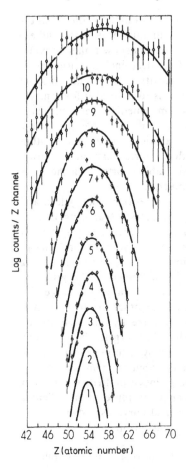

Fig. 6.26. Charge distributions of the Xe-
-like fragment for the system ^{136}Xe (1130 MeV)
+ ^{209}Bi and different total kinetic energy
windows. The energy loss increases from near
zero for curve 1 to approximately 300 MeV
for curve 11. (From Sc 76)

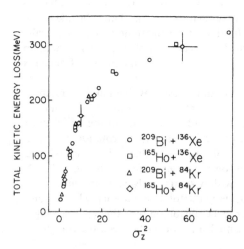

Fig. 6.27. Correlation between total kinetic energy loss and variance,
σ_z^2, of the charge distribution for
projectile-like fragments in various
Kr- and Xe-induced reactions. (From
Hu 76)

might be an important basic mechanism for energy and angular momentum dissipation.

A number of questions remain open at this stage, and are the object of continued investigation. In particular, one should like to reach a detailed microscopic understanding of the processes responsible for the very high rates of kinetic energy loss (of order 10^{23} MeV/s) which must be assumed during the initial phase of a typical deep-inelastic collision. Closely related to this problem are the questions of how the total excitation energy is shared initially between the fragments, and of how fast an equilibrium distribution of excitation energies is approached. The latter will presumably correspond to equal nuclear temperature of the fragments, and hence to a partition of excitation energies roughly in proportion to the fragment masses. Another aspect of the same problem concerns the total angular momentum transfer from relative to intrinsic motion, and the partition of intrinsic angular momenta between the fragments. Finally a knowledge of the spin alignment and (or) spin polarization of the fragments would provide an especially sensitive test of classical collision models, and of the concept of deflection to "negative angles."

Experimental work related to these questions invariably involves the observation of either gamma rays or light particles in coincidence with heavy fragments. We first consider briefly some experiments where evidence for a "fast" emission of light charged particles in deep-inelastic collisions was obtained. The systems investigated include $^{16}O + ^{27}Al$ ($E_{Lab} = 65$ MeV; Ha 77), $^{16}O + ^{58}Ni$ ($E_{Lab} = 92$ and 96 MeV; Ho 77a), $^{16}O + ^{197}Au$ and ^{208}Pb ($E_{Lab} = 140$ and 310 MeV; Ge 77), $^{32}S + ^{197}Au$ ($E_{Lab} = 373$ MeV; Ga 78), and $^{86}Kr + ^{197}Au$ ($E_{Lab} = 724$ MeV; Mi 78). Angular correlations between projectile-like fragments and α particles were found to exhibit significant forward-peaking in most cases. This effect was interpreted in terms of pre-equilibrium decay of the composite system, presumably resulting from the existence of a localized region of high excitation density near the point of contact ("hot pot"; see We 78). It seems likely that further work along the same lines will provide interesting insights into the mechanisms of energy and angular momentum dissipation in deep-inelastic collisions.

Pre-equilbirium processes, however, in general account only for a small fraction of the total light-particle emission at the energies considered here. The yield of evaporated particles, on the other hand, gives information on the excitation energy remaining in the heavy fragments after scission, and thus on the question to what extent thermal equilibrium is reached during the period of contact. One experimental method used in this context involves coincident detection of the two heavy fragments, with detailed identification of at least one partner by energy, charge, and mass. Deviations of the observed kinematic correlations from those expected for strict two-body decay are then exploited to estimate the evaporated mass as a function of primary mass, and the evaporated charge for symmetric partitions. This approach has been followed in studies of the systems $^{40}Ar + ^{58}Ni$ [Ba 78a] and $^{40}Ar + Ag$ [Ca 78]. In another type of experiment, performed with the system $^{86}Kr + ^{124}Sn$, only one fragment was detected at two different bombarding energies [Pl 78]. In this case the primary excitation of the observed fragments was inferred from the measured change in average final mass number for a given value of Z with bombarding energy. The results of these various studies indicate a partition of excitation energies in proportion to the fragment masses, and hence approximately equal nuclear temperatures of the primary fragments in deep-inelastic collisions.

A more direct way of investigating this problem is by detecting the evaporated particles in coincidence with heavy fragments. In particular, neutron emission has been studied for the systems ^{63}Cu + ^{197}Au [Pe 77b], ^{86}Kr + ^{166}Er [Ey 78] and ^{132}Xe + ^{197}Au [Go 78]. The fact that the two heavy primary fragments are produced with high relative velocity is very helpful in this context, as it allows the neutron yields associated with each fragment to be separated kinematically. The measured neutron multiplicities are approximately proportional to fragment mass and energy loss for wide ranges of these parameters. Thus it appears that thermal equilibrium between the primary fragments is reached even for peripheral collisions which do not lead to full relaxation of the relative kinetic energy. The energy spectra and angular distributions of the emitted neutrons were found to be consistent with statistical evaporation from fully accelerated fragments, with no evidence for fast neutron emission from the composite systems.

The angular momentum transfer in a collision can be studied by measuring the multiplicity of gamma rays which are emitted by the fragments. For this purpose either one or several gamma detectors are operated in coincidence with a detector for heavy fragments. The relationship between fragment spin and gamma multiplicity depends on the angular momentum carried away by particle evaporation (usually assumed negligible or small) and on the average multipolarity and relative alignment of the gamma rays (usually taken as predominantly "stretched" E2). In some cases, gamma-multiplicity measurements have been calibrated empirically in terms of fragment spin. This is done by comparison with fusion reactions, leading to the same products at similar excitation energy, where the primary angular momentum distribution is known. It should be noted, however, that the gamma multiplicity depends on composition (A,Z), excitation energy, and angular momentum, and that calibration procedures taking all these effects into account are not generally available. Another limitation of the method, as it has been applied so far, lies in the fact that the combined multiplicity of gamma quanta emitted by both fragments is obtained. Therefore the individual fragment spins remain undetermined. A summary of gamma-multiplicity measurements performed up to 1976 has been given by Galin [Ga 76b]. More recently, a number of groups have performed detailed studies of gamma multiplicity as a function of fragment charge and Q value, using projectiles from ^{20}Ne to ^{86}Kr [Gl 77, Al 78, Ch 78, Na 78, Ol 78]. The gross features of the results are summarized in the following (see Figs. 6.28 and 6.29).

In the "quasi-elastic" and moderately inelastic domains, for energy losses up to about 50% of the deep-inelastic values, the gamma multiplicity increases approximately linearly with energy loss, and with the number of transferred

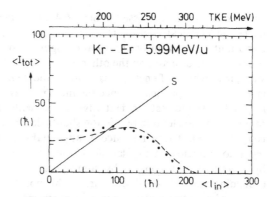

Fig. 6.28. Total intrinsic angular momentum in inelastic scattering of krypton from erbium, as deduced from gamma-multiplicity measurements. Plotted along the abscissa is the incident angular momentum (bottom scale) or, alternatively, the total kinetic energy loss (top scale). The sticking limit for the initial fragmentation is indicated by the straight line, and the dashed curve gives results of a calculation by Wolschin and Nörenberg. (From Ol 78)

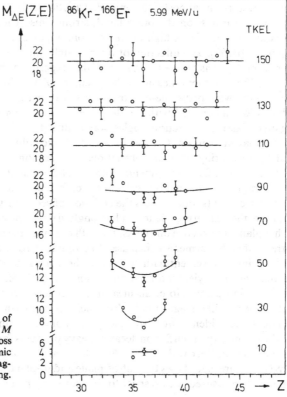

Fig. 6.29. Dependence of the gamma multiplicity M on total kinetic energy loss (TKEL, in MeV) and atomic number Z of the light fragment in Kr + Er scattering. (From Ol 78)

charges. As will be discussed in Sect. 6.4, this is the qualitative behaviour expected in classical friction models, and points to nucleon transfer as an important mechanism of energy and angular momentum dissipation. In the deep-inelastic domain, on the other hand, the multiplicity remains fairly constant, irrespective of energy loss or mass and charge transfer. This behaviour is not expected classically, since the most negative Q-values presumably correspond to the lowest incident angular momenta, and therefore the angular momentum transferable by friction should decrease as the deep-inelastic limit is approached. The persistence of appreciable spins at the lowest Q values seems to indicate that angular momentum is generated in the fragments either by thermal fluctuations or by some peculiar collective mode of vibration ("bending mode"). We shall return to this point in Sect. 6.4.

In order to test scattering models more specifically, it would be desirable to measure not only the total fragment spin, but also its alignment or polarization. In a classical model, of course, the fragment spins should be completely polarized along the normal to the scattering plane. Moreover, the direction of the spin vector should depend on whether the particles are deflected to positive or negative scattering angles. Unfortunately, the angular correlation of unresolved gamma radiation is not very sensitive to the spin alignment, since apparently summing over many transitions with different multipole orders results in almost isotropic distributions. In a study of discrete gamma transitions, on the other hand, strong anisotropies were found and taken as evidence for a high degree of spin alignment of the deep-inelastic fragments [Bi 77]. The direction of the residual spins, following scattering of ^{40}Ar from Ag, was studied in a remarkable experiment by Trautmann et al. [Tr 77]. These authors measured the circular polarization of gamma rays in coincidence with projectile-like fragments. From the observed sign of the polarization, associated with the deep-inelastic component, the existence of negative classical deflection angles was deduced.

Angular momentum transfer in deep-inelastic collisions may also be studied by measuring angular correlations between one heavy fragment and light particles or fission fragments emitted in the sequential decay of the associated fragment. In the classical limit, i.e. for complete polarization of the primary fragments, one expects the emission of secondary particles to be isotropic in the scattering plane, and strongly decreasing towards the normal to that plane. In practice one measures both out-of-plane and in-plane angular correlations; the former are then used to estimate the degree of spin alignment of the primary fragment with respect to the normal to the scattering plane, whereas the latter yield information on (anisotropic) spin components within the scattering plane. Ho et al. measured angular correlations of α particles with projectile-like fragments in deep-inelastic scattering of ^{16}O from ^{58}Ni, and obtained evidence for a strong alignment of the residual target-like nuclei [Ho 77a]. The sequential fission decay of heavy products formed in deep-inelastic collisions of ^{86}Kr with ^{209}Bi at 610 MeV was investigated by Dyer et al. [Dy 77]. Figure 6.30 shows the out-of-plane and in-plane correlation of fission fragments, measured with respect to the scattering plane of the Kr-like frag-

Fig. 6.30. In-plane and out-of-plane angular correlation between krypton-like primary fragments and secondary products from sequential fission of the bismuth-like fragment in Kr + Bi scattering. The curves represent theoretical predictions for different angular momenta of the fissioning nucleus. (From Dy 77)

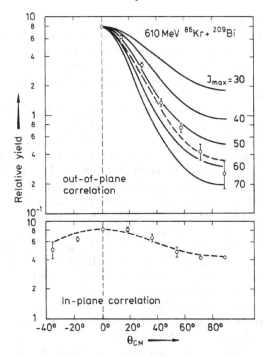

ments. An angular momentum of the fissioning nucleus of about 55 $\hbar$ was deduced from the out-of-plane correlation, consistent with the classical "sticking limit" for the appropriate range of incident angular momenta. A subsequent, more rigorous analysis of the same data by Back and Bjornholm yielded a somewhat lower spin value of about 40 $\hbar$, with rather large uncertainties, however [Ba 78b]. The observed anisotropy of the in-plane correlation indicates a preferential orientation of the in-plane spin projection perpendicular to the recoil axis.

We conclude this section by noting a definite trend towards more detailed and more comprehensive investigations. It appears that, in order to analyse fully the dynamic structure of deep-inelastic collisions, simultaneous studies of all open decay channels with the highest possible resolution will be needed.

6.4 Models of Deep-Inelastic Collisions

The experimental results on deep-inelastic collisions have stimulated tremendous theoretical activity. A variety of methods and concepts, appropriate to classical and statistical mechanics, fluid dynamics, and thermodynamics, have been applied in addition to the traditional methods of quantal nuclear theory. A common feature of these models is that at some stage the detailed, quantum-mechanical description of the many-body system is abandoned in favour of performing statistical averages over certain microscopic degrees of freedom, while treating certain macroscopic degrees of freedom classically. An adequate presentation of the various approaches is clearly beyond the scope of this book. Instead, we merely give a—necessarily incomplete—list of references, followed by a somewhat more detailed discussion of a few selected models.

Among the more ambitious approaches, we first mention the time-dependent Hartree–Fock method developed, in particular, by the Oak Ridge group [Bo 76a,b, Cu 76a,b, Ko 77, Ma 77]. Statistical theories of transport phenomena in nuclear collisions were worked out by Nörenberg and collaborators [Nö 75, Nö 76a,b, Ay 76a,b, Ay 78, Sc 78c]; by Hofmann and Siemens [Ho 76a,b, Ho 77]; and by Weidenmüller and collaborators [Ko 76, Ag 77b, Ag 78]. On a more phenomenological level, but again based on microscopic, statistical arguments, we have the diffusion model of Moretto and Sventek [Mo 75c, Mo 76b, Sv 76], and the "one-body dissipation" model of Swiatecki, Randrup, and collaborators [Bl 77, Bl 78, Ra 78a,b]. In the latter approach the energy loss from relative to intrinsic motion is attributed to the interaction of individual nucleons with the mean field produced by the sum of all nucleons in the system. In a simplified picture this interaction may be viewed as mediated by collisions between nucleons and a moving "container wall", or by the passage of nucleons from one fragment to the other through a "window" [Sw 72a].

Other models concentrate on the macroscopic, collective degrees of freedom. Wong and collaborators have formulated foundations of "nuclear hydrodynamics" [Wo 75]. The coupling between relative and vibrational motion of the fragments was treated by Broglia and collaborators in the framework of the semiclassical coupled channels method [Br 74, Br 76b, Br 78a,b].

The combination of macroscopic and microscopic approaches (i.e. the incorporation of shell effects into a macroscopic description based on the liquid-drop model) was briefly discussed in Sect. 2.2. The method has been extended to the calculation of inertial parameters and applied to the dynamics of fission and heavy-ion collisions. As an example we mention the theory of fragmentation developed by Greiner and collaborators [Fi 74, Zo 75, Ya 76]. Here mass fragmentation is introduced as a collective coordinate which is treated quantum-mechanically by solving a time-dependent Schrödinger equation.

The macroscopic-microscopic method has been very successful when applied to low-energy fission phenomena; its relevance to deep-inelastic collisions is less obvious, however, since the typical excitation energies involved (of the order of 50 - 100 MeV) will tend to reduce effects due to shell structure and coherent collective motion. In addition, energy dissipation into intrinsic degrees of freedom is not explicitly included and must therefore be added using some phenomenological prescription. Nevertheless a detailed comparison of the predictions with experiment should be rewarding since shell effects, even if slight, may be a decisive factor in collisions leading to the formation of rare products such as superheavy elements [Gu 77].

Finally there is a large group of models which describe nucleus–nucleus collisions in completely macroscopic and classical terms. In some cases these models have been based on elaborate parametrizations of the nuclear shape and detailed prescriptions for the internal flow of nuclear matter [Ni 69, Ni 74, Da 76]. In other cases (to be discussed in more detail later), the colliding nuclei have been pictured simply as rigid, nondeformable spheres. The transfer of energy and angular momentum from collective (relative) to intrinsic motion is taken into account by including phenomenological "friction" forces in the classical equations of motion. In this way, cross sections, angular distributions, and energy and angular momentum loss are predicted.

At this point it is important to note that any model which correctly incorporates the geometry of trajectory motion, and provides for the possibility of dissipative phenomena, leading to fluctuations in the dynamical variables, will at least qualitatively fit the gross features of the existing data. It is therefore extremely difficult to draw conclusions about the validity of specific model assumptions, until much more detailed comparisons with experiment become possible than have been published in the past. In the following we take an empirical point of view, concentrating on those phenomenological models which were shown to be useful for quantitative analyses of experimental data. Thus we neglect the more fundamental approaches which may ultimately lead to an understanding of the microscopic structure of deep inelastic collisions, but have—in our judgement—not yet been tested critically by comparison with observation.

This leaves essentially the classical scattering models and the so-called diffusion models. Each of these two classes of models focusses attention on some particular aspect of the collision problem: the former on the dynamics of trajectory motion and the resulting distribution of the products in phase space, and the latter on the redistribution of mass, charge, and energy during the collision. Clearly these two aspects are coupled to some extent in real collisions, and this coupling imposes restrictions on the validity of the simple models employed so far. Finally we shall briefly discuss possible improvements of the models and ask to what extent the existing phenomenological analyses can provide an improved understanding of the physical mechanisms involved.

The various classical scattering models are characterized by three basic ingredients, namely:
a) the collective degrees of freedom (q_i) which are treated explicitly, and their associated inertial parameters;
b) the potential energy $V(q_i)$ including the nucleus–nucleus potential; and
c) the assumed dissipative forces (friction) which remove energy from the collective degrees of freedom.

Once these points are settled, it is straightforward in principle to compute the time evolution of the collective variables by solving the generalized

Lagrange equations

$$\frac{d}{dt}\frac{\partial \mathscr{L}}{\partial \dot{q}_i} - \frac{\partial \mathscr{L}}{\partial q_i} = -\frac{\partial \mathscr{F}}{\partial \dot{q}_i}.$$ (6.16)

Here $\mathscr{L}(q_i, \dot{q}_i) = T(q_i, \dot{q}_i) - V(q_i)$ is the Lagrangian of the system, and

$$\mathscr{F}(q_i, \dot{q}_i) = -\frac{1}{2}\frac{d}{dt}[T(q_i, \dot{q}_i) + V(q_i)]$$

is Rayleigh's "dissipation function". In practice solving the coupled differential equations (6.16) is usually a complicated numerical problem, and therefore one retains only the minimum number of degrees of freedom q_i which is considered absolutely essential.

The question as to which degrees of freedom should be included in a description of nuclear fission and nucleus–nucleus collisions has been discussed by Swiatecki [Sw 72a] and by Swiatecki and Björnholm [Sw 72b]. Attention was focussed on three coordinates, namely a separation coordinate α_2, a mass asymmetry coordinate α_3, and a "neck forming coordinate" α_4. If one is primarily interested in collisions of initially separated nuclei, it seems more appropriate to replace the coordinates α_2, α_4 by the centre-to-centre distance r and one or several coordinates describing simple deformations of the two fragments. The latter degrees of freedom come into play especially during the final stages of a deep-inelastic collision. This follows from dynamical arguments and from the observation that the final fragment kinetic energy is often well below the entrance-channel Coulomb barrier (see Sect. 6.2). We therefore expect that the assumption of a "frozen" shape of the fragments (disregarding deformations) may be a reasonable approximation for calculating the ingoing part of the trajectory, but cannot give reliable results for the outgoing part. Some authors have tried to simulate the effects of deformation without explicit consideration of deformation coordinates by introducing friction forces acting only at the turning point and during separation [Bo 75], or by adding a (nonconservative) deformation potential for the outgoing part of the trajectory only [Si 76].

A further complication arises from the fact that nucleus–nucleus collisions usually involve significant angular momenta. During the collision a part of the initial orbital angular momentum may be transferred to intrinsic rotation of either fragment. Hence additional coordinates are required to specify the angular position of the internuclear axis and the intrinsic axes of the fragments.

Next we consider specific examples of classical scattering models. The first calculations of this type were published by Gross and Kalinowski [Gr 73, Gr 74a, Gr 74a,b, De 76a, Gr 78]. The equations of motion used by these authors were given in Cartesian coordinates as

$$\mu \frac{d^2 x_i}{dt^2} = -\frac{\partial V}{\partial x_i} - \sum_k C_{ik}\frac{dx_k}{dt},$$ (6.17)

where the subscripts i, k are equal to 1 for the radial, and 2 for the tangential, component. The quantities C_{ik} are the components of the position-dependent friction tensor ($C_{12} = C_{21} = 0$). Intrinsic angular momenta, shape changes, and mass transfer are neglected, since only the relative position vector enters. In order to facilitate the comparison with other models we transform (6.17) to spherical polar coordinates and substitute the orbital angular momentum L for the angular coordinate Θ. This yields, with $C_{11} = C_r(r)$ and $C_{22} = C_L(r)$,

$$\mu \frac{d^2 r}{dt^2} = - \frac{\partial}{\partial r}\left[V(r) + \frac{\hbar^2 L^2}{2\mu r^2} \right] - C_r(r) \frac{dr}{dt}, \qquad (6.18)$$

$$\mu \frac{dL}{dt} = \qquad\qquad - C_L(r)L. \qquad (6.19)$$

In practical applications of this model, different assumptions about the potential $V(r)$ and the friction coefficients $C_r(r)$, $C_L(r)$ were employed. The most extensive calculations were performed with a potential obtained by folding the nucleon density distribution of one nucleus with a standard optical model potential for nucleon scattering from the other nucleus. The friction coefficients were parametrized as

$$C_r(r) = C_1 f(r); \quad C_L(r) = C_2 f(r), \qquad (6.20)$$

with universal, adjustable constants C_1, C_2. In the original version of the model [Gr 73, Gr 74a] the shape function $f(r)$ was taken as a simple Fermi function centred at a distance 2.7 fm outside the half-density distance (i.e. close to the strong absorption distance as deduced from elastic scattering; see Sect. 3.3). Improved results were obtained, especially for heavy scattering systems, when the strength of the friction forces was coupled to the fragment sizes by putting $f(r)$ equal to the square of the gradient of the nuclear potential [Gr 74b]. With this ansatz for the friction shape-function, extensive analyses of fusion cross-sections and inelastic scattering data were performed, leading to the following best fit values for the friction parameters C_1, C_2:

$$C_1 = 4 \times 10^{-23}\text{s MeV}^{-1}; \quad C_2 = 1 \times 10^{-25}\text{s MeV}^{-1}. \qquad (6.21)$$

Gross and Kalinowski were able to reproduce with this model the main qualitative features of deep-inelastic scattering. At the same time, a very good quantitative fit to experimental fusion cross sections for many systems and bombarding energies was achieved (see Sect. 7.4). Obvious limitations of the model are the neglect of intrinsic angular momenta, mass transfer, and fragment deformation. The assumption of "frozen" spherical fragment shapes necessarily yields too small energy losses in deep-inelastic collisions, since final kinetic energies below the Coulomb barrier—as observed experimentally and explained by stretching of the fragments prior to scission—cannot be reproduced with plausible friction forces. Further, as in all classical scattering models published so far, a unique deflection angle, final energy, and mass partition are implied for a given initial angular momentum. This leads to certain unrealistic features, like the divergence of the differential cross section at rainbow angles where $d\Theta/dL = 0$ [see (6.6)]. Nevertheless, the apparent overall success of the model is remarkable, considering its simplicity.

Several points require further comment, however. The single-folding nuclear potential employed by Gross and Kalinowski is unusually deep by comparison with phenomenological analyses of elastic scattering data, and with theoretical considerations based on nuclear matter and liquid-drop concepts (see Sect. 7.4). Another conspicuous feature of the model is the extremely anisotropic friction tensor, with a large radial and small tangential component. Such a large anisotropy is not consistent with models which relate friction either to nucleon transfer across the neck or to nucleon–nucleon collisions in the overlap region, nor can it explain the substantial transfers of angular momentum which have been observed experimentally [Dy 77, Gl 77, Ol 78].

These apparent discrepancies may be understood by noting that the success of a classical scattering model in reproducing certain gross features of experimental data does not necessarily imply an independent physical significance of the conservative and friction forces entering the model. The calculations of Gross and Kalinowski with different potentials and friction shape functions, and other evidence to be discussed in Chap. 7, show that changes in the potential can be compensated to some extent by associated changes in friction. Moreover, restrictions in configuration space will be reflected in the potential and friction needed to fit experimental data. This indicates that one actually deals with effective forces which may only work well in the context of a particular model. The "validity" of the model assumptions can then be judged only by the amount of experimental data that can be reproduced.

Next we turn to a scattering model discussed by Tsang [Ts 74]. Here the shape and mass asymmetry degrees of freedom are frozen, as in the previous model, but the intrinsic rotation of each fragment is explicitly treated assuming rigid moments of inertia. The angular coordinates and intrinsic velocity distributions are illustrated schematically in Fig. 6.31. The tangential component of the friction force now transfers angular momentum from orbital to intrinsic motion, but explicitly conserves total angular momentum. The following ansatz is used for the (total) friction force:

$$\boldsymbol{F} = - k \int \rho_1 \rho_2 \, \boldsymbol{u}_{12} d^3 \boldsymbol{r} \, . \tag{6.22}$$

Here k is a constant, ρ_1 and ρ_2 are the local nucleon densities due to fragment 1 and 2, respectively, and $\boldsymbol{u}_{12}$ is the associated local velocity difference. This form of the friction force may be interpreted as arising from an interaction of extremely short range between two penetrating classical fluids or solids. We note that the force is completely specified in terms of one parameter, k; hence no freedom is left to adjust independently the strengths of "radial" and "tangential" friction.

Fig. 6.31. Schematic representation of rotational degrees of freedom and tangential friction force in nucleus–nucleus collisions

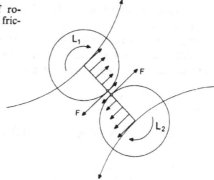

Equation (6.22) can be used to calculate the dissipation function and hence the equations of motion. This yields [Ts 74]

$$\mu \frac{d^2r}{dt^2} = -\frac{\partial}{\partial r}\left[V(r) + \frac{\hbar^2 L^2}{2\mu r^2}\right] - C_r(r)\frac{dr}{dt}, \tag{6.23}$$

$$\frac{dL}{dt} = -\frac{dL_1}{dt} - \frac{dL_2}{dt}, \tag{6.24}$$

$$\frac{dL_1}{dt} = C_r(r)[(\omega r - \omega_1 g_1 - \omega_2 g_2)g_1 - b^2(\omega_1 - \omega_2)], \tag{6.25}$$

$$\frac{dL_2}{dt} = C_r(r)[(\omega r - \omega_1 g_1 - \omega_2 g_2)g_2 + b^2(\omega_1 - \omega_2)], \tag{6.26}$$

where we have substituted

$$C_r(r) = k \int \rho_1 \rho_2 \, d^3\mathbf{r} . \tag{6.27}$$

Obviously (6.23) is identical with (6.18), and (6.24) expresses angular momentum conservation. The time dependence of the intrinsic angular momenta is governed by (6.25) and (6.26), where g_1, g_2 are the distances between the centre-of-mass of the overlap region and the centres of fragments 1 and 2, respectively, and b is an effective radius of gyration of the overlap region around its centre of mass. The first term in the square bracket on the right-hand side of (6.25) and (6.26) is proportional to the relative surface velocity of the two fragments, and the resulting force is therefore attributed to "sliding friction". Conversely, the second term is proportional to the difference in angular velocities and therefore arises from "rolling friction". It should be noted that sliding friction produces rolling of the two surfaces on each other, while rolling friction produces rigid rotation of the system as a whole ("sticking"). The parameter b^2 is of the order of the product of a nuclear radius with the thickness of the nuclear surface, and may be thought of as describing the stiffness of the neck between the two fragments. The resulting strength of the rolling friction is about one order of magnitude less than that of the sliding friction [Ts 74].

For practical purposes, (6.24–26) may be simplified somewhat by eliminating redundant variables. Confining our attention to small surface separations of the two fragments, we set $g_i \approx R_i$ and write ($R_{12} = R_1 + R_2$)

$$L = \mu r^2 \omega \approx \mu R_{12}^2 \omega , \tag{6.28}$$

$$L_i = \mathscr{I}_i \omega_i = \frac{2}{5} M_i R_i^2 \omega_i \quad (i = 1,2) , \tag{6.29}$$

$$L^+ = L_1 + L_2 = L_0 - L , \tag{6.30}$$

$$L^- = \frac{g_2 L_1 - g_1 L_2}{g_1 + g_2} \approx \frac{R_2 L_1 - R_1 L_2}{R_{12}} . \tag{6.31}$$

This leads to the following approximate equations for L and L^-:

$$\mu \frac{dL}{dt} = -C_L(r) \left[\frac{7}{2} \left(L - \frac{5}{7} L_0 \right) - \frac{5}{2} \frac{R_{12}}{M_1 + M_2} \left(\frac{M_2}{R_1} - \frac{M_1}{R_2} \right) L^- \right], \tag{6.32}$$

$$\mu \frac{dL^-}{dt} = -\tilde{C}_L(r) \left[\frac{5}{2} \frac{R_{12}}{M_1 + M_2} \left(\frac{M_2}{R_1} - \frac{M_1}{R_2} \right) (L_0 - L) \right.$$
$$\left. + \frac{5}{2} \frac{R_{12}^2}{M_1 + M_2} \left(\frac{M_1}{R_2^2} + \frac{M_2}{R_1^2} \right) L^- \right]. \tag{6.33}$$

Here we have introduced the coefficients $C_L(r)$ and $\tilde{C}_L(r)$ to describe sliding and rolling friction, respectively, in order to allow for a more general friction force than that given by (6.22). If (6.22) holds, then we have

$$C_L(r) = C_r(r); \quad \tilde{C}_L(r) = \frac{b^2}{R_{12}^2} C_r(r) . \tag{6.34}$$

We note that the scattering problem is now formulated in terms of the three coupled differential equations (6.23), (6.32), and (6.33) for the functions $r(t)$, $L(t)$, and $L^-(t)$. A further simplification can be achieved by neglecting the rolling friction [$\tilde{C}_L(r) \equiv 0$]. This leads to $L^- \equiv 0$ and [Ba 74]

$$\mu \frac{dL}{dt} = -\frac{7}{2} C_L(r) \left(L - \frac{5}{7} L_0 \right) . \tag{6.35}$$

In the same approximation, the total energy loss from the relative degrees of freedom (r, L) due to friction is given by

$$\frac{dE}{dt} = -C_r(r) \left(\frac{dr}{dt} \right)^2 - \frac{7}{2} C_L(r) \frac{\hbar^2 L \left(L - \frac{5}{7} L_0 \right)}{\mu^2 r^2} . \tag{6.36}$$

It is now straightforward to deduce the final angular momenta for the limiting cases of "rolling" and "sticking" (see also Table 6.4). In the rolling limit (negligible rolling friction, but strong sliding friction) we have $L^- \equiv 0$, and from (6.31) and (6.35)

$$
L = \frac{5}{7} L_0; \quad L_i = \frac{2}{7} \frac{R_i}{R_{12}} L_0 \quad (i = 1, 2) .
\tag{6.37}
$$

In this case, the final relative angular momentum is reduced by the factor 5/7 compared with its initial value, independently of the fragment mass ratio. The sticking limit (strong sliding and rolling friction) is obtained by setting the right-hand sides of (6.32) and (6.33) equal to zero, and re-introducing L_1, L_2 by means of (6.30) and (6.31). The same result is deduced more easily from (6.25) and (6.26), however, by noting that the sticking limit corresponds to $\omega_1 = \omega_2 = \omega$. It follows that the final angular momenta L, L_1, L_2 must be proportional to the appropriate moments of inertia as given by (6.28) and (6.29). Thus we obtain

$$
L = \frac{\mu R_{12}^2}{\mu R_{12}^2 + \mathscr{I}_1 + \mathscr{I}_2} L_0; \quad L_i = \frac{\mathscr{I}_i}{\mu R_{12}^2 + \mathscr{I}_1 + \mathscr{I}_2} L_0 \quad (i = 1,2). \tag{6.38}
$$

For symmetric fragmentations $(M_1, R_1 = M_2, R_2)$, (6.38) and (6.37) yield identical results. In this case rolling and sticking are physically not distinguishable. An important consequence of this result is that experiments designed to differentiate between different models of angular momentum coupling in deep-inelastic collisions should be performed with highly asymmetric systems.

Classical scattering models similar to those outlined above have been developed by a number of authors [Ba 74, Sp 74, Bo 74b, Da 74, Se 75]. Most of these models include internal rotation, but not deformation of the fragments among the dynamical variables. As mentioned previously, effects of the latter degrees of freedom have been simulated by assuming additional friction forces [Bo 75] or potentials [Si 76], acting on the outgoing part of the trajectory only. A model which explicitly takes fragment deformation into account was proposed by Deubler and Dietrich [De 75, De 77]. Using potentials consistent with nuclear-matter considerations (in contrast to the deeper folding

Table 6.4. Final angular momentum of relative motion (L) and intrinsic angular momenta (L_1, L_2) for different model assumptions about angular momentum transfer ($L_0 =$ initial angular momentum; $\mathscr{I}_1 = \frac{2}{5} M_1 R_1^2$, $\mathscr{I}_2 = \frac{2}{5} M_2 R_2^2$, $\mathscr{I}_{12} = \mu R_{12}^2$)

Model	L/L_0	L_1/L_0	L_2/L_0
No loss	1	0	0
"Rolling"	$\dfrac{5}{7}$	$\dfrac{2}{7} \dfrac{R_1}{R_{12}}$	$\dfrac{2}{7} \dfrac{R_2}{R_{12}}$
"Sticking"	$\dfrac{\mathscr{I}_{12}}{\mathscr{I}_{12} + \mathscr{I}_1 + \mathscr{I}_2}$	$\dfrac{\mathscr{I}_1}{\mathscr{I}_{12} + \mathscr{I}_1 + \mathscr{I}_2}$	$\dfrac{\mathscr{I}_2}{\mathscr{I}_{12} + \mathscr{I}_1 + \mathscr{I}_2}$

potentials) and moderately strong radial friction forces, the latter authors calculated deflection functions which exhibit two rainbow angles, a "Coulomb rainbow" occuring at larger angular momenta, and a "nuclear rainbow" occuring at smaller angular momenta (compare cases b and c of Fig. 6.15). This is in contrast to the calculations of Gross and Kalinowski, where usually only a single rainbow angle appears, followed by a transition to orbiting for smaller angular momenta.

One might ask to what extent the existing model calculations allow the strength of friction in nucleus–nucleus collisions to be deduced. Unfortunately, the use of different parametrizations and form factors for the friction forces makes a direct quantitative comparison of different models difficult. Nevertheless it appears that C_r values of the order of 10^{-21} MeV s fm^{-2} are required by the experimental data obtained with heavy projectiles (Ar, Kr) and target nuclei. From an analysis of the Dubna data for ^{40}Ar + ^{232}Th, based on (6.27), $k = 2 \times 10^{-20}$ MeV s fm was deduced [Si 76].

These numbers may be compared with simple theoretical estimates. Albrecht and Stocker discussed a gas-kinetic model in which the friction force arises from momentum transfer by nucleon–nucleon collisions in the overlap region [Al 77]. Their result, $k = 1 \times 10^{-20} - 5 \times 10^{-20}$ MeV s fm in general and 2.6×10^{-20} MeV s fm for ^{40}Ar + ^{232}Th, agrees well with the empirical value quoted above.

In an alternative approach, friction is related to the flow of nucleons which pass from one fragment to the other upon contact. The associated flow of collective momentum is assumed to be thermalized in the acceptor nucleus and hence constitutes a friction force. This effect, also known as the "window effect", was first discussed by Swiatecki [Sw 72a]. Quantitative estimates of the resulting radial and tangential friction coefficients were obtained by Bass [Ba 76b] and by Randrup [Ra 78a]. Using geometrical arguments (also known as the "proximity theorem"; see Sect. 7.4), the total flux of nucleons $I(s)$ between two spherical nuclei can be deduced from the flux per unit area $j(s)$ between two static, semi-infinite regions filled with nuclear matter and bounded by plane surfaces at separation s:

$$I(s) = 2\pi \frac{R_1 R_2}{R_{12}} \int_0^\infty j(s + x)dx = 2\pi \frac{R_1 R_2}{R_{12}} i(s) . \tag{6.39}$$

Here $i(s)$ is a universal function of the surface separation coordinate $s = r - R_{12}$. The resulting radial and tangential friction forces are (with m the nucleon mass)

$$F_r(s) = - mI(s)u_r = - 2\pi m \frac{R_1 R_2}{R_{12}} i(s)u_r, \tag{6.40}$$

$$F_L(s) = - \frac{m}{2} I(s)u_L = - \pi m \frac{R_1 R_2}{R_{12}} i(s)u_L, \tag{6.41}$$

where u_r, u_L are local, collective components of the nucleon velocity at the fragment surfaces which must be small compared to the average thermal velocity in the neck region. We note that the force given by (6.40) and (6.41) is not isotropic; the radial component of the friction tensor is enhanced by a factor of 2 due to the correlation between radial nucleon flux and the associated radial momentum.

The functions $j(s)$, $i(s)$ can be estimated simply for the special case of touching surfaces ($s = 0$); this yields [Ba 76b]

$$C_r(s = 0) = 2C_t(s = 0) \approx \frac{9\pi}{32}\left(\frac{3}{\pi}\right)^{2/3}\frac{R_1 R_2}{R_{12}}\frac{\hbar}{r_0^3}$$

$$= 1.3 \times 10^{-21}\text{MeV s fm}^{-2} \text{ for } A_1, A_2 = 125, r_0 = 1.07 \text{ fm}. \tag{6.42}$$

Thus again the correct order of magnitude is obtained. We consider this interpretation of the energy loss in deep-inelastic collisions especially satisfying, as it is conceptually simple and provides a natural link between friction and nucleon transfer. There is strong experimental evidence for a close relationship between these phenomena, as has been discussed in Sect. 6.3 (see Figs. 6.24, 6.26, 6.27). It should be noted, however, that a number of other mechanisms have been suggested in the literature as responsible for the energy loss in deep-inelastic collisions. These include the excitation of particle–hole states by the time-dependent single-particle potential [Gr 75, Ay 76a,b, Ho 76a] and the coupling of the radial motion to giant multipole excitations [Br 74, Br 76b]. An important—and still open—question, in this context, is whether the energy lost from the relative motion is promptly dissipated in particle degrees of freedom, or temporarily stored in some collective mode to be thermalized during a later stage of the collision. We have emphasized here the former aspect, and have shown that simple statistical considerations based on nucleon transfer can explain at least an important fraction of the observed energy loss. This conclusion is supported by several recent analyses of experimental data [Oe 78, Sv 78, Br 87b, Sc 78]; further quantitative details are given later in this section.

The complete function $i(s)$ [see (6.39–41)], and hence the dependence of the coefficients of transfer-induced friction on surface separation, has been calculated by Randrup in the Thomas–Fermi approximation [Ra 78a]. The results are available in tabulated and analytical form, and are usually referred to by the term "proximity friction". A remarkable feature of these calculations is that they provide a parameter-free prediction of friction forces. Thus systematic comparisons with experimental data should provide a meaningful test of the model assumptions.

Another class of models—appropriately termed "diffusion models"—concentrates on the statistical aspects of transport phenomena which arise from the interaction of two nuclei in close contact. For reasons of simplicity, detailed considerations of trajectory motion and the resulting time dependence of the interaction are usually avoided in these models. In addition, "memory effects" reflecting the previous history of the system are neglected (Markoff approximation). Thus one arrives at a set of "master equations" for the probability distribution $P(x,t)$ as a function of the time t and variables x which describe characteristic macroscopic properties like mass or charge partition between the fragments. The conditions under which master equations can be derived from a microscopic, quantum-mechanical formulation of the problem have been discussed extensively by Nörenberg, and we refer the reader to his work for further details [Nö 75, Nö 76b].

In the following we consider the special case of a single (discrete) variable x, as has been done in most of the practical applications of diffusion models published so far; the extension of the formalism to more than one variable is straightforward in principle, but tedious in practice. The master equations can then be written as [Nö 74]

$$\frac{\partial}{\partial t} P(x, t) = \sum_{x'} w(x, x', t) [\rho(x)P(x', t) - \rho(x')P(x, t)], \tag{6.43}$$

where the summation extends over all possible (discrete) values of the variable x'. The function $\rho(x)$ gives the statistical weight for each value of x, as measured by the corresponding number of microscopic states, and $w(x,x',t) = w(x',x,t)$ is the microscopic transition probability per unit time between states in the subsets x and x'.

In most of the practical applications, the transition probabilities $w(x,x',t)$ are non-zero only for values of x' in the vicinity of x. The set of coupled equations (6.43) can then be simplified by treating x,x' as continuous variables and expanding the functions ρ and P with respect to x' around $x' = x$. Retaining terms up to second order in $(x' - x)$ yields the Fokker–Planck equation [Nö 74]:

$$\frac{\partial}{\partial t} P(x, t) = -\frac{\partial}{\partial x}[v(x, t)P(x,t)] + \frac{\partial^2}{\partial x^2}[D(x, t)P(x, t)]. \tag{6.44}$$

Thus we have a differential equation describing the evolution of $P(x,t)$ in terms of the "drift coefficient"

$$v(x, t) = 2\mu_2(x, t)\frac{\partial}{\partial x}\rho(x, t) + \rho(x, t)\frac{\partial}{\partial x}\mu_2(x, t), \tag{6.45}$$

and the "diffusion coefficient"

$$D(x, t) = \mu_2(x, t)\rho(x, t), \tag{6.46}$$

with

$$\mu_2(x, t) = \frac{1}{2} \int_{-\infty}^{+\infty} w(x, x', t)(x' - x)^2 \, dx'. \tag{6.47}$$

These equations are well known in statistical mechanics and thermodynamics where they describe a broad range of diffusion phenomena. As mentioned earlier, applications to nuclear collisions have usually involved the assumption of a quasi-static contact configuration with a definite lifetime τ, which may also be interpreted as an effective interaction time. This means that the functions $\mu_2(x,t)$ and $\rho(x,t)$, and hence also $v(x,t)$ and $D(x,t)$, are assumed to be time independent during an interval of length τ, and zero both before and after that time interval. It is further assumed that the distribution $P(x,\tau)$ can be measured by observing products at a certain centre-of-mass angle $\Theta(\tau)$ or with a certain energy loss $\Delta E(\tau)$ as discussed in Sect. 6.2.

The solution of (6.44) is particularly simple if the coefficients v and D are not only independent of time, but also independent of x. If, in addition, the variable x has a sharp value x_0 for $t = 0$ (i.e. $P(x,0)$ is given by a delta-function in $(x - x_0)$ the solution for $t = \tau$ is

$$P(x, \tau) = \frac{1}{\sqrt{4\pi D\tau}} \exp\left(-\frac{(x - x_0 - v\tau)^2}{4D\tau}\right). \tag{6.48}$$

The average and the variance of x are in this case given by the equations

$$\langle x \rangle = \int_{-\infty}^{+\infty} xP(x, \tau)\,dx = x_0 + v\tau, \tag{6.49}$$

$$\langle (x - \langle x \rangle)^2 \rangle = \int_{-\infty}^{+\infty} (x - x_0 - v\tau)^2 P(x, \tau)\,dx = 2D\tau, \tag{6.50}$$

which clearly exhibit the significance of the coefficients v and D. From (6.45) and (6.46) we deduce, for constant v and D,

$$v = \left(\frac{1}{\rho}\frac{\partial \rho}{\partial x}\right)_{x_0} D \approx -\frac{1}{T}\left(\frac{\partial V_L}{\partial x}\right)_{x_0} D, \tag{6.51}$$

where T is a nuclear temperature, $V_L(x)$ is the sum of the potential and centrifugal energies of the two touching nuclei, and the density of intrinsic levels ρ is approximated by the expression [Mo 75c]

$$\rho(E, x) = \bar{\rho}(E)\exp\left(-\frac{V_L(x)}{T}\right). \tag{6.52}$$

Equation (6.51) defines the relationship between the drift and diffusion coefficients, and may be recognized as the Einstein relation describing the Brownian motion of a particle in thermal equilibrium with a heat bath at temperature T. The drift coefficient v is seen to be proportional to the appropriate potential gradient, or "driving force".

It should be noted that the approximation of constant coefficients v and D restricts the range of interaction times for which physical systems behave according to (6.48–50). In the limit of large times, i.e. when equilibrium has been reached, the probability distribution $P(x,t)$ becomes stationary and proportional to the density of microscopic states ρ. This follows directly from the master equations (6.43).

In the practical application of diffusion models to nuclear collisions, several cases may be distinguished. In one type of analysis, the dependence of the mean value and the variance of the mass or charge distribution on the centre-of-mass angle is studied. The latter is taken as a measure of the interaction time τ using (6.9), and from the resulting linear relationship between mean value or variance and τ [see (6.49), (6.50)] the corresponding drift coefficient or diffusion coefficient is deduced. A well-known example is the analysis by Nörenberg of the Dubna results for $^{40}Ar + ^{232}Th$ which is illustrated in Figs. 6.32 and 6.33. Similar analyses have been performed for a

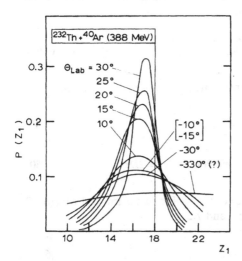

Fig. 6. 32. Normalized distributions of product charges for the system $^{40}Ar + ^{232}Th$ and different laboratory angles. (From Nö 74; experimental data from Ar 73b)

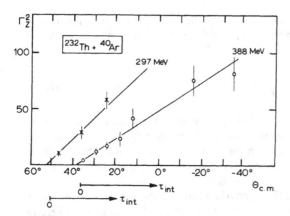

Fig. 6.33.   Square of the width (FWHM) of the charge distribution as a function of centre-of-mass angle for the system $^{40}Ar + ^{232}Th$ and two bombarding energies. The arrows below the figure show the origin and direction of the interaction time scale for each energy. (From Nö 74; experimental data from Ar 73b)

number of other systems, involving projectiles from ^{16}O to ^{86}Kr [Nö 76b, Wo 76]. The method works best where the scattering angle varies strongly over a limited range of angular momenta (i.e. for the class I systems of Sect. 6.2). Obvious limitations of the method arise from the schematic treatment of the time dependence of the interaction, and from uncertainties in the assumed relationship between (effective) interaction time and scattering angle.

For class II systems, where the projectile-like products are strongly focussed into a narrow range of angles, other methods of analysis have been devised. In general one starts by deducing an empirical deflection function $\Theta(L)$ from the experimental reaction cross section as a function of energy loss [Sc 77a, Sc 78] or scattering angle [Wo 78a]. In a second step, angular-momentum-dependent interaction times $\tau(L)$ are calculated using a schematic scattering model as discussed in Sect. 6.2. Finally, drift and diffusion coefficients are determined by assuming the validity of (6.48) and fitting experimental charge distributions integrated over all angles. The latter may be taken either for a specified energy loss [and hence L; Sc 77a, Sc 78] or integrated over final energies [Wo 78a].

Explicit reference to a time scale may be avoided by correlating directly the variance with the mean value of a mass or charge distribution. A linear relationship, independent of time, is expected between these quantities on the basis of (6.49) and (6.50). The corresponding slope can be used to calculate the ratio of the drift coefficient to the diffusion coefficient, which may then be compared to the ratio predicted theoretically with the help of (6.51). This approach was applied by Ngô et al. to experimental mass distributions for the system ^{63}Cu $+$ ^{197}Au at two different bombarding energies, and good agreement between experiment and theoretical expectation was obtained [Ng 76].

An extended version of the diffusion model has been developed by Moretto and Sventek, and applied to the calculation of angle-integrated Z distributions and of angular distributions for products of specified Z [Mo 75c, Sv 76, Mo 76b]. In a first step, these authors take the master equations (6.43) to compute time-dependent populations $P(Z,t)$, assuming a constant strength of interaction (note that Z here refers to one of the fragments). The following ansatz is used for the transition probabilities:

$$w(Z, Z', t) = \frac{\kappa f}{[\rho(Z)\rho(Z')]^{1/2}} \qquad (Z' = Z \pm 1) , \tag{6.53}$$

where κ is an adjustable constant (of the order of 10^{20} s^{-1} fm^{-2}) and

$$f = 2\pi d \frac{R_1 R_2}{R_{12}} \qquad (d = 1.0 \text{ fm}) \tag{6.54}$$

is the area of the "window" between the touching fragments [compare (6.39)]. The diffusion and drift coefficients of the equivalent Fokker–Planck equation are given approximately by

$$D_Z = \frac{1}{3}\kappa f \; ; \quad v_Z = -\frac{1}{3}\kappa f \frac{1}{T}\frac{\partial V_L}{\partial Z} . \tag{6.55}$$

The evolution of the Z distribution $P(Z,t)$ is governed by the properties of the "ridge line potential" $V_L(Z)$ in the vicinity of the "injection point" (see our qualitative discussion of this point in Sect. 6.3, and Mo 76b). This is demonstrated in Fig. 6.34 which shows results of calculations by Sventek and Moretto for the system ^{86}Kr + ^{197}Au [Sv 76]. In Fig. 6.34 b the initially asymmetric Z distribution is seen to broaden and to evolve towards symmetry within a few times 10^{-21}s. It is interesting to note that apparently a small fraction of the total intensity also passes over the potential energy maximum to end up in compound-nucleus formation.

In a second step, Moretto and Sventek calculate angle-integrated Z distributions from the equation

$$\frac{d\sigma}{dZ} = 2\pi\lambda^2 \int_{\tau=0}^{\infty} d\tau \int_{L=0}^{\infty} \Pi(L,\tau)P_L(Z,\tau)R(L)L \, dL , \tag{6.56}$$

where $\Pi(L,\tau)$ is the probability distribution of interaction times τ associated with initial angular momentum L, and $R(L)$ is the probability for re-emission of two fragments after a deep-inelastic collision. The latter quantity is taken as unity in a certain angular momentum window, determined by the total cross section for deep-inelastic scattering and fusion, and zero outside.

Finally, in the same spirit, angular distributions for products of specified Z are obtained from

$$\frac{\partial^2\sigma}{\partial\Omega\partial Z} = \lambda^2 \int_{\tau=0}^{\infty} d\tau \sum_L \frac{\Pi(L,\tau)P_L(Z,\tau)R(L)L}{\sin\Theta \frac{\partial}{\partial L}\Theta(L,\tau,Z)} , \tag{6.57}$$

where the sum under the integral sign is extended over those (discrete, but in general noninteger) values of L which produce scattering to the angle Θ for given values of τ and Z. The deflection function $\Theta(L,\tau,Z)$ is approximated by a sum of contributions due to Coulomb deflection on the ingoing and outgoing parts of the trajectory [see (6.12)] and a contribution due to rigid rotation of the intermediate complex (with appropriately averaged moment of inertia) during its lifetime τ. Using (6.56) and (6.57) with simple analytic expressions for the probability distribution $\Pi(L,\tau)$, Moretto and Sventek were able to reproduce semiquantitatively a large body of experimental data on Z distributions and angular distributions in deep-inelastic scattering [Mo 75c, Sv 76, Mo 76b].

We now turn to a discussion of the magnitudes of diffusion coefficients which have been deduced either from analyses of experimental data or from theoretical considerations. In general, mass diffusion coefficients have been derived from measured charge diffusion coefficients assuming the following relationship [Nö 76b]:

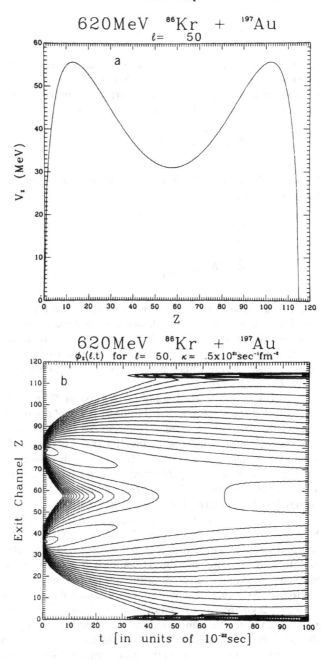

Fig. 6.34.   a Potential energy as a function of Z of one fragment, for the intermediate complex formed by ^{86}Kr bombardment of ^{197}Au at $E_{Lab} = 620$ MeV, $L = 50$; **b** Corresponding contour plot of probability distribution $P(Z, t)$. (From Sv 76)

$$D_A = \left(\frac{\bar{A}}{\bar{Z}}\right)^2 D_Z,$$ (6.58)

where $\bar{A}$ and $\bar{Z}$ are the mass and charge number, respectively, of the compound system. It should be noted that (6.58) refers to strongly correlated diffusion of neutrons and protons, corresponding to a frozen ratio Z/A. In this case the variances of Z, N, and A are related by

$$\sigma_Z^2 + \sigma_N^2 = \frac{\bar{Z}^2 + \bar{N}^2}{\bar{A}^2}\sigma_A^2 < \sigma_A^2.$$ (6.59)

A completely independent diffusion of protons and neutrons, on the other hand, would imply

$$D_A = \left(\frac{\bar{A}}{\bar{Z}}\right) D_Z,$$ (6.60)

$$\sigma_Z^2 + \sigma_N^2 = \frac{\bar{Z} + \bar{N}}{\bar{A}}\sigma_A^2 = \sigma_A^2.$$ (6.61)

As discussed in Sect. 6.3, there is ample evidence that the ratio Z/A reaches its equilibrium distribution much faster than the mass asymmetry degree of freedom. Therefore (6.58) and (6.59) are expected to be more realistic in general than (6.60) and (6.61). This view is supported by experimental data for systems where both mass and charge distributions have been measured [Ja 75, Ga 77, Ba 77].

A recent, more detailed study of the system ^{40}Ar + ^{208}Pb has shown, however, that the combined distribution in A and Z approaches the situation described by (6.59) with increasing energy loss, but is better represented by (6.61) in the limit of small energy loss [Ba 78c]. This result clearly reflects the finite equilibrium width of the distribution with respect to Z/N, and the different time scales of equilibration for the charge-to-mass ratio and the mass partition, respectively. It shows, moreover, that analyses of experimental data based on (6.58) may be subject to error.

Equation (6.59) illustrates that the distribution of a variable (here Z or N) may be narrower than expected from the total number of independent diffusion steps (as reflected by σ_A^2) in the absence of correlations. Conversely, the total number of nucleon jumps would be underestimated if the correlation between proton and neutron diffusion was not taken into account [as in (6.60) and (6.61)]. As pointed out by Randrup, additional correlations—with similar consequences—are due to "Pauli blocking" which arises from the fact that nucleons can only be transferred into unoccupied states [Ra 78b]. Hence, if all states were occupied, transfers would have to occur simultaneously in opposite directions, and would thus not contribute to the widths of observable distributions. Quite generally, limitations in the available phase space will induce correlations among the dynamic variables. Consequently the total number of diffusion steps will be larger than estimated from observed distributions by applying simple random-walk arguments. This effect has been largely disregarded in the analysis of experimental data so far.

Absolute values of mass or charge diffusion coefficients have been deduced from experimental data for argon-, krypton-, and xenon-induced reactions by several authors [Nö 74, Nö 76b, Mo 76b, Wo 76, Sc 78b, Wo 78a]. The results vary somewhat with the particular sets of data considered and the method of analysis; typical values are about 1×10^{22} s^{-1} for D_Z, and correspondingly larger for D_A [see (6.58)]. In principle, the more refined methods yield angular-momentum-dependent diffusion coefficients [Sc 78, Wo 78a]; in practice, however, this dependence was found to be weak over large ranges of angular momentum.

Nörenberg and collaborators have performed microscopic calculations of transport coefficients in a statistical single-particle model. The resulting mass diffusion coefficients, which contain one adjustable parameter, agree within about a factor of two with "experimental values" for many different systems [Nö 76b, Ay 76a,b, Wo 78a]. A simple estimate of mass diffusion coefficients—without adjustable parameters—may also be derived for nuclei touching at their half-density radii from the "window model". This model forms the basis of the diffusion model of Moretto and Sventek, and has also been discussed here in the context of energy dissipation. From simple statistical considerations, and neglecting correlation effects such as "Pauli blocking", we can relate the mass diffusion coefficient to the total nucleon flux $I(s)$ between two nuclei at separation s [see (6.39)]:

$$D_A(s) = \frac{1}{2} I(s) = 2\pi \frac{R_1 R_2}{R_{12}} i(s) . \tag{6.62}$$

For $s = 0$ this leads to [see (6.42)]:

$$D_A(0) \approx \frac{9\pi}{64}\left(\frac{3}{\pi}\right)^{2/3} \frac{R_1 R_2}{R_{12}} \frac{\hbar}{mr_0^3} . \tag{6.63}$$

Equation (6.63) yields $D_A(0) \approx 3 \times 10^{22} - 6 \times 10^{22}$ s^{-1} in agreement with empirical values.

A comparison of (6.40), (6.41), and (6.62) suggests that for transfer-induced friction, the friction and diffusion coefficients should be connected by the following simple relationship, independently of the radial separation s:

$$\frac{C_r}{D_A} = \frac{2C_L}{D_A} = 2m . \tag{6.64}$$

This is precisely the kind of relationship needed to explain the correlation between energy loss and charge distribution observed in krypton- and xenon--induced reactions [Hu 76; see Sect. 6.3 and Fig. 6.27]. A quantitative comparison with the available experimental data indicates that actual ratios of energy loss to mass dispersion are systematically somewhat larger (typically by factors of 2–5) than implied by (6.64). This could be partly due to the neglect of correlation effects in (6.64), which would tend to increase the number of nucleon transfers and hence the total energy loss for a given mass dispersion, and partly due to the presence of energy loss mechanisms other than nucleon transfer. It seems clear, nevertheless, that the diffusive motion of nucleons across the neck is responsible for a major part of the energy loss.

The diffusion models, discussed here in the context of mass and charge transport, have been generalized to include the relative motion as a diffusive process in configuration space [Ho 76a,b, Nö 76b]. The result is a spreading of the probability distribution in phase space superimposed on a drift given by the classical equations of motion. Among other consequences, this leads to the disappearance of the unphysical divergence of differential cross sections at rainbow angles [Ho 76b]. Since the corresponding drift and diffusion coefficients can in general not be approximated by constants, the theory becomes mathematically rather complex.

The dissipation of angular momentum in heavy-ion collisions was described in the framework of transport theory by Wolschin and Nörenberg [Wo 78b]. It was shown that appreciable angular momenta are generated in the fragments due to statistical fluctuations, especially for the lower partial waves which are associated with large energy losses and long interaction times. This effect can explain the plateau in the measured gamma multiplicities in the region of large energy loss (see Fig. 6.28). Figure 6.35 shows the calculated dependence of the total fragment spin (integrated over energy loss) on the charge of the projectile-like fragment for the system $^{86}Kr + {}^{166}Er$ at 5.99 MeV per nucleon. The full curve, including fluctuations, is seen to reproduce the experimental spin values deduced from gamma multiplicity data of Olmi et al. [Ol 78; see also Fig. 6.29]. The conspicuous dip at $Z = 36$ is due to the fact that Z values near the projectile charge are populated preferentially in peripheral collisions which correspond to small interaction times.

To summarize our discussion of simple models of deep-inelastic scattering,

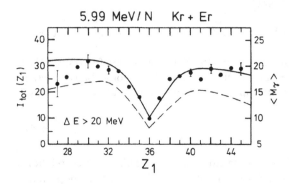

Fig. 6.35. Energy-integrated total intrinsic angular momentum I_{tot} (left scale) and corresponding gamma multiplicity $\langle M_\gamma \rangle$ (right scale) as a function of the atomic number of the light fragment in Kr + Er scattering. The full and broken curves represent theoretical values calculated including and neglecting fluctuations, respectively, and the points are experimental values from Ol 78. (From Wo 78)

we can state that macroscopic and statistical concepts have been successfully applied to nuclear systems comprising no more than about 100 particles. The gross features of the data have been explained as due to the action of (phenomenological) conservative and dissipative forces, combined with mass and charge transfer by diffusion of nucleons across the neck. Last, but not least, the analysis of correlations between mass transfer and energy loss has begun to provide insight into the microscopic properties of the dissipative forces. The next step should now be the development and application of a unified model which simultaneously describes the collective motion in configuration space (taking all relevant degrees of freedom into account) and microscopic transport phenomena. The phenomenological coefficients entering such a model could then be determined from an analysis of experimental data, for comparison with the predictions of more fundamental microscopic theories.

References Chapter 6

(For CA 76, HE 69, HE 74, MU 73, NA 74 see Appendix D)

Ag 77a Agarwal, S. et al.: Nucl. Phys. *A293*, 230 (1977)
Ag 77b Agassi, D.; Ko, C.M.; Weidenmüller, H.A.: Ann. Phys. (N.Y.) *107*, 140 (1977)
Ag 78 Agassi, D.; Ko, C.M.; Weidenmüller H.A.: Phys. Rev. *C18*, 223 (1978)
Al 77 Albrecht, K.; Stocker, W.: Nucl. Phys. *A278*, 95 (1977)
Al 78 Aleonard, M.M. et al.: Phys. Rev. Lett. *40*, 622 (1978)
Ar 69a Artukh, A.G. et al.: Nucl. Phys. *A137*, 348 (1969)
Ar 69b Artukh, A.G. et al.: HE 69 p. 140
Ar 70 Artukh, A.G. et al.: Phys. Lett. *32B*, 43 (1970)
Ar 71a Artukh, A.G. et al.: Nucl. Phys. *A160*, 511 (1971)
Ar 71b Artukh, A.G. et al.: Nucl. Phys. *A168*, 321 (1971)
Ar 71c Artukh, A.G. et al.: Nucl. Phys. *A176*, 284 (1971)
Ar 73a Artukh, A.G. et al.: Nucl. Phys. *A211*, 299 (1973)
Ar 73b Artukh, A.G. et al.: Nucl. Phys. *A215*, 91 (1973)
Ar 76 Artukh, A.G. et al.: Dubna Report E7-9974 (1976)
Ay 76a Ayik, S.; Schürmann, B.; Nörenberg, W.: Z. Phys. *A277*, 299 (1976)
Ay 76b Ayik, S.; Schürmann, B.; Nörenberg, W.: Z. Phys. *A279*, 145 (1976)
Ay 78 Ayik, S.; Schürmann, B.; Nörenberg, W.: Z. Phys. *A286*, 271 (1978)
Ba 74 Bass, R.: Nucl. Phys. *A231*, 45 (1974)
Ba 75a Barette, J. et al.: Z. Phys. *A274*, 121 (1975)
Ba 75b Bass, R.; v. Czarnecki, J.; Zitzmann, R.: Nucl. Instr. Methods *130*, 125 (1975)
Ba 76a Babinet, R. et al.: Nucl. Phys. *A258*, 172 (1976)
Ba 76b Bass, R.: CA 76 p. 147
Ba 77 Barette, J. et al.: GSI-Report J-1-77, p. 50 (1977)
Ba 78a Babinet, R. et al.: Nucl. Phys. *A296*, 160 (1978)
Ba 78b Back, B.B.; Bjornholm, S.: Nucl. Phys. *A302*, 343 (1978)
Ba 78c Bass, R. et al.: GSI-Report J-1-78, p. 28 (1978)
Bi 75 Bimbot, R. et al.: Nucl. Phys. *A248*, 377 (1975)
Bi 77 Van Bibber, K. et al.: Phys. Rev. Lett. *38*, 334 (1977)
Bl 77 Blocki, J. et al.: Ann. Phys. (N.Y.) *105*, 427 (1977)
Bl 78 Blocki, J. et al.: Ann. Phys. (N.Y.) *113*, 330 (1978)
Bo 71 Bondorf, J.P. et al.: J. Phys (Paris) *32*, C6-145 (1971)
Bo 73 Bondorf, J.P.; Nörenberg, W.: Phys. Lett. *44B*, 487 (1973)
Bo 74a Bohr, A.; Mottelson, B.R.: Phys. Scr. *10A*, 13 (1974)
Bo 74b Bondorf, J.P.; Sobel, M.I.; Sperber, D.: Phys. Rep. *15*, 83 (1974)
Bo 75 Bondorf, J.P. et al.: Phys. Rev. *C11*, 1265 (1975)
Bo 76a Bonche, P.; Koonin, S.E.; Negele, J.W.: Phys. Rev. *C13*, 1226 (1976)
Bo 76b Bonche, P.: J. Phys. *37*, C5-213 (1976)
Br 61 Britt, H.C.; Quinton, A.R.: Phys. Rev. *124*, 877 (1961)

Br 74   Broglia, R.; Dasso, C.; Winther, A.: Phys. Lett. *53B*, 301 (1974)
Br 76a Braun-Munzinger, P. et al.: Phys. Rev. Lett. *36*, 849 (1976)
Br 76b Broglia, R.A.; Dasso, C.H.; Winther, A.: Phys. Lett. *61B*, 113 (1976)
Br 78a Broglia, R.A. et al.: Phys. Lett. *73B*, 405 (1978)
Br 78b Broglia, R.A. et al.: Phys. Rev. Lett. *41*, 25 (1978)
Ca 78   Cauvin, B. et al.: Nucl. Phys. *A301*, 511 (1978)
Ch 78   Christensen, P.R. et al.: Phys. Rev. Lett. *40*, 1245 (1978)
Co 76   Cormier, T.M. et al.: Phys. Rev. *C13*, 682 (1976)
Cu 76a Cusson, R.Y.; Maruhn, J.: Phys. Lett. *62B*, 134 (1976)
Cu 76b Cusson, R.Y.; Smith, R.K.; Maruhn, J.: Phys. Rev. Lett. *36*, 1166 (1976)
Da 74   Davis, R.H.: Phys. Rev. *C9*, 2411 (1974)
Da 76   Davies, K.T.R.; Sierk, A.J.; Nix, J.R.: Phys. Rev. *C13*, 2385 (1976)
De 75   Deubler, H.H.; Dietrich, K.: Phys. Lett. *56B*, 241 (1975)
De 76a De, J.N.; Gross, D.H.E.; Kalinowski, H.: Z. Phys. *A227*, 385 (1976)
De 77   Deubler, H.H.; Dietrich, K.: Nucl. Phys. *A277*, 493 (1977)
Dy 77   Dyer, P. et al.: Phys. Rev. Lett. *39*, 392 (1977)
Eb 71   Eberhard, K.A.; Richter, A.: In *Proc. International Conference on Statistical Properties of Nuclei*, Albany 1971, ed. by J. Garg. New York: Plenum Press.
Eg 76   Eggers, R. et al.: Phys. Rev. Lett. *37*, 324 (1976)
Ey 73   Eyal, Y. et al.: Phys. Rev. *C8*, 1109 (1973)
Ey 78   Eyal, Y. et al.: Phys. Rev. Lett. *41*, 625 (1978)
Fi 74   Fink, H.J. et al.: NA 74 Vol.2 p. 21
Ga 70   Galin, J. et al.: Nucl. Phys. *A159*, 461 (1970)
Ga 75a Galin, J. et al.: Nucl. Phys. *A255*, 472 (1975)
Ga 75b Gatty, B. et al.: Nucl. Phys. *A253*, 511 (1975)
Ga 75c Gatty, B. et al.: Z. Phys. *A273*, 65 (1975)
Ga 76a Galin, J. et al.: Z. Phys. *A278*, 347 (1976)
Ga 76b Galin, J.: J. Phys. *37*, C5-83 (1976)
Ga 77   Galin, J. et al.: Z. Phys. *A283*, 173 (1977)
Ga 78   Gamp, A. et al.: Phys. Lett. *74B*, 215 (1978)
Ge 77   Gelbke, C.K. et al.: Phys. Lett. *71B*, 83 (1977)
Gl 77   Glässel, P. et al.: Phys. Rev. Lett. *38*, 331 (1977)
Go 78   Gould, C.R. et al.: Z. Phys. *A284*, 353 (1978)
Gr 70   Gridnev, G.F.; Volkov, V.V.; Wilczynski, J.: Nucl. Phys. *A142*, 385 (1970)
Gr 73   Gross, D.H.E.; Kalinowski, H.; Beck, R.: MU 73 Vol.1 p. 394
Gr 74a Gross, D.H.E.; Kalinowski, H.: Phys. Lett. *48B*, 302 (1974)
Gr 74b Gross, D.H.E.; Kalinowski, H.; De, J.N.: HE 74 p. 194
Gr 75   Gross, D.H.E.: Nucl. Phys. *A240*, 472 (1975)
Gr 78   Gross, D.H.E.; Kalinowski, H.: Phys. Rep. *45*, 175 (1978)
Gu 77   Gupta, R.K.; Sandulescu, A.; Greiner, W.: Phys. Lett. *67B*, 257 (1977)
Ha 74   Hanappe, F. et al.: Phys. Rev. Lett. *32*, 738 (1974)
Ha 77   Harris, J.W. et al.: Phys. Rev. Lett. *38*, 1460 (1977)
Hi 77   Hildenbrand, K.D. et al.: Phys. Rev. Lett. *39*, 1065 (1977)
Ho 76a Hofmann, H.; Siemens, P.J.: Nucl. Phys. *A257*, 165 (1976)
Ho 76b Hofmann, H. Ngô, C.: Phys. Lett. *65B*, 97 (1976)
Ho 77a Ho, H. et al.: Z. Phys. *A283*, 235 (1977)
Ho 77b Hofmann, H.; Siemens, P.J.: Nucl. Phys. *A275*, 464 (1977)
Hu 75   Huizenga, J.R.: Nukleonika *20*, 291 (1975)
Hu 76   Huizenga, J.R. et al.: Phys. Rev. Lett. *37*, 885 (1976)
Ja 75   Jacmart, J.C. et al.: Nucl. Phys. *A242*, 175 (1975)
Ka 59   Kaufmann, R.; Wolfgang, R.: Phys. Rev. Lett. *3*, 232 (1959)
Ka 61   Kaufmann, R. Wolfgang, R.: Phys. Rev. *121*, 192 (1961)
Ko 76   Ko, C.M.; Pirner, H.J.; Weidenmüller, H.A.: Phys. Lett. *62B*, 248 (1976)
Ko 77   Koonin, S.E. et al.: Phys. Rev. *C15*, 1539 (1977)
Kr 74   Kratz, J.V.; Norris, A.E.; Seaborg, G.T.: Phys. Rev. Lett. *33*, 502 (1974)
Kr 76   Kratz, J.V. et al.: Phys. Rev. *C3*, 2347 (1976)
Kr 77   Kratz, J.V. et al.: Phys. Rev. Lett. *39*, 984 (1977)
Le 69   Lefort, M. et al.: HE 69 p. 181
Le 73   Lefort, M. et al.: Nucl. Phys. *A216*, 166 (1973)
Le 75   Lefort, M. Orsay Report IPNO-RC-7507 (1975)
Le 76   Lefort, M.: Rep. Prog. Phys. *39*, 129 (1976)

Le 78 Lefort, M.; Ngô, C.: Ann. Phys. (Paris) *3*, 5 (1978)
Ma 77 Maruhn-Rezwani, V.; Davis, K.T.R.; Koonin, S.E.: Phy. Lett. *67B*, 134 (1977)
Mi 78 Miller, J.M. et al.: Phys. Rev. Lett. *40*, 100 (1978)
Mo 75a Moretto, L.G. et al.: Nucl. Phys. *A255*, 491 (1975)
Mo 75b Moretto, L.G. et al.: Phys. Lett. *B58*, 31 (1975)
Mo 75c Moretto, L.G.; Sventek, J.S.: Phys. Lett. *B58*, 26 (1975)
Mo 76a Moretto, L.G. et al.: Nucl. Phys. *A259*, 173 (1976)
Mo 76b Moretto, L.G.; Schmitt, R.: J. Phys. (Paris) *37*, C5-109 (1976)
Mo 76c Moretto, L.G. et al.: Phys. Rev. Lett. *36*, 1069 (1976)
Na 76 Natowitz, J.B.; Namboodiri, M.N.; Chulik, E.T.: Phys. Rev. *C13*, 171 (1976)
Na 77 Natowitz, J.B. et al.: Nucl. Phys. *A277*, 477 (1977)
Na 78 Natowitz, J.B. et al.: Phys. Rev. Lett. *40*, 751 (1978)
Ng 76 Ngô, C. et al.: Nucl. Phys. *A267*, 181 (1976)
Ng 77 Ngô, C. et al.: Z. Phys. *A283*, 161 (1977)
Ni 69 Nix, J.R.: Nucl. Phys. *A130*, 241 (1969)
Ni 74 Nix, J.R.; Sierk, A.J.: Phys. Scr. *10A*, 94 (1974)
Nö 74 Nörenberg, W.: Phys. Lett. *52B*, 289 (1974)
Nö 75 Nörenberg, W.: Z. Phys. *A274*, 241 (1975)
Nö 76a Nörenberg, W.: Z. Phys. *A276*, 84 (1976)
Nö 76b Nörenberg, W.: J. Phys. (Paris) *37*, C5-141 (1976)
Oe 78 Oeschler, B. et al.: Phys. Rev. *C18*, 239 (1978)
Ol 78 Olmi, A. et al.: Phys. Rev. Lett. *41*, 688 (1978)
Ot 76 Otto, R.J. et al.: Phys. Rev. Lett. *36*, 135 (1976)
Ot 78 Otto, R.J. et al.: Z. Phys. *A287*, 97 (1978)
Pe 73 Péter, J. et al.: MU 73 p. 611
Pe 75 Péter, J.; Ngô, C.; Tamain, B.: Nucl. Phys. *A250*, 351 (1975)
Pe 76 Péter, J.: Orsay Report IPNO-RC-7602 (1976)
Pe 77a Péter, J. et al.: Nucl. Phys. *A279*, 110 (1977)
Pe 77b Péter, J. et al.: Z. Phys. *A283*, 413 (1977)
Pl 78 Plasil, F. et al.: Phys. Rev. Lett. *40*, 1164 (1978)
Ra 78a Randrup, J.: Ann. Phys. (N.Y.) *112*, 356 (1978)
Ra 78b Randrup, J.: Nucl. Phys. *A307*, 319 (1978)
Re 62 Read, J.B.J.; Ladenbauer-Bellis, I.-B.; Wolfgang, R.: Phys. Rev. *127*, 1722 (1962)
Ro 77 Roynette, J.C. et al.: Phys. Lett. *67B*, 395 (1977)
Ru 77 Russo, P. et al.: Phys. Lett. *67B*, 155 (1977)
Sa 77 Sann, H. et al.: GSI-Report P-5-77 (1977)
Sc 76 Schröder, W.U. et al.: Phys. Rev. Lett. *36*, 514 (1976)
Sc 77a Schröder W.U. et al.: Phys. Rev. *C 16*, 623 (1977)
Sc 77b Schröder, W.U. Huizenga, J.R.: Ann. Rev. Nucl. Sci. *27*, 465 (1977)
Sc 78a Schädel, M. et al.: Phys. Rev. Lett. *41*, 469 (1978)
Sc 78b Schroeder, W.U. et al.: Phys. Rep. *45*, 301 (1978)
Sc 78c Schürmann, B.; Nörenberg, W.; Simbel, M.: Z. Phys. *A286*, 263 (1978)
Se 75 Seglie, E.; Sperber, D.; Sherman, A.: Phys. Rev. *C11*, 1227 (1975)
Si 76 Siwek-Wilczynska, K.; Wilczynski, J.: Nucl. Phys. *A264*, 115 (1976)
Sp 74 Sperber, D.: Phys. Scr. *10A*, 115 (1974)
St 67 Strudler, P.M.; Preis, I.L.; Wolfgang, R.: Phys. Rev. *154*, 1126 (1967)
Sv 76 Sventek, J.S.; Moretto, L.G.: Phys. Lett. *65B*, 326 (1976)
Sv 78 Sventek, J.S.; Moretto, L.G.: Phys. Rev. Lett. *40*, 697 (1978)
Sw 72a Swiatecki, W.J.: J. Phys. (Paris) *33*, C5-45 (1972)
Sw 72b Swiatecki, W.J.; Björnholm, S.: Phys. Rep. *4*, 325 (1972)
Ta 75 Tamain B. et al.: Nucl. Phys. *A252*, 187 (1975)
Ta 76 Tamain B. et al.: Phys. Rev. Lett. *36*, 18 (1976)
Tr 77 Trautmann, W. et al.: Phys. Rev. Lett. *39*, 1062 (1977)
Ts 74 Tsang, C.F.: Phys. Scr. *10A*, 90 (1974)
Va 76a Vandenbosch, R.; Webb, M.P.; Thomas, T.D.: Phys. Rev. Lett. *36*, 459 (1976)
        Phys. Rev. *C14*, 143 (1976)
Va 76b Vandenbosch, R. et al.: Nucl. Phys. *A269*, 210 (1976)
Vo 69 Volkov, V.V. et al.: Nucl. Phys. *A126*, 1 (1969)
Vo 74a Volkov, V.V.: NA 74 p. 363
Vo 74b Volkov, V.V.: HE 74 p. 253
We 78 Weiner, R.; Weström, M.: Nucl. Phys. *A286*, 282 (1977)

Wi 67   Wilczynski, J.; Volkov, V.V.; Decowski, P.: J. Nucl. Phys. (USSR) 5, 942 (1967)
Wi 73   Wilczynski, J.: Phys. Lett. 47B, 484 (1973)
Wo 74   Wolf, K.L. et al.: Phys. Rev. Lett. 33, 1105 (1974)
Wo 75   Wong, C.Y.; Maruhn, J.A.; Welton, T.A.: Nucl. Phys. A253, 469 (1975)
Wo 76   Wolf, K.L. et al.: CA 76 p. 176
Wo 78a Wolschin, G.; Nörenberg, W.: Z. Phys. A284, 209 (1978)
Wo 78b Wolschin, G.; Nörenberg, W.: Phys. Rev. Lett. 41, 691 (1978)
Ya 76   Yamaji, S. et al.: Z. Phys. A278, 69 (1976)
Zo 75   Zohni, O. et al.: Z. Phys. A275, 235 (1975)

7. Complete Fusion

7.1 General Considerations

The term "complete fusion" and its physical implications have been discussed extensively in the literature. From an experimental point of view one would like to have observable criteria by which complete fusion may be recognized. However, as discussed in Chap. 6, there is no unambiguous way of infering a sequence of intermediate states from observable features. This leads us to adopt a schematic definition of complete fusion which may be expressed symbolically by writing.

$$(A_1, Z_1) + (A_2, Z_2) \rightarrow (A_1+A_2, Z_1+Z_2)_{E_{ex},J} . \tag{7.1}$$

Here the right-hand side stands for a state of the system which is completely characterized by its total mass, charge, energy, and angular momentum and has reached equilibrium with respect to all other (internal) degrees of freedom. We note that this definition of complete fusion is synonymous with the classical concept of compound-nucleus formation as introduced by Niels Bohr. It clearly refers to an idealized situation which is never fully realized in practice due to the finite lifetime of the compound nucleus, especially at high excitation energies.

A number of more or less direct criteria may be used to assign complete fusion in the sense defined above. The most convincing evidence is provided by the observation of reaction products with masses near that of the compound system ("evaporation residues"). Weaker arguments rely on symmetry properties of angular distributions, or on the independence of decay properties on the entrance channel at a given excitation energy and angular momentum. A more detailed discussion of these points may be found in Sect. 6.1.

The possibility of inducing complete fusion of heavy nuclei has been a particularly fascinating aspect of heavy-ion research for many years. Large-scale efforts in the field have been motivated by the desire to synthesize new nuclear species beyond the known regions of stable or radioactive nuclides. Predictions of an "island of stability" near $A = 300$, $Z = 114$—where shell effects are believed to stabilize the nuclear ground states against decay by spontaneous fission—have especially contributed to this activity. We shall briefly return to the prospects and problems of superheavy production by fusion reactions in Sec. 7.4

Another unique aspect of fusion with heavy projectiles ($A \gtrsim 40$) is associated with the high angular momenta—in excess of 100 $\hbar$—which can be imparted to the compound nucleus. These offer the possibility to study otherwise unaccessable nuclear states in the vicinity of the "γ-rast line", which marks the limit of nuclear stability with respect to angular momentum at a given excitation energy (see Figs. 8.1, 8.3).

Next we consider the energy balance in nuclear fusion. The excitation energy of the compound nucleus, E_{ex}, at a given centre-of-mass bombarding energy E_{CM} can be written as

$$E_{ex} = E_{CM} + Q_{fu} = E_{CM} + c^2(M_1 + M_2 - M_{1+2}), \qquad (7.2)$$

where Q_{fu} represents the gain in binding energy associated with formation of the compound nucleus in its ground state, and M_1, M_2, and M_{1+2} are the ground-state masses of projectile, target, and compound nucleus, respectively. At a given bombarding energy, the Q value for compound-nucleus formation (with excitation energy E_{ex}) is

$$Q = Q_{fu} - E_{ex} = -E_{CM} . \qquad (7.3)$$

This shows that compound-nucleus formation represents the limit of a completely inelastic process, where all of the relative kinetic energy in the incident channel is absorbed.

An important question is what range of compound-nucleus excitation energies can be reached at bombarding energies above the "fusion barrier" B_{fu}. The latter is defined here as the barrier height of the effective two-body potential at zero angular momentum, and should not be confused with the "interaction barrier" B_{int} discussed in Chap. 3 (see Figs. 3.10, 7.1). In Table 7.1, we have collected values of Q_{fu} and corresponding excitation energies at $E_{CM} = B_{fu}$ for a number of representative target–projectile combinations. We note that, except for comparatively light systems with $A_1 + A_2 < 100$, the quantity Q_{fu} assumes negative values, and hence fusion is an endothermic process. The reason is that, in order to form a heavy compound nucleus, more energy must be supplied to overcome the Coulomb repulsion than is gained from nuclear binding forces.

Comparing now the excitation energies reached on top of the fusion barrier ($E_{ex} = Q_{fu} + B_{fu}$), we find values around 40–50 MeV for a considerable range of projectile and compound nucleus masses ($A_1 \lesssim 80$, $A_1 + A_2 < 300$). For these systems it is clearly impossible to produce "cold" compound nuclei with excitation energies significantly below 40 MeV. Consequently the primary compound nuclei will decay either by light-particle evaporation or by fission, and the relative probabilities for these processes will determine the nature of the final reaction products.

For heavier systems the excitation energy at the fusion barrier is reduced,

Table 7.1. Q values ($Q_{fu}{}^a$), potential barriers ($B_{fu}{}^b$), and minimum excitation energies ($E_{ex} = B_{fu} + Q_{fu}$) in complete fusion

Projectile	^{14}N			^{40}Ar			^{86}Kr			^{136}Xe		
Target	Q_{fu} [MeV]	B_{fu} [MeV]	E_{ex} [MeV]	Q_{fu} [MeV]	B_{fu} [MeV]	E_{ex} [MeV]	Q_{fu} [MeV]	B_{fu} [MeV]	E_{ex} [MeV]	Q_{fu} [MeV]	B_{fu} [MeV]	E_{ex} [MeV]
^{12}C	15.1	7.0	22.1	20.4	16.2	36.6	4.8	29.5	34.3	−9.0	41.4	32.4
^{60}Ni	3.6	27.9	31.5	−14.3	65.3	51.0	−71.9	120.3	48.4	−125.9	170.4	44.5
^{120}Sn	−2.9	45.2	42.3	−60.2	106.7	46.5	−165.3	198.1	32.8	−272	282	10
^{172}Yb	−17.2	59.9	42.7	−93.9	142.2	48.3	−250	265	15	−388	379	−8^c
^{208}Pb	−35.4	68.0	32.6	−128.6	161.9	33.3	−299	303	4	−463	434	−29^c
^{238}U	−27.0	74.6	47.6	−133	178	45	−326	333	7	−500	478	−22^c

^{a} Values with decimal point are based on the mass compilation of Wapstra and Gove [Wa 71]; values without decimal point are estimates taken or extrapolated from calculations given in Jo 67, Se 70, Ts 70, Fi 72.

^{b} S-wave potential barrier for spherical nuclei as calculated with the empirical nucleus–nucleus potential of (7.35), (7.50) and (7.51).

^{c} Negative values of E_{ex} imply that compound nucleus formation is energetically not possible at the potential barrier. In this case additional energy equal to $-E_{ex}$ must be supplied to form a (hypothetical) compound nucleus in its ground state.

and eventually the fusion barrier—defined as a maximum of the effective two-body potential—disappears altogether (see Fig. 7.1). This does not imply, however, that very heavy compound nuclei can be formed at zero excitation with appropriately chosen bombarding energies ($E_{CM} \approx -Q_{fu}$). On the contrary, compound-nucleus formation in these heavy systems is severely restricted for dynamical reasons. The experimental evidence concerning this question and its interpretation will be considered in detail in Sect. 7.3 and 7.4.

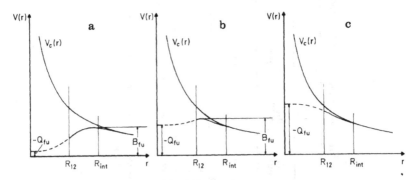

Fig. 7.1. Schematic nucleus–nucleus potentials $V(r)$ for a typical light **(a)**, intermediate **(b)**, and heavy **(c)** system [$V_C(r)$: Coulomb potential, B_{fu}: fusion barrier, Q_{fu}: Q value for complete fusion, R_{int}: interaction distance, R_{12}: half-density distance]

A quantitative analysis of complete fusion cross sections must explicitly specify angular momentum. The classical cross section (for spinless particles) as a function of angular momentum can be written as [see (3.36)]

$$\frac{d\sigma}{dL} = 2\pi \lambdabar^2 L . \tag{7.4}$$

It is intuitively plausible, and consistent with the predictions of simple classical scattering models (see sect .6.4 and 7.4), that complete fusion—if it exists—will be associated wiht the lowest partial waves up to a limiting value L_{fu}. Classically the fusion probability will be unity for $L < L_{fu}$ and zero for $L > L_{fu}$; this "sharp cut-off approximation" yields for the fusion cross section

$$\sigma_{fu} = 2\pi \lambdabar^2 \int_0^{L_{fu}} L dL = \pi \lambdabar^2 L_{fu}^2 , \tag{7.5}$$

and for the average angular momentum $\bar{L}_{cn}$ of the compound nucleus

$$\bar{L}_{cn} = \frac{2}{3} L_{fu} . \tag{7.6}$$

In reality we expect a transition region in L space where the fusion probability drops from values near unity to practically zero, as shown qualitatively by the broken curve in Fig. 7.2. It seems therefore reasonable to interpret

Fig. 7.2. Angular momentum spectrum of the total reaction cross section and decomposition according to complete fusion (σ_{fu}), deep-inelastic (σ_{in}), and quasi-elastic (σ_{qe}) scattering

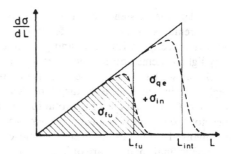

Fig. 7.2. Angular momentum spectrum of the total reaction cross section and decomposition according to complete fusion (σ_{fu}), deep-inelastic (σ_{in}), and quasi-elastic (σ_{qe}) scattering

the quantity L_{fu} as an effective angular momentum limit, defined in relation to the (experimental) fusion cross section by (7.5). Similar arguments apply also to the total reaction cross section σ_R and to the corresponding angular momentum limit L_{int} [see (3.37)]. The decomposition of the total angular momentum spectrum into contributions leading to complete fusion on one side, and to quasi-elastic or deep-inelastic interactions on the other side, is shown schematically in Fig. 7.2. Evidently the ratio of the fusion cross section to the total reaction cross section can be expressed as

$$\frac{\sigma_{fu}}{\sigma_R} = \frac{L_{fu}^2}{L_{int}^2} = P_{fu} \,, \tag{7.7}$$

where we have introduced an overall fusion probability, P_{fu}, averaged over angular momentum.

We note that both L_{fu} and L_{int} depend on bombarding energy for a given target–projectile combination. The relevant experimental evidence is discussed in Sect. 7.3, and model predictions of L_{fu} are presented and compared with experiment in Sect. 7.4. Anticipating to some extent the contents of these later sections we show schematically in Fig. 7.3 the dependence of L_{fu} and L_{int} on bombarding energy for a moderately heavy system. We observe, in

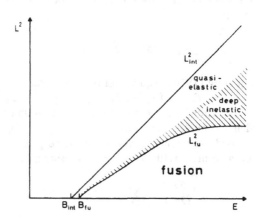

Fig. 7.3. Plot of squared angular momentum (L^2) versus energy (E), showing schematically regions where different reaction mechanisms dominate

particular, the existence of an intermediate energy region where the ratio σ_{fu}/σ_R remains fairly constant for a given system. The same ratio varies, however, strongly with target and projectile mass, from practically unity for very light systems and low energies to zero for very heavy systems.

So far we have considered the fusion cross section in the classical limit, neglecting intrinsic spins of target nucleus and projectile. This is normally adequate for discussions of the total fusion cross section at bombarding energies not too close to the fusion barrier, where high orbital angular momenta are involved. At bombarding energies close to the barrier, or when transitions to specific final channels are of interest, a more detailed description is usually required, however. Denoting by I_1, I_2 the intrinsic spins of target and projectile (assumed nonidentical), the cross section for formation of the compound nucleus with total angular momentum J can be written as

$$\sigma_J = \pi \lambda^2 \frac{(2J+1)}{(2I_1+1)(2I_2+1)} \sum_{IS} T_{ISJ}, \tag{7.8}$$

where S is the "channel spin" which results from coupling the intrinsic spins of target and projectile and is restricted by the condition

$$|I_2 - I_1| \leq S \leq I_2 + I_1. \tag{7.9}$$

The sum on the right-hand side of (7.8) runs over all combinations l, S which are consistent with a given value of J, and the quantities T_{ISJ} are "transmission coefficients" which describe the probability of compound nucleus formation with total angular momentum J for an incident partial wave characterized by l, S. It should be noted that these coefficients are different from the transmission coefficients calculated with the optical model [in this book denoted as $(1 - |S_l|^2)$ or $(1 - A_l^2)$, see (1.30)] which refer to the total reaction cross section (see also Fig. 2.9).

The coefficients T_{ISJ} will not depend on S or J if the target–projectile interaction is spin independent, as is normally assumed. In this case ($T_{ISJ} \equiv T_l$), summing (7.8) over J yields the familiar result for the total fusion cross section

$$\sigma_{fu} = \sum_J \sigma_J = \pi \lambda^2 \sum_{l=0}^{\infty} (2l+1) T_l. \tag{7.10}$$

Equation (7.10) is equivalent to (7.5) in the limit of large values of L_{fu}, and assuming $T_l = 1$ for $l < L_{fu}$, $T_l = 0$ for $l > L_{fu}$.

Equations (7.8) and (7.10) must be modified if target and projectile are identical nuclei, since wave functions with proper exchange symmetry must be used. The relevant arguments were given in Sect. 1.3 and will not be repeated here. With $I_1 = I_2 = I$ we obtain, instead of (7.8) and (7.10),

$$\sigma_J = \pi \lambda^2 \frac{2J+1}{(2I+1)^2} \sum_{l,S} [1 + (-1)^{l+S}] T_{ISJ}, \tag{7.11}$$

$$\sigma_{fu} = \sum_J \sigma_J = \pi \lambda^2 \sum_{l=0}^{\infty} \left(1 + \frac{(-1)^{2l+l}}{2l+1}\right)(2l+1)T_l, \tag{7.12}$$

where the coefficients T_{lSJ}, T_l are calculated for distinguishable nuclei with identical properties.

Finally we consider the cross section for a transition from the entrance channel α (composed of target nucleus and projectile in their respective ground states) via the corresponding compound nucleus to a specific exit channel β. The latter may be defined just by the masses and charges of the final fragments, or, in addition, by specifying their internal quantum states. The compound nucleus cross section is given quite generally by

$$\sigma_{\alpha\beta} = \sum_J \sigma_J(\alpha) \frac{\Gamma_J(\beta)}{\Gamma_J}. \tag{7.13}$$

Here $\sigma_J(\alpha)$ denotes the fusion cross section in channel α, $\Gamma_J(\beta)$ the decay width of the compound nucleus into channel β, and Γ_J its total decay width (at angular momentum J). Equation (7.13) is a consequence of our assumption of an equilibrated compound nucleus, which implies independence of compound-nucleus formation and decay. It is very important to realize, however, that such independence holds separately for each angular momentum and not for the cross section as a whole. Therefore identical decay properties can only be expected for entrance channels populating identical or similar angular momentum spectra in the same compound nucleus at the same excitation energy, a condition which is not easily met in practice with heavy projectiles of significantly different mass and charge. In this respect, reactions induced by heavy projectiles are quite different from reactions induced by light projectiles, where only limited angular momenta contribute which have little or no effect on compound nucleus decay.

7.2 Sub-Barrier Fusion

At bombarding energies below the Coulomb barrier, complete fusion is classically forbidden, but can nevertheless occur due to quantum-mechanical barrier penetration. This process has been studied extensively with target nuclei and projectiles up to and including ^{16}O. The relevant fusion barriers are between 5 and 10 MeV, and fusion reactions have been detected at bombarding energies down to about 40 % of the barrier with cross sections as low as 10^{-32} cm². Little is known, on the other hand, about sub-barrier fusion of heavier target nuclei or projectiles.

There are several reasons why the interest in this area has concentrated on light systems. First of all such experiments have so far been technically feasible only for light systems, where the barriers are within reach of smaller accelerators; there beams of the required quality and intensity have been available without undue restrictions in running time. For heavier systems not only higher bombarding energies are necessary, but also the cross sections are expected to decrease more rapidly with decreasing energy below the barrier, and this places more stringent requirements on beam energy definition and stability. In addition background problems arising from the presence of low Z target contaminants and problems associated with the identification of the reaction yield become much more serious.

The physical motivation for studying low-energy fusion cross sections has been largely provided by their astrophysical implications. In order to test theoretical models of nucleosynthesis during stellar evolution, one needs fusion excitation functions down to energies well below those which have so far been reached experimentally. The results are therefore customarily analysed with the aim of extrapolating to lower energies. We shall not discuss the problems associated with such extrapolations or the consequences for nuclear astrophysics in this book, but refer the reader to the relevant literature (see, for example, Bu 57, Ba 71, Cu 76). It is interesting to note, however, that the systems of special interest in the astrophysical context are $^{12}C + {}^{12}C$, $^{12}C + {}^{16}O$, and $^{16}O + {}^{16}O$ which are known to exhibit resonance phenomena both below and above the barrier in contrast to other systems. These resonances and their possible interpretation are discussed in Chap. 2, while in this section we shall be mainly concerned with the nonresonant part of the cross section. Unfortunately the presence and magnitude of resonant contributions to the fusion cross section are sometimes difficult to judge and this introduces uncertainties into the analysis and extrapolation of the data. A well-known example is the $^{12}C + {}^{12}C$ system where pronounced resonance structure persists to the lowest energies studied [Ma 73, Hi 77b, Ke 77].

Another important aspect of sub-barrier fusion, which is related to the astrophysical applications but of more general significance, is the possibility to deduce information on the nucleus–nucleus potential. At a given incident energy and angular momentum the probability of penetration through the potential barrier will depend both on the height and on the shape and position of the barrier. One may therefore hope to obtain, by measuring sub-barrier fusion excitation functions with sufficient precision over a sufficiently large energy range, a rather detailed mapping of the potential barrier and hence of the nuclear part of the potential. In practice, the analysis of sub-barrier fusion in terms of a potential is not completely unambiguous, however, as will be discussed in more detail below.

Measurements of sub-barrier fusion cross sections are experimentally difficult because of the low cross sections. The problem is to achieve high sensitivity together with good discrimination against background radiation arising from low-Z target contaminants, especially hydrogen. In principle, compound-nucleus formation can be observed by detecting any of the following decay products: heavy evaporation residues, light evaporation products (mainly α particles, protons, and neutrons), gamma rays emitted in the decay of bound excited states of the residual nuclei, and finally delayed beta or gamma radiation from the decay of radioactive products. The first and the last of these possibilities are practically ruled out for light systems at low bombarding energies, the former for technical reasons (low energies of the heavy recoils and high background from Rutherford scattering), and the latter because most of the evaporation residues are stabile nuclei. This leaves the observation of light charged particles or of prompt gamma rays as the most useful techniques. In charged-particle work, problems arise from the fact that usually several particles are emitted sequentially and the corresponding multiplicities are not accurately known. In gamma-ray work, on the other hand, events where particle evaporation proceeds directly to the ground states of the residual nuclei cannot be detected. Additional losses or uncertainties may exist depending on what type of gamma detector is used. Large--volume NaI counters provide high efficiency (at the expense of energy resolution) but may loose counts due to accidental summing. Germanium counters largely avoid this problem due to their smaller size and efficiency, and allow individual transitions to be resolved, thus providing superior background discrimination and the possibility to differentiate between different product nuclei. A disadvantage with germanium counters is that weak transitions cannot be analysed, and therefore the total cross sections are deduced from the yield of a few prominent transitions connecting low-lying states. This is not a serious handicap, however, since the observed transitions typically account for 60 - 80 % of the total yield of a given product (including direct ground-state production) and this percentage can be fairly reliably estimated from statistical evaporation calculations. Consequently much of the recent work on sub-barrier fusion cross sections has been performed with the gamma-ray method using germanium counters (see Table 7.2). The advantages of this technique have been summarized by Switkowski et al. [Sw 76]; the use of NaI detectors is discussed, for example, in [Cu 76]. Table 7.2 gives a list of experimental papers on sub-barrier fusion cross sections for light systems.

As an example we show in Fig. 7.4, taken from [Cu 76], experimental fusion cross sections for the system $^{12}C + ^{16}O$ and centre-of-mass bombarding energies between 4 and 12 MeV. The cross section is seen to rise over this energy range by more than seven orders of magnitude. In order to remove partly the strong energy dependence caused by barrier penetration, it is customary to plot instead of the cross section the "nuclear cross-section factor" $S(E)$ defined by [Bu 57]

$$S(E) = E\,\sigma_{fu}(E)\exp\left(2\pi n\right), \tag{7.14}$$

where n, as before, is the Sommerfeld parameter (1.13). The factor $\exp\left(2\pi n\right)$ is essentially the ratio of probability densities in the asymptotic region and at the origin for Coulomb scattering of point charges; it compensates for that part of the barrier penetrability which is independent of nuclear properties (including the nuclear radii). The effect of the transformation (7.14) on the data shown in Fig. 7.4 is demonstrated in Fig. 7.5. In contrast to the cross section, the S factor (not to be confused with the spectroscopic factor for a transfer reaction introduced in Chaps. 4 and 5) increases with decreasing energy and tends to become nearly constant at the lowest energies measured. In the energy region between the fusion barrier, which is near 8 MeV (see

Table 7.2. Measurements of sub-barrier fusion cross sections for light systems $(A_1, A_2 \leq 16)$.

System	E_{CM} [MeV]	Method[a]	References
$^{10}B + {}^{10}B$	1.8 — 3.7	(1)	Hi 76a
$^{10}B + {}^{11}B$	1.6 — 3.9	(1)	
$^{11}B + {}^{11}B$	1.6 — 3.7	(1)	
$^{10,11}B + {}^{12}C$	2.0 — 7.5	(2)	Da 76a
$^{10}B + {}^{12}C$	2.1 — 5.4	(1)	Hi 77a
$^{11}B + {}^{12}C$	2.1 — 6.0	(1)	
$^{12}C + {}^{12}C$	5.0 — 12.5	(1),(3)	Vo 64
$^{12}C + {}^{12}C$	3.2 — 6.2	(3)	Pa 69
$^{12}C + {}^{12}C$	2.5 — 5.0	(3)	Ma 73
$^{12}C + {}^{12}C$	3.5 — 7.5	(1)	Sp 74
$^{12}C + {}^{12}C$	2.5 — 6.2	(2)	Ke 77
$^{12}C + {}^{12}C$	2.5 — 5.9	(2)	Hi 77b
$^{10}B + {}^{14}N$	2.6 — 6.0	(1)	Hi 77a
$^{12}C + {}^{13}C$	3.1 — 11.9	(2)	Da 76b
$^{12}C + {}^{14}N$	5.5 — 14.8	(1),(3)	Ku 64
$^{12}C + {}^{16}O$	6.4 — 15.5	(1),(3)	
$^{12}C + {}^{14}N$	6.0 — 16.0	(2),(3)	Ol 74
$^{12}C + {}^{14}N$	3.6 — 9.2	(2)	Sw 77
$^{14}N + {}^{16}O$	5.6 — 12.6	(2)	
$^{14}N + {}^{14}N$	4.7 — 11.0	(1),(2)	Sw 76
$^{12}C + {}^{16}O$	4.6 — 9.2	(2)	
$^{12}C + {}^{16}O$	4.5 — 8.5	(3)	Pa 71
$^{12}C + {}^{16}O$	4.0 — 11.8	(1)	Ču 76
$^{12}C + {}^{16}O$	3.9 — 12.0	(2)	Ch 77
$^{12}C + {}^{16,17,18}O$	7.0 — 14.0	(4)	Ey 76
$^{16}O + {}^{16}O$	6.8 — 11.9	(1),(3),(5)	Sp 74

[a] Measurement of (1) gamma rays with NaI detector; (2) gamma rays with Ge(Li) detector; (3) light charged particles (p,α); (4) heavy evaporation residues; (5) delayed activity.

Table 7.5), and about 50 % of the barrier, the S factor varies only by about one order of magnitude, whereas the fusion cross section varies by more than six orders of magnitude. At the same time, resonance structure becomes apparent which is hidden in Fig. 7.4 in the steep rise of the cross section. Above the fusion barrier, the S factor decreases exponentially with increasing energy over a range of several MeV, as shown by the linear portion in the semilogarithmic plot of Fig. 7.5. Some authors have taken this latter part of the S-factor curve as a reference, against which changes in the sub-barrier region are measured, by introducing [Pa 69, Ma 73, Ke 77]

$$\tilde{S}(E) = E\sigma_{fu}(E)\exp(2\pi n + gE),\tag{7.15}$$

where g is a parameter which depends on nuclear properties (the "size parameter", equal to 1.14 MeV^{-1} for the linear part of Fig. 7.5.).

It is interesting to compare the system $^{12}C + {}^{16}O$ with another system of the same total mass, $^{14}N + {}^{14}N$. From a macroscopic point of view, one expects nearly identical properties since the charge product, reduced mass, and dimensions are nearly the same. Microscopic differences arise, however, from the presence of valence nucleons in the unfilled $1p\frac{1}{2}$ shell in ^{14}N and the more

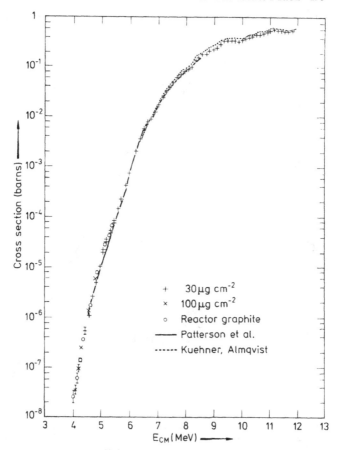

Fig. 7.4. Fusion cross section as a function of centre-of-mass bombarding energy for the system $^{12}C + ^{16}O$. Experimental points, measured by Čujec and Barnes with different targets, are compared with the results of other authors. (From Ču 76)

tightly bound structure of the α-particle nuclei ^{12}C and ^{16}O. For the $^{14}N + ^{14}N$ system the compound nucleus is more highly excited at corresponding bombarding energies (see Table 2.3), and the ground state Q values for nucleon transfer are close to zero. As a result, sizable cross sections for sub--barrier nucleon transfer are observed in the $^{14}N + ^{14}N$ system (see Sect. 5.3) in contrast to $^{12}C + ^{16}O$, where the transfer Q values are much less favourable.

Figure 7.6 shows results of Switkowski et al. for fusion and one-neutron transfer in the system $^{14}N + ^{14}N$, plotted as the corresponding S factors versus centre-of-mass bombarding energy. Two points are especially noteworthy: firstly, the transfer cross section (including the proton-transfer cross section which is not shown in Fig. 7.6, and is of similar magnitude as the neutron-transfer cross section) is considerably smaller than the fusion cross

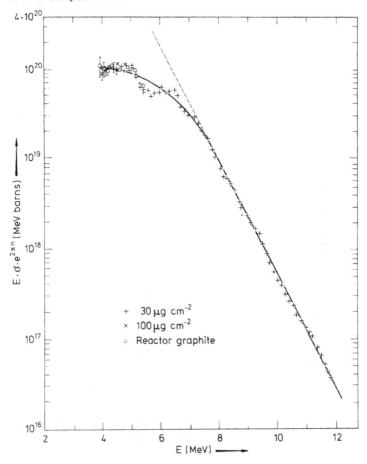

Fig. 7.5. The "nuclear cross-section factor" $S = E\sigma\exp(2\pi n)$ as a function of energy for the results shown in Fig. 7.4. (From Ču 76)

section at the barrier, but rises more steeply towards smaller energies. Consequently, at sufficiently low energies, transfer becomes more probable than fusion and eventually dominates the reaction cross section. This can be understood qualitatively by noting that transfer at low energies occurs by nucleon tunnelling at large core–core separations, whereas fusion requires tunnelling of the heavy cores through their mutual potential barrier.

A second point of interest in Fig. 7.6 is the absence of structure in the $^{14}N + {^{14}N}$ fusion cross section, whereas structure is clearly present in the $^{12}C + {^{16}O}$ cross section shown in Fig. 7.5. As discussed in Chap. 2, this difference is probably due to a stronger coupling between the elastic channel and surface reaction channels in $^{14}N + {^{14}N}$. Such a coupling, which is also reflected in the imaginary part of the optical potential, tends to remove the

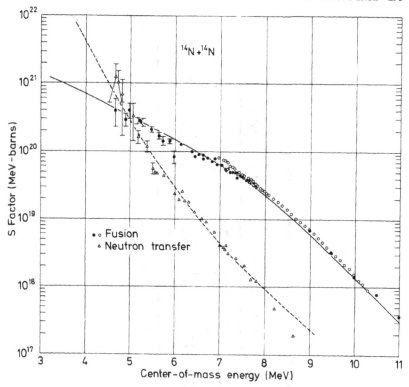

Fig. 7.6. S factors for fusion and neutron transfer in the system $^{14}N + ^{14}N$ as a function of centre-of-mass bombarding energy. The solid and dashed curves represent calculations with the ingoing-wave boundary-condition model and with a semiclassical model of sub-Coulomb transfer, respectively. (From Sw 76; the transfer data are from Hi 65)

system from the elastic channel, and hence destroys its coherence. It must be emphasized, however, that this type of interaction neither implies nor prevents fusion. As a result the fusion cross section becomes smooth and structureless, but is not affected in average magnitude compared with macroscopic estimates.

We now turn to the analysis of sub-barrier fusion cross sections. A very successful approach has been the application of the standard optical model (introduced in Sect. 2.2) where the fusion cross section is identified with the total reaction cross section. The latter assumption seems justified for most of the light systems, except in special cases like $^{14}N + ^{14}N$ at low energies. Representative sets of optical model parameters are given in Table 7.3. One set in particular [labelled (1) in Table 7.3], originally deduced by Reeves from a fit to $^{12}C + ^{12}C$ cross sections [Re 66], was subsequently found to reproduce the fusion cross sections for many light systems quite well (see Fig. 7.8). An impressive example is shown in Fig. 7.7 where a perfect fit to the experimental fusion cross section for the system $^{12}C + ^{10}B$ has been achieved

Table 7.3. Parameters of optical model potentials used in analyses of sub-barrier fusion cross sections for light systems (for further explanation see text)

Label	System	V [MeV]	r_V [fm]	a_V [fm]	W [MeV]	r_W [fm]	a_W [fm]	b [MeV][a]	c [fm^{-2}][a]	References
(1)	Many	50	1.26	0.44	10	1.215	0.45			Re 66
(2)	^{12}C + ^{10,11}B	200	0.989	0.52	30	0.989	0.52			Hi 77a
	^{10,11}B + ^{10,11}B									Hi 76a
(3)	^{12}C + ^{12}C	13.0	1.354	0.55	0.22E	1.354	0.55	100	0.100	Mi 73
(4)	^{12}C + ^{16}O	10.2	1.362	0.50	0.14E	1.362	0.50	100	0.156	
(5)	^{16}O + ^{16}O	8.2	1.488	0.35	0.123E	1.488	0.35	100	0.190	

[a] For definition see (7.16)

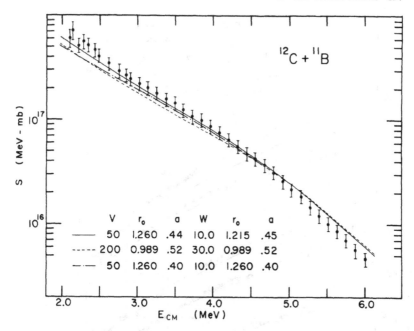

Fig. 7.7. S factor for the system $^{12}C + {}^{10}B$ as a function of centre-of-mass bombarding energy. The experimental points are compared with optical model calculations based on three different parameter sets. (From Hi 77a)

over a large energy range (corresponding to a variation in cross section by six orders of magnitude) with three different parameter sets [Hi 77a]. It should be noted that the same parameter sets will not fit the data for other systems equally well, but in many cases (except for systems exhibiting resonance structure, see below), minor parameter changes were found to result in an adequate reproduction of the measured cross sections [Hi 76a, Da 76a,b, Hi 77a].

A comparison of fusion cross sections for nine different systems, ranging from $^{12}C + {}^{12}C$ to $^{16}O + {}^{16}O$, with optical model calculations using the same standard parameter set [number (1) of Table 7.3, except that $r_0 = 1.27$ fm] was presented by Stockstad et al. [St 76b] and is shown in Fig. 7.8. While the data are well reproduced on the average, significant discrepancies are evident for some systems. These discrepancies were interpreted by the authors as evidence for microscopic effects of nuclear structure, assumed to change rapidly with the proton and neutron number of the interacting nuclei. On the other hand, it seems likely that some of the discrepancies could be removed by a suitable systematic variation of the model parameters with nuclear size—leading to an increase of the cross section for the heavier systems and intermediate energies, and a general decrease at the lowest energies.

A detailed analysis of the systems $^{12}C + {}^{12}C$, $^{12}C + {}^{16}O$, and $^{16}O + {}^{16}O$ was performed by Michaud and Vogt [Mi 72, Mi 73]. It was found that the

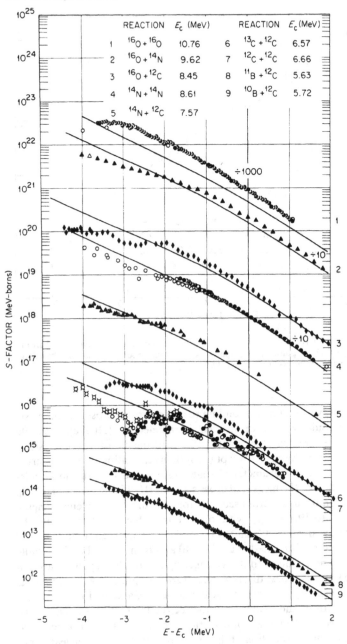

Fig. 7.8. Fusion cross sections for nine different systems, presented in terms of the S factor as a function of energy difference relative to the Coulomb barrier. The solid curves have been calculated with the optical model using the same standard parameter set. (From St 76b)

resonant behaviour peculiar to these systems could be obtained qualitatively
with optical potentials of Woods–Saxon shape by choosing a sufficiently small
value for the imaginary part. However, the use of such potentials necessitated
unreasonable parameter changes from one system to another. This difficulty
could be avoided by adding a soft repulsive core, which was derived from an
α-particle model of the interacting nuclei, to the real part of the potential:

$$V(r) = V_{\mathrm{C}}(r) + V_{\mathrm{ws}}(r) + b \exp\left(-cr^2\right). \tag{7.16}$$

Here $V_{\mathrm{C}}(r)$ is the Coulomb potential and $V_{\mathrm{ws}}(r)$ the real part of a standard
Woods–Saxon potential (attractive), while the last term represents the repul-
sive core. The best-fit parameters of the Woods–Saxon potentials and of the
repulsive core (b,c) for the three systems in question [Mi 73] are given in
Table 7.3.

It was further observed that absorption by the outer tail of the imaginary
part of the potential, at distances beyond the barrier, was essential for producing the correct magnitude of the fusion cross section at low energies. This
phenomenon was termed "absorption under the barrier" [Mi 72]; it produces
a marked increase of the calculated S factor at low energies. The energy where
absorption under the barrier becomes effective depends sensitively on the
diffuseness of the imaginary part of the potential. Hence an extrapolation of
the S factor to lower energies—within the optical model with Woods–Saxon
potentials—is affected dramatically by the value chosen for this parameter,
which in turn is not well determined by the data for higher energies [Cu 76].

However, the physical significance of absorption under the barrier is not
clear. The calculations imply that any interaction removing the system from
the elastic channel automatically leads to fusion, even if it occurs at large
distances beyond the barrier position. Such a picture strongly contradicts intuitive concepts of fusion, and alternative models have therefore been applied
which do not allow for absorption at large distances. These are the "ingoing-
-wave boundary condition" model [Ch 77, Sw 77], the "equivalent square-well
model" [Mi 70, Mi 72, Fo 75a] and the "barrier-penetration model" [Cu 76].

In the ingoing-wave boundary condition model, a real potential is used
together with the requirement that the wave function has ingoing character at
a suitably chosen contact radius (for a more detailed discussion see Sect. 3.3).
Hence the transition into the compound nucleus occurs at a well-defined distance inside the potential barrier. This model has been used successfully to
analyse sub-barrier fusion cross sections for $^{12}\mathrm{C} + ^{12}\mathrm{C}$, $^{12}\mathrm{C} + ^{16}\mathrm{O}$, and
$^{16}\mathrm{O} + ^{16}\mathrm{O}$ [Ch 77] and for $^{14}\mathrm{N} + ^{14}\mathrm{N}$ [Sw 77; see Fig. 7.6]. With potentials
deduced from fits to elastic scattering data, the average (nonresonant) be-
haviour of the fusion cross sections was well reproduced except for $^{16}\mathrm{O} +$
$^{16}\mathrm{O}$, where a readjustment of parameters was found necessary.

As shown by Michaud et al. [Mi 70], an optical model potential of Woods–
-Saxon shape can be replaced by an equivalent (complex) square-well potential
which yields practically the same reaction cross section over an extended
energy range. More precisely, this holds provided the transmission coefficients

calculated for the complex square well are corrected by application of "reflection factors", which remove unphysical effects due to reflections at the square-well boundary, and thus produce an overall increase of the reaction cross section. The equivalent square-well model has been used by Fowler et al. to obtain excellent fits to experimental S factors for a number of astrophysically interesting systems [Fo 75a].

An even simpler approach was followed by Čujec and Barnes in an analysis of their results for the system $^{12}C + ^{16}O$ [Ču 76]. These authors used a real potential $V(r)$, calculated by a folding procedure [Br 74], and derived transmission coefficients from the approximate relationship

$$T_l(E) = \exp\left(-\frac{2}{\hbar} \int_{r_1}^{r_2} \{2\mu[V(r) - E]\}^{1/2} \, dr\right), \tag{7.17}$$

where r_1 and r_2 are the inner and outer classical turning points, respectively. Equation (7.17) is based on the JWKB approximation which is expected to become poor close to the barrier. Therefore for energies in the vicinity of the barrier, the potential was approximated by an inverted parabola for which the transmission coefficients are given by [Hi 53]

$$T_l = \left[1 + \exp\left(+2\pi \frac{B_l - E}{\hbar\omega_l}\right)\right]^{-1}. \tag{7.18}$$

Here B_l is the effective barrier height for the partial wave with angular momentum l, and the "oscillator frequency" is

$$\omega_l = \left\{\frac{1}{\mu} \frac{d^2}{dr^2}\left[V(r) + \frac{l(l+1)\hbar^2}{2\mu r^2}\right]\right\}_{r=r_B}^{1/2}. \tag{7.19}$$

This model has been referred to as "barrier-penetration model". Its main merit lies in the fact that it contains no free parameters except those entering the real potential. The resulting fit to the average behaviour of the $^{12}C + ^{16}O$ sub-barrier fusion cross section is of similar quality to those obtained with the previously discussed models [Ču 76]. The success of the models in which absorption into the compound nucleus is confined to a well-defined radial region supports the view that absorption under the barrier is not needed to explain the existing data. It seems likely, then, that this should also be true for still lower energies, where the relevant absorptive region would move to even larger distances with imaginary potentials of standard diffuseness. We conclude that the optical model is not applicable to extrapolations of the fusion cross section to very low energies, unless provision is made to limit the range of the imaginary potential.

It remains to examine the question to what extent excitation functions for sub-barrier fusion reflect details of the real part of the nucleus–nucleus potential. With potentials of Woods–Saxon shape we have the familiar problem of parameter ambiguities as demonstrated in Fig. 7.7. The success of the equivalent square-well model, on the other hand, indicates that for each system

two, or at most three, significant parameters are adequate to reproduce sub-
-barrier fusion over an extended energy range. This is similar to the situation
in elastic scattering above the barrier, were the important quantities are a
characteristic distance and the potential at that distance [Ch 76].

The question which properties of the potential are probed in sub-barrier
fusion experiments may be studied by comparing explicitly different potentials
which are known or expected to reproduce the data. This has been done in
Fig. 7.9 for the systems $^{12}C + {}^{10}B$ and $^{12}C + {}^{16}O$. In each case we show the
standard Woods–Saxon potential, set (1) of Table 7.3, and a semi-empirical
potential deduced from an analysis of fusion cross sections above the barrier
[Ba 77b; see also Sect. 7.4, (7.50), (7.51)]. Included are further another
Woods–Saxon potential for $^{12}C + {}^{10}B$ [set (2) of Table 7.3] which is known
to reproduce the data (see Fig. 7.7) and the repulsive core potential of Michaud for $^{12}C + {}^{16}O$. Table 7.4 lists barrier heights, barrier positions, and
distances where the potential has dropped to 50 % of the barrier height (at
the inside of the barrier), as calculated with the same potentials for different
systems.

An inspection of Fig. 7.9 and Table 7.4 indicates that the quantities most

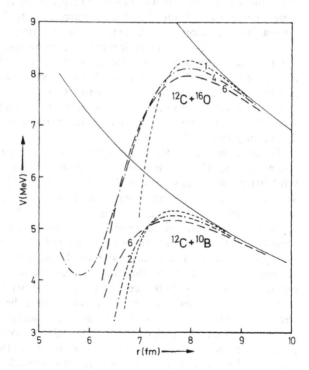

Fig. 7.9. Comparison of different real potentials which fit experimental fusion data for
the systems $^{12}C + {}^{10}B$ and $^{12}C + {}^{16}O$. The numbers given at the curves are explained in
Tables 7.4 and 7.5. The continuous curves without numbers are the pure Coulomb potentials

reliably determined are the barrier heights and barrier positions. This is not unexpected since the same quantities enter decisively the magnitude and derivative of the cross section above the barrier, as shown by the classical formula (7.25).

Barrier penetration arguments suggest that the sub-barrier cross section should be sensitive, in addition, to the curvature of the potential near the top of the barrier and its rate of decrease towards smaller distances. It appears, however, that the detailed shape of the potential inside the barrier is not simply related to the fusion cross section in general. For example, the different potentials which fit the $^{12}C + ^{10}B$ system behave differently inside the barrier. This lack of sensitivity may be partly due to the dominant influence of the broad outer part of the barrier, and partly due to the onset of absorptive processes near the inner edge of the barrier. As a consequence of these effects, the system may reach somewhere within the barrier a point of no return such that the radial dependence of the potential for smaller distances becomes irrelevant.

An interesting departure from this pattern is observed, however, for the system $^{12}C + ^{16}O$ where the two potentials marked 4 and 6 in Fig. 7.9 coincide rather closely down to a distance 1.5 fm inside the barrier. The former of these potentials was deduced by Michaud from a fit to the gross structure in the near-barrier fusion cross section, whereas the latter was determined from an analysis of higher-energy fusion data for many systems. A similar correspondence of potentials exists for the system $^{12}C + ^{12}C$ (not included in Fig. 7.9) which, like $^{12}C + ^{16}O$, is also known to be "weakly absorbing". We may tentatively conclude that the systems $^{12}C + ^{12}C$ and $C^{12} + ^{16}O$ exhibit more transparency than others in the sense that the sub-barrier fusion cross section for these systems is sensitive to the shape of the potential down to significantly smaller distances.

So far, we have considered only light systems and approximately spherical fragments. The effect of the static deformation of a heavy target nucleus on sub-barrier fusion was studied by Stokstad et al. [St 78]. These authors used activation techniques to measure precise cross sections for the fusion of ^{16}O with a sequence of even samarium isotopes, from ^{148}Sm ($\beta_2 \approx 0$) to ^{154}Sm ($\beta_2 \approx 0.3$). The results are shown in Fig. 7.10 and reveal a drastic variation—roughly by a factor 20—of the cross section with mass number at the lowest energy investigated. Above the barrier, on the other hand, the cross sections become almost identical. As shown by the authors, this behaviour cannot be explained simply as a consequence of the proportionality of nuclear radii to $A^{1/3}$ and the associated variation in Coulomb barrier height. Instead it gives clear evidence for a direct and strong influence of nuclear deformation.

This effect can be understood qualitatively by noting that the nuclear potential depends sensitively on the local separation of the fragment surfaces in the vicinity of the centre-to-centre line. Thus the nuclear interaction will be stronger, and hence result in a lower Coulomb barrier, when a prolate target nucleus is hit by a spherical projectile near its poles, then when it is hit near its equator (see Fig. 7.32). The overall fusion cross section corresponds to an

Table 7.4. Barrier parameters for light nucleus–nucleus systems

System	B_{fu} [MeV]				r_B [fm]				$r(V = \frac{1}{2}B_{fu})$ [fm]			
	(1)[a]	(2)	(3)–(5)	(6)[b]	(1)	(2)	(3)–(5)	(6)	(1)	(2)	(3)–(5)	(6)
$^{12}C + ^{10}B$	5.35	5.26		5.17	7.7	7.7		7.7	6.6	6.4		6.0
$+ ^{12}C$	6.34		6.20	6.11	7.8		7.8	7.8	6.7		—	6.1
$^{14}N + ^{10}B$	6.16	6.11		5.95	7.8	7.7		7.8	6.7	6.4		6.1
$+ ^{12}C$	7.30			7.04	7.9			7.9	6.8			6.2
$+ ^{14}N$	8.41			8.10	8.0			8.0	6.9			6.3
$^{16}O + ^{12}C$	8.25		8.11	7.95	8.0		8.0	8.0	6.9		—	6.3
$+ ^{14}N$	9.51			9.16	8.1			8.1	6.9			6.4
$+ ^{16}O$	10.76		10.40	10.35	8.1		8.5	8.1	7.0		—	6.5

[a] Labels (1) to (5) correspond to the potentials given in Table 7.3

[b] Semi-empirical potential derived from a classical analysis of fusion cross sections above the barrier [Ba 77b]; see (7.35), (7.50), and (7.51).

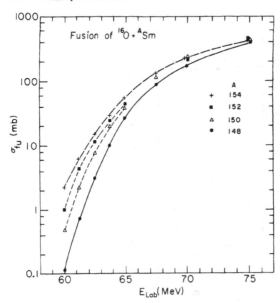

Fig. 7.10. Excitation functions for fusion of ^{16}O with different samarium isotopes. (From St 78)

average taken over all possible orientations of the nuclear symmetry axis. At low energies, where barrier penetration is important, the relationship between effective barrier height and cross section will be highly nonlinear and hence the cross section will be dominated by contributions from those orientations which give the largest cross sections. The net effect is then a strong increase of the cross section with increasing deformation. A quantitative estimate of the dependence of the nuclear potential on deformation is given in Sect. 7.5.

To summarize this section we note that a large body of data on sub-barrier fusion of light systems has been accumulated. In addition, interesting evidence has been obtained concerning the influence of nuclear deformation for heavier systems. The main features of the data can be understood or reproduced with simple phenomenological approaches. A satisfactory quantitative theory which would allow to extrapolate to lower energies or to heavier systems does not yet exist, however.

7.3 Fusion Above the Barrier: Experimental Evidence

In this section we discuss measurements of complete fusion cross sections performed at bombarding energies above the barrier, and the most important qualitative conclusions which emerge from these measurements. A quantitative discussion of the results follows in Sect. 7.4.

We begin with a brief review of experimental methods and their historical development. As mentioned in the preceding section, fusion cross sections can be measured in principle by detecting any of the (prompt or delayed) decay products of the primary compound nucleus. For technical reasons, the methods most successful in sub-barrier studies of light systems—observation of prompt gamma rays or light charged particles—are much less useful for the higher energies and heavier systems considered in this section. This is primarily a consequence of the complex decay patterns of highly excited heavy compound nuclei which are usually not known in sufficient detail. For the same reason measurements of radioactive decay products are of limited applicability in quantitative measurements of the fusion cross section (but may be an excellent tool for studies of specific aspects of the reaction mechanism, as will be shown later). On the other hand, the larger compound-nucleus momenta reached at higher bombarding energies, combined with near-geometrical cross sections, offer the possibility for a direct observation of heavy recoil nuclei ("evaporation residues")at small angles with respect to the beam direction. Measurements of this type were originally performed using mica track detectors, and later with counter telescopes. The latter method has emerged as the most powerful technique for a precise quantitative determination of complete fusion cross sections; more details are given below.

Heavy compound nuclei $(A_1 + A_2 > 150)$ may decay with significant probability by fission. The fusion cross section is then given by the sum of the cross sections for evaporation residue formation (σ_{er}) and for fusion followed by fission ("fusion-fission", σ_{ff}). In general both cross sections must be determined separately. For very fissile systems $(A_1 + A_2 > 240)$, however, practically all compound nuclei decay ultimately by fission, and the fusion cross section may then be identified approximately with the cross section for fusion-fission.

It should be recalled, at this point, that the observation of "fission fragments"—or, more precisely, of a nearly symmetric component in the mass distribution of reaction products—does not by itself permit an unambiguous assignment of the reaction mechanism. Such fragments may in general originate from any of three possible processes:

i) Fission decay (either directly, or following the evaporation of light particles) of a primary compound nucleus, formed by amalgamation of projectile and target nucleus;
ii) Emission of two heavy fragments of similar mass following a deep-inelastic collision (without intermediate formation of a compound nucleus);
iii) Inelastic scattering or transfer followed by sequential fission of the excited heavy product.

These processes are discussed more fully in Chap. 6. Here we are concerned with the question to what extent component (i)—corresponding to fusion-fission—can be separated from components (ii) and (iii), which are irrelevant for the fusion problem.

As noted before, an unambiguous distinction between processes (i) and (ii) is not possible from experimental data alone, and hence we have to invoke indirect arguments. These may be based on angular distributions (permitting estimates of the composite system's lifetime), calculated potential energy surfaces (indicating the tendency of the tangent system to evolve towards symmetry or asymmetry), or comparisons between different entrance channels leading to the same or similar compound systems (as discussed in detail later in this section). For the lighter compound systems up to $A_1 + A_2 \approx 120$, the yield of "fission-like products" usually contributes only a very small fraction to the total reaction cross section. The available evidence indicates that in this mass region most or all of the observed symmetric yields should be attributed to process (ii). For heavy compound systems $(A_1 + A_2 \gtrsim 180)$, on the other hand, the symmetric yields seem to be dominated by process (i) for extremely asymmetric entrance channels $(A_1 : A_2 < 0.2)$ and by process (ii) for nearly symmetric entrance channels $(A_1 : A_2 > 0.5)$. There is also a broad range of intermediate situations—especially for argon-induced reactions—where the origin and composition of the symmetric component are not well established.

In contrast to processes (i) and (ii), where two heavy fragments are emitted, process (iii) corresponds to a three-body final state. It differs from the former processes by the fact that the two fission fragments do not carry the full momentum of the incident projectile. Hence process (iii) can be distinguished experimentally from processes (i) and (ii). A simple and elegant method of discrimination was introduced by Sikkeland and Viola [Si 62, Si 63, Si 68]. These authors measured angular correlations between the two fission fragments emitted following heavy-ion bombardment of ^{238}U and other heavy target nuclei. One counter was placed at a fixed angle of 90° with respect to the beam direction, while the angular position of the other counter was varied both within (ψ) and perpendicular to (ζ) the reaction plane defined by the first counter. Figure 7.11 shows a contour plot of the angular correlation in the ψ–ζ plane, obtained by bombarding a ^{238}U target with 10 MeV/u ^{16}O ions. Two regions of enhanced yield are clearly distinguishable: one centred at a smaller value of $\psi(64°)$ with a narrow distribution in ζ, and one centred at a larger value of $\psi(80°)$ with a much wider distribution in ζ. Events in the former region were shown to arise from fission following full momentum transfer to the composite system, whereas events in the latter region were attributed to sequential fission following inelastic collisions with partial momentum and mass transfer. Similar data were obtained with other projectiles incident on ^{238}U; a summary of angular correlations as a function of ψ ($\zeta = 0$) is shown in Fig. 7.12. The relative yield of fission events following incomplete momentum transfer is seen to increase strongly with increasing projectile mass. The following empirical relationship was deduced between the cross sections for fission following complete momentum transfer (σ_{cf}) and fission following incomplete momentum transfer (σ_{if}), for projectiles with mass number A_1 and an energy of 10 MeV/u incident on a ^{238}U target:

$$\frac{\sigma_{cf}}{\sigma_{cf} + \sigma_{if}} = (1 + 0.03 \, A_1)^{-1} . \tag{7.20}$$

We note that the sum $\sigma_{cf} + \sigma_{if}$ can be identified approximately with the total reaction cross section since, for the systems and energies considered

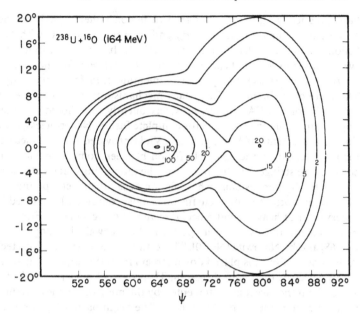

Fig. 7.11. Contour diagram of fission-fragment angular correlation observed in the ^{16}O bombardment of ^{238}U. For further explanation, see text. (From Si 63)

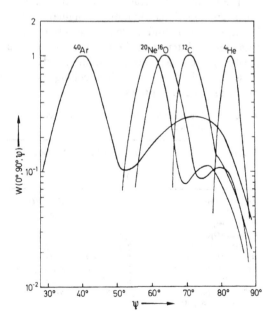

Fig. 7.12. In-plane fission-fragment angular correlation functions ($\zeta = 0$) observed in the bombardment of ^{238}U with different projectiles at 10.4 MeV/u. The lower peaks to the right correspond to fission following incomplete momentum transfer. (From Si 68)

here, practically all interactions lead to fission either directly or indirectly.

Originally the cross sections σ_{cf} and σ_{if} were associated with "complete fusion" and "incomplete fusion", respectively. It should be emphasized, however, that the experiment can only distinguish between full momentum transfer (two-body decay) and partial momentum transfer (three-body decay). Therefore contributions of a surface mechanism to σ_{cf} cannot be excluded, especially for the ^{40}Ar projectile. Consequently, σ_{cf} should be taken as an upper limit to the cross section for complete fusion. Nevertheless, the correlation measurements of Sikkeland and Viola provided the first clear evidence for a limitation on the complete fusion between complex nuclei, which becomes more stringent as the masses of the interacting fragments increase.

For the lighter systems, where fission decay of the compound nucleus is of minor importance, complete fusion is most conveniently detected by direct observation of heavy recoil nulcei ("evaporation residues"). First systematic measurements of this type were performed by Kowalski, Jodogne, and Miller [Ko 68] and by Natowitz [Na 70]. These authors used an ingenious technique: A thin sheet of mica is placed down-stream from the target at an angle of 45° with respect to the beam direction. Heavy recoil nuclei ejected at forward angles from the target are intercepted by the mica where they produce tracks which can be made visible by etching. The beam particles and reaction products of comparable or smaller mass, on the other hand, do not produce sufficient damage in the mica to be detectable. The method thus depends critically on the mass difference between detected and background particles, and on the existence of a well-defined threshold of the track detector with respect to mass (charge).

The results of the track detector measurements provided a first survey of fusion cross sections and limiting angular momenta over a large range of projectile and compound nucleus masses [Ko 68, Na 70], and thus stimulated a great deal of interest in the fusion problem. Due to inherent limitations in flexibility and quantitative reliability, and as a consequence of technical progress with electronic counting techniques, the method was later replaced by the use of counter telescopes.

Table 7.5. Measurements of fusion cross sections at bombarding energies above the barrier

System	Bombarding energy [MeV/u]	Method[a]	Deduced cross section[b]	References
^{11}B + ^{159}Tb	10.5	Mica, f(tot)	σ_{er}, σ_{ff}	Ze 73
^{12}C + ^{158}Gd	10.5	$E/\Delta E$		Ze 74
^{16}O + ^{154}Sm	8.6			Ko 74
^{20}Ne + ^{150}Nd	7.2			
^{12}C, ^{14}N, ^{15}N + ^{12}C	2.0 − 4.7	$E/\Delta E$	σ_{er}	Co 76
^{12}C, ^{18}O, ^{19}F + ^{12}C	1.3 − 5.2	$E/\Delta E$	σ_{er}	Sp 76b
^{12}C + Al, Ti, Cu, Ni	3.8 − 15.0	Mica, $E/\Delta E$	σ_{er}	Na 72
^{12}C + ^{27}Al	8.3, 15.0	$E/\Delta E$	σ_{er}	Na 73
^{12}C, ^{16}O, ^{20}Ne + Cu, Ag, Au, Bi	≤ 10.5	Mica	σ_{er}	Na 70

Table 7.5.   (*Continued*)

System	Bombarding energy [Mev/u]	Method[a]	Deduced cross section[b]	References
$^{12}C + {}^{152}Sm$	3.3 — 5.3	in beam γ	σ_{er}	Br 75
$^{16}O + {}^{148,150}Nd$	3.1 — 5.3			
$^{18}O + {}^{148}Nd$	2.8 — 4.5			
$^{12}C, {}^{14}N, {}^{16}O, {}^{20}Ne$ $+ {}^{238}U$	≤ 10.4	$f(tot), f(corr)$	σ_{ff}	Si 62, Vi 62 Si
$^{14}N + {}^{12}C$	3.1 — 12.7	$E/\Delta E$	σ_{er}	St 76a
$^{14}N + Al, Cr, Ni$	11.2, 18.7	$E/\Delta E$	σ_{er}	Na 75
$^{16}O + {}^{12}C$	1.9 — 3.9	$E/\Delta E$	σ_{er}	Sp 76a
$^{16}O + {}^{12}C$	2.5 — 4.1	in beam γ	σ_{er}	Ko 76
$^{16}O + {}^{12}C, {}^{16}O$	3.75	$E/\Delta E/t$	σ_{er}	We 76
$^{16}O + {}^{27}Al$	3.1 — 5.0	E/t	σ_{er}	Ba 77a
$^{16}O + {}^{27}Al, {}^{28}Si$	1.6 — 3.1	in beam γ	σ_{er}	Da 75
$^{16,17,18}O + {}^{27}Al$	1.2 — 2.5	$E/\Delta E$	σ_{er}	Ei 77
$^{16}O, {}^{20}Ne, {}^{32}S + {}^{27}Al$	≤ 10.5	$E/\Delta E$	σ_{er}	Ko 75
$^{16}O + {}^{59}Co$	≤ 10.5	Mica	σ_{er}	Ko 68
$^{20}Ne + {}^{27}Al$				
$^{16}O + {}^{134}Ba$	3.9 — 7.7	$act(\gamma)$	σ_{er}	De 77a
$^{40}Ar + {}^{110}Pd$	3.5 — 5.9			
$^{16}O + {}^{197}Au, {}^{209}Bi$	≤ 10.4	$f(tot)$	σ_{ff}	Si 64
$^{19}F + {}^{12}C$	2.6 — 4.0	E/t	σ_{er}	Pü 75
$^{20}Ne + {}^{150}Nd$	6.4 — 8.6	in beam γ	σ_{er}	Sa 76
$^{20}Ne + {}^{150}Nd$	8.75	$E/\Delta E$	σ_{er}	Ha 78
$^{20}Ne + {}^{235}U$	8.75, 12.6	$f(corr.)$	σ_{ff}	Vi 76
$^{32}S + {}^{24}Mg {}^{27}Al, {}^{40}Ca$	2.0 — 4.1	$E/\Delta E$	σ_{er}	Gu 73b
$^{32}S + {}^{115}In$	10.5	$E/\Delta E$	σ_{er}, σ_{ff}	Lu 76
$^{35}Cl + {}^{48}Ti$	3.0	$E/t; act(\gamma)$	σ_{er}	Hi 76b
$^{35}Cl + {}^{27}Al, {}^{48}Ti$	2.0 — 4.9	$E/\Delta E$	σ_{er}, σ_{ff}	Sc 75
$^{54,56}Fe, {}^{58,60,62,64}Ni$				Sc 76
$^{90}Zr, {}^{116,124}Sn$				
$^{40}Ar + {}^{58}Ni, {}^{109}Ag, {}^{121}Sb$	4.2 — 8.4	$E/\Delta E$	σ_{er}, σ_{ff}	Gu 74
$^{84}Kr + {}^{65}Cu$	5.9, 7.2			Br 76
$^{40}Ar + Ni, Ge, Ag, Sb$	4.2 — 7.4	$E/\Delta E$	σ_{er}	Ga 75
$^{40}Ar + {}^{118,121}Sb$	3.8 — 7.0	$act(\alpha)$	σ_{er}	Ga 74c
$^{40}Ar + Mo, Sb, Ho, U$	4.0 — 7.5	$f(corr.)$	σ_{ff}	Ta 75
$^{40}Ar + {}^{197}Au$	4.6 — 6.2	$f(tot)$	σ_{ff}	Ng 77
$^{40}Ar + {}^{238}U$	10.4	$f(corr.)$	σ_{ff}	Si 68
$^{52}Cr + {}^{56}Fe$	4.0, 5.1	$E/\Delta E$	σ_{er}	Ag 76
$^{63}Cu + Ni, Ag$	5.5			

[a] Mica: observation of heavy recoils with mica track detectors; $E/\Delta E$, E/t, $E/\Delta E/t$: observation of heavy recoils with counter telescopes measuring residual energy (E), energy loss (ΔE), time of flight (t); $f(corr.)$: Measurement of fission fragment angular correlations; $f(tot)$: Measurement of total yield of fission fragments; in beam γ: observation of prompt gamma radiation; $act(\alpha)$: observation of delayed α activitiy; $act(\gamma)$: observation of gamma radiation from radioactive products.
[b] σ_{er}: cross section for evaporation residue formation; σ_{ff}: fusion-fission cross section (note that in general $\sigma_{fu} = \sigma_{er} + \sigma_{ff}$).

Early measurements of evaporation residues from ^{32}S- and ^{40}Ar-induced reactions with a counter telescope were performed by Gutbrod and collaborators [Gu 73a, Gu 74]. These authors used a gas proportional counter for ΔE measurements together with a surface barrier counter for measurements of the residual energy $E - \Delta E$. The combination of ΔE and E yields (approximately) the atomic number of the detected product and hence allows one to discriminate between evaporation residues and other reaction products, especially scattered projectiles. Other versions of the method make use of a solid-state detector [Na 72, Ch 73, Na 73] or an ionization chamber [Fo 76b, Ga 75] to measure ΔE, or employ time-of-flight techniques to deduce the product mass from a combined measurement of particle velocity and kinetic energy (see Table 7.5).

A disadvantage of the counter-telescope method is that very small angles, typically below about 3° in the laboratory system, are not accessible due to very high rates of elastically scattered projectiles. Therefore, the measurement of fusion cross sections by this method relies on the fact that compound nuclei—originally moving along the beam direction—are deflected to angles $\gtrsim$ 3° by the recoil due to evaporated light particles, or by multiple scattering in the target. With not too heavy projectiles, up to about ^{40}Ar, a major fraction of the evaporation residues is usually emitted at angles > 3°; hence it is possible to correct the measured yields for the unobserved emission at smaller angles. With projectiles heavier than ^{40}Ar, the unobserved intensity may become larger than the observed intensity, thus causing large uncertainties in

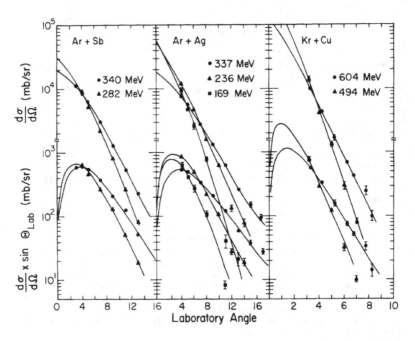

Fig. 7.13. Angular distributions of evaporation residues from Ar- and Kr-induced fusion reactions. Curves are drawn through the experimental points and extrapolated to smaller angles. The upper curves are differential cross sections per unit solid angle, and the lower curves are multiplied by sin ϑ in order to show the relative yield per unit angle. In the Kr-bombardments, the extrapolated part of the cross section is much larger than the observed part, while in the Ar bombardments the extrapolated and observed parts are comparable. (From Br 76)

the corresponding correction. The situation is illustrated in Fig. 7.13, which shows angular distributions of evaporation residues as observed in various ^{40}Ar- and ^{84}Kr-induced reactions [Br 76].

In the following we give some representative examples of fusion excitation functions measured with counter telescopes. Figure 7.14 shows results for three typical light systems, ^{12}C + ^{12}C, ^{18}O + ^{12}C, and ^{19}F + ^{12}C, as obtained by Sperr et al. at Argonne [Sp 76b]. One remarkable feature is the oscillatory structure in the ^{12}C + ^{12}C data which is, however, not of interest here. We note that with increasing energy, the fusion cross sections reach maxima and then slowly decrease again. This behaviour is in contrast to that of the total reaction cross section for these systems, which is known to increase monotonically with energy. Quantitative comparisons with optical model calculations of the reaction cross section lead to the conclusion that, in each of the cases shown, the fusion cross section is close to the reaction cross section in the vicinity of the fusion barrier, but accounts only for about half of that cross section near the upper end of the investigated energy range.

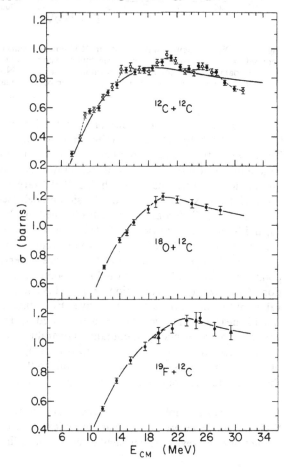

Fig. 7.14. Fusion excitation functions for the systems ^{12}C + ^{12}C, ^{18}O + ^{12}C, and ^{19}F + ^{12}C as deduced from counter-telescope measurements of evaporation residues (from Sp 76b; the triangles in the case ^{19}F + ^{12}C are from Pü 75) The full curves are fits to the data with a phenomenological model due to Glas and Mosel (Gl 74b)

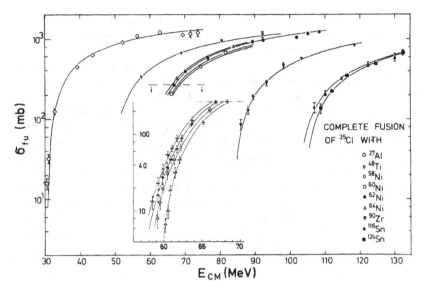

Fig. 7.15.   Fusion excitation functions for ^{35}Cl bombardments of different target nuclei, as deduced from counter-telescope measurements of evaporation residues. The full curves are fits to the data assuming constant fusion radii (see Sect. 7.4.) Note the systematic variation of fusion barrier and cross section with neutron number of the target. (From Sc 76)

Figure 7.15 shows the results of an extensive set of measurements performed by Scobel et al. at Rochester and Brookhaven with ^{35}Cl projectiles [Sc 76]. From these data, fusion barriers and barrier radii were deduced for a large number of target isotopes. It was shown that these parameters depend systematically on the mass and charge of the target nuclei, in approximate agreement with simple macroscopic fusion models (see Table 7.7).

Finally, Fig. 7.16 combines results obtained by Gauvin et al. at Orsay [Ga 75] and by Britt et al. at Berkeley [Br 76] for the system ^{40}Ar + ^{121}Sb. In this case, appreciable yields of reaction products corresponding to a (nearly) symmetric division of the compound system are observed. Two types of data points are shown in Fig. 7.16: the lower points give the measured cross sections for heavy recoil production (σ_{er}), and the upper points give maximum values of the fusion cross section, derived by adding to σ_{er} the total cross section for nearly symmetric division as measured by Tamain et al. [Ta 75]. We note that the latter may well contain comparable contributions from fusion--fission [process (i)] and quasi-fission [process (ii)] as indicated by an analysis of the neighbouring system ^{40}Ar + ^{109}Ag due to Britt et al. [Br 76]. The curves shown in Fig. 7.16 represent various theoretical predictions as explained in the caption.

Additional examples of fusion cross section measurements are given in Sect. 7.4, where we present comparisons of experimental data with theoretical predictions. Surveys of experimental results, including numerical tables with

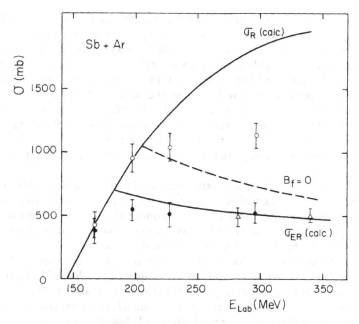

Fig. 7.16. Fusion excitation functions for the system ^{40}Ar + ^{121}Sb. The lower points are cross sections for evaporation-residue formation (full circles from Ga 75, open triangles from Br 76) and the upper points are obtained by adding the fission cross sections from Ta 75. The full curves give theoretical predictions of the total reaction cross section (upper) and the cross section for evaporation residue formation (lower). The dashed curve results from the assumption that all angular momenta contribute to the cross section for which the calculated fission barrier of the compound nucleus (B_f) is larger than zero. (From Ga 75)

fusion cross sections and limiting angular momenta, have been presented by Lefort and others [Le 74a,b, Ga 74a, Pl 74a, Gr 74a,b, Le 75, Le 76a,b, Le 77]. In Table 7.5 we give a list of published experimental work on fusion cross sections; further information may be found in the references quoted there. In the following we try to summarize qualitatively the experimental evidence on fusion cross sections. We consider first the "fusion probability" (P_{fu}), defined here as the ratio of the cross section for complete fusion to the total reaction cross section (7.7). We note that the latter quantity is usually not measured directly but either estimated on the basis of systematics [using either the classical formula (3.39), or the optical model] or deduced from elastic scattering data (see Sect. 3.3). Both procedures should be accurate to within a few percent, if applied carefully.

For light systems ($A_1, A_2 \lesssim 40$), the fusion probability is close to unity at bombarding energies near the fusion barrier. At much higher energies, of the order of two to five times the barrier, the fusion probability is found to decrease to values of the order of 50%. For systems involving at least one heavy partner (A_1 or $A_2 > 100$), on the other hand, surface reactions are usually significant—if not dominant—already near the interaction barrier. The fusion probability of these heavier systems in general depends only weakly on bombarding energy between about 1.1 and 2 times the barrier. For a given target nucleus or projectile it decreases strongly with increasing mass number of the other fragment [see (7.20)]. In particular, using Kr projectiles, significant fusion probabilities have been observed with target masses up to $A_2 \approx 160$, but not for very heavy target nuclei ($A_2 \gtrsim 200$). This trend clearly limits the possibilities to produce very heavy nuclei ($A > 300$) by fusion reactions. In our judgement, no unambiguous evidence for the (temporary) production of such superheavy compound nuclei has yet been presented, in spite of extensive experimental efforts.

Comparing different systems which lead to the same or neighbouring compound nuclei, one finds a tendency for the fusion probability to increase with increasing mass asymmetry in the entrance channel. Again this tendency is most obvious in the case of argon- and krypton-induced reactions [Ga 74d, Pe 75, Br 76; see also Sect. 6.2 and the discussion below].

Another way to discuss fusion cross sections is in terms of the limiting angular momentum for fusion, L_{fu}. This quantity, as defined by (7.5), does not involve the noncompound part of the cross section, and can thus be compared directly with predictions of theoretical models of compound nucleus formation. By definition L_{fu} is practically zero below the fusion barrier, and starts to increase with energy above the barrier, remaining, however, in general below the maximum angular momentum L_{int}. Actual numerical values of L_{fu} depend strongly not only on the bombarding energy, but also on the reduced mass and charge product of the interacting fragments. High values of L_{fu}—in excess of 100—have been reported in particular for argon bombardments of medium and heavy target nuclei [$A_2 > 100$; Si 68, Gu 74, Ga 74c, Ga 75, Ta 75, Br 76]. Due to uncertainties in the interpretation of the observed fission yields, these numbers must be taken with some caution as discussed earlier. It seems clear, nevertheless, that compound nuclei with angular momenta of the order of 100 $\hbar$ have been produced.

The question whether the angular momentum limit L_{fu} is determined by properties of the entrance channel or of the compound nucleus has been discussed extensively in the literature. It is not obvious a priori, of course, that either one or the other influence should alone be decisive, since in principle the transition probability from the entrance channel into the compound nucleus will depend both on the initial and on the final state of the system. In other words, compound-nucleus formation will in general occur only if the initial conditions are favourable *and* if appropriate states are available in the compound nucleus. If, however, one of these conditions is satisfied trivially in a specific situation, then the other one will become the limiting factor for compound nucleus formation.

Important evidence concerning this question has been obtained in experiments involving the same compound system, but different entrance channels. A list giving details of relevant published work is presented in Table 7.6. The first experiment of this type was performed by Zebelman, Miller, and collaborators, who studied the formation of the compound nucleus ^{170}Yb at 107 MeV excitation with ^{11}B, ^{12}C, ^{16}O, and ^{20}Ne projectiles [Ze 73, Ze 74, Ko 74]. Compound-nucleus formation was detected by observing the yields of evaporation residues and fission fragments (the latter contribute, however, only between 0.6 and 6.1 percent to the measured fusion cross sections and are thus not essential for the interpretation of the results). A clear variation of L_{fu} with projectile mass was observed (see Table 7.6); it was concluded that fusion in this case is limited by the entrance channel and not by the compound nucleus. This conclusion is supported by the results of a number of similar experiments performed by other groups involving, however, different compound nuclei and bombarding energies. As an example we show in Fig. 7.17 the correlation between limiting angular momentum and excitation energy for the compound nucleus ^{150}Gd, when produced either by ^{16}O bombardment of ^{134}Ba, or by ^{40}Ar bombardment of ^{110}Pd [De 77a]. Here the fusion cross sections (σ_{er}) were deduced from measurements of delayed activities by gamma spectroscopy. Again the two entrance channels exhibit a clearly different behaviour with respect to L_{fu} as a function of excitation energy. We can

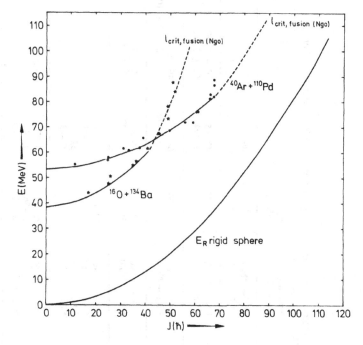

Fig. 7.17. Limiting angular momenta (J) for production of the compound nucleus ^{150}Gd through the entrance channels ^{16}O + ^{134}Ba and ^{40}Ar + ^{110}Pd as a function of excitation energy (E). The points are deduced from measured cross sections for evaporation-residue formation and show clearly a dependence on the entrance channel. The upper curves are theoretical predictions of the fusion cross section for the two entrance channels, and the lower curve represents the rotational energy of a rigid sphere. (From De 77a)

Table 7.6. Studies of the same compound nucleus via different entrance channels

Compound nucleus	Entrance channel	E_{ex} [MeV]	L_{int}[a]	L_{fu} (er)[b]	L_{fu} (er + ff)[c]	Observed quantity[d]	References
^{108}Sn	^{16}O + ^{92}Mo	148	89	55 ± 5		σ_{er}, σ_f	Le 77
	^{52}Cr + ^{56}Fe	100	94	52 ± 5			
^{117}Te	^{14}N + ^{103}Rh	71, 107	43, 65	40 ± 5, 52 ± 5		α-ang. dist.	Ga 74b
	^{40}Ar + ^{77}Se		56, 94	52 ± 3, 70 ± 5			
^{149}Tb	^{40}Ar + ^{109}Ag	71–194	68–170	49–71	49–103	σ_{er}, σ_f	Gu 74, Br 76
	^{84}Kr + ^{65}Cu	137, 185	156, 192	92 ± 20, 77 ± 10	128 ± 15, 150 ± 9		
^{150}Gd	^{16}O + ^{134}Ba	38–92	0–67	50 ± 3°		σ_{er}	De 77a,b
	^{40}Ar + ^{110}Pd	55–125	31–122	66 ± 4°			
^{158}Er	^{16}O + ^{142}Nd	50–110	37–81			σ_{xn}	Ga 74d, Ca 75, Le 76b, Le 77
	^{40}Ar + ^{118}Sn		37–117				
	^{84}Kr + ^{74}Ge		52–141				
^{159}Tm	^{63}Cu + ^{96}Zr		47–137				
^{164}Er	^{12}C + ^{152}Sm	35–49	0–27	0–26		σ_{er}, γ-mult.	Br 75, Ha 75
	^{16}O + ^{148}Nd	42–56	0–34	0–31			Br 77
	^{40}Ar + ^{124}Sn	60–120	61–126	35–65			

Table 7.6. (*Continued*)

Compound nucleus	Entrance channel	E_{ex} [MeV]	L_{int} [a]	L_{fu} (er) [b]	L_{fu} (er + ff) [c]	Observed quantity [d]	References
^{166}Er	^{4}He + ^{162}Dy	16–25	0–13	0–13		σ_{er}, γ-mult.	Br 75, Ha 75
	^{7}Li + ^{159}Tb	31–44	0–18	0–18			
	^{16}O + ^{150}Nd	42–61	0–38	0–34			
	^{18}O + ^{148}Nd	42–58	0–36	0–30			
^{170}Yb	^{11}B + ^{159}Tb	107	63	40 ± 3	40 ± 3	σ_{er}, σ_f	Ze 73, Ze 74, Ko 74
	^{12}C + ^{158}Gd		68	46 ± 4	46 ± 4		
	^{16}O + ^{154}Sm		75	57 ± 4	58 ± 4		
	^{20}Ne + ^{150}Nd		79	68 ± 6	70 ± 6		
^{181}Re	^{12}C + ^{169}Tm	60–100	40–66			σ_f	Si 64
	^{16}O + ^{165}Ho	60–130	41–91				
	^{22}Ne + ^{159}Tb	60–170	43–127				

[a] Calculated classically using (3.39), (3.40) and an empirical nucleus–nucleus potential.
[b] Limiting angular momentum for evaporation residue formation.
[c] Limiting angular momentum for complete fusion, including fusion-fission.
[d] σ_{er}: cross section for evaporation residue formation;
σ_f: total cross section for emission of fission and quasi-fission fragments;
σ_{xn}: cross section for (HI-, xn) reactions;
α-ang. dist.: angular distribution of evaporated α particles;
γ-mult.: γ-multiplicity distribution.
[e] Taken at $E_{ex} = 80$ MeV (see Fig. 7.17).

summarize the available evidence by stating that the fusion cross section appears to be limited by entrance-channel dynamics at moderate bombarding energies and for not too heavy systems, i.e. under conditions where compound nucleus formation can be clearly identified.

The approach used in these studies—to produce the same compound system via different entrance channels—has been very fruitful in providing insight not only into the question of L_{fu}, but also into other aspects of the reaction mechanism. By definition the decay properties of a given compound nucleus, with specified excitation energy and angular momentum distribution, should not depend on previous history. Hence any differences in absolute or relative yields of various reaction products (corrected for phase space) must reflect specific entrance-channel effects, like pre-equilibrium emission of particles or different angular momentum distributions of the compound nucleus. In the following we give some examples.

In a number of cases the cross sections for fission decay (or symmetric fragmentation) of a given compound system were measured using different target–projectile combinations. Quite different results were obtained when producing the compound system ^{108}Sn either via ^{16}O $+$ ^{92}Mo or via ^{52}Cr $+$ ^{56}Fe: in the former case, the yield of fission-like products accounted for less than 10^{-3}, and in the latter case for approximately 60% of the total reaction cross section [Le 77]. Although the two numbers are not strictly comparable due to the different excitation energies involved (see Table 7.7), they clearly indicate that compound nucleus fission is unimportant in this mass region, and that strong yields of symmetric products must therefore arise from quasi--fission. This result is in sharp contrast to observations made in studies of very heavy compound systems. As discussed in Sect. 6.2, a comparison of the systems ^{40}Ar $+$ ^{238}U and ^{84}Kr $+$ ^{186}W (corresponding to similar compound systems) shows that symmetric fragmentations are present with the more asymmetric entrance channel, but absent with the more symmetric entrance channel [Pe 75]. This strongly suggests that the symmetric products in the argon--induced reaction arise at least partly from fusion-fission.

7.4   Fusion Above the Barrier: Models and Limitations

In this section we discuss to what extent experimental fusion cross sections can be understood quantitatively in terms of simple models. A first crude classification of the various approaches is obtained by distinguishing models based on entrance channel dynamics, and models concerned with properties of the compound nucleus.

Much of the early discussion of the fusion problem was dominated by compound-nucleus arguments. A macroscopic approach was followed in the model of the "rotating liquid drop" developed by Cohen, Plasil, and Swiatecki [Co 63, Co 74, Pl 74b]. These authors studied the deformation of a nucleus under the influence of angular momentum, assuming rigid rotation and shapes of axial symmtery. The results were expressed in terms of two dimensionless parameters x and y, defined by

$$x = \frac{E_C}{2E_s} \approx \frac{Z^2}{50A}, \tag{7.21}$$

$$y = \frac{E_{Rot}}{E_s} \approx \frac{2L^2}{A^{7/3}}. \tag{7.22}$$

Here E_C, E_s, and E_{Rot} are the Coulomb, surface and rotational energies, respectively, of a spherical drop with mass number A and charge number Z. The parameter x is usually referred to as "fissility parameter". For $x = 1$ and zero angular momentum ($y = 0$), the system becomes unstable against prompt fission, i.e. the fission barrier B_f vanishes. For $x < 1$, on the other hand, B_f decreases with increasing angular momentum L and vanishes at a finite value of L. The latter is shown as a function of compound nucleus mass number A in Fig. 7.18 and may be interpreted as an upper limit for L_{fu}, the limiting angular momentum for fusion. This follows from the argument that in the absence of a fission barrier, the lifetime of the composite system is of the order

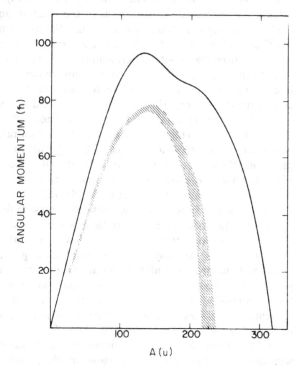

Fig. 7.18. Liquid-drop model prediction of the angular momentum at which the fission barrier of beta-stable nuclei with mass number A vanishes (solid curve). The hatched area indicates the angular momentum, at which the fission barrier is equal to the neutron separation energy. (From Pl 74b)

of the vibrational period (see Table 6.1) and hence a compound nucleus in the classical sense—i.e. with a lifetime much larger than characteristic nucleonic periods—should not exist. Indeed most of the measured fusion cross sections are well below the limits predicted in this manner and thus point to some other mechanism restricting fusion. However, in a number of cases—involving, in particular, ^{40}Ar projectiles and intermediate or heavy target nuclei (see Fig. 7.16)—experimentally deduced values of L_{fu} exceed the limits corresponding to $B_f = 0$ [Si 68, Na 75, Ga 75a, Br 76, Le 76 b]. Whether these discrepancies arise from a defect of the model calculation or from problems associated with the interpretation of the experimental data is not yet clear.

Further restrictions on the fusion cross section may be imposed by the requirement that levels must exist in the compound nucleus at appropriate excitation energy E_{ex} and angular momentum L. Considering the E_{ex} versus $J(\approx L)$ plane it follows that compound nuclei can only be produced to the left of the "y-rast line" [Gr 67b] which connects the levels of lowest excitation energy for each spin value.

A very crude first-order estimate of the y-rast line may be obtained macroscopically by identifying the y-rast energy with the rotational energy of a rigidly rotating sphere with angular momentum L (see Fig. 7.17). More refined calculations include deformation, as in the rotating liquid-drop model, or apply microscopic methods based on the Fermi-gas model or the single--particle shell model [Gr 67a, b; Hi 69]. Again, in most cases where comparisons have been made with experimental fusion cross sections, it was found that compound-nucleus formation was not limited by the nonexistence of levels in the compound nucleus [see, for example, Na 72]. This conclusion is consistent with the experimental results discussed in Sect. 7.3 which clearly demonstrate the importance of entrance-channel effects.

A first attempt to relate fusion cross sections to entrance channel dynamics was made by Kalinkin and Petkov [Ka 64], but was not successful in reproducing the experimentally observed systematics (see, for example, Na 70, Na 72). Later the problem was analysed semiquantitatively by Swiatecki [Sw 70] and by Swiatecki and Bjornholm [Sw 72], with particular emphasis on heavy compound systems such as those of interest in fission and superheavy element production. These authors discussed the basic ingredients of a quantitative description: the choice of a few, essential degrees of freedom ("elongation" α_2, mass asymmetry α_3, and "necking" α_4), and of the corresponding functions describing potential energy, inertia, and dissipation. Important concepts like the "critical asymmetry" and the "misalignment of the fusion and fission valleys" (in the α_2–α_4 plane) were introduced, and the consequences of friction for the possible production of superheavy compound nuclei at low excitation were considered (see Sect. 7.5).

A breakthrough towards a quantitative understanding of fusion cross sections for the lighter systems occured in 1973, when several authors proposed classical scattering models of the general type discussed in Sect. 6.4 [Ba 73a,b; Ba 74a,b; Gr 74a,b; Wi 73]. The basic assumptions common to these models can be summarized as follows:

(i) Deformation and mass transfer are neglected, i.e. the sizes and shapes of the fragments are considered "frozen".

(ii) The relative motion is treated classically assuming phenomenological (conservative and nonconservative) two-body forces.

(iii) For fusion to occur, the system must penetrate to distances where strong dissipative forces remove energy from the relative motion and hence cause trapping in an attractive region of the effective two-body potential.

In order to understand the success of this simple approach to the fusion problem one must realize that these conditions are only needed for the ingoing part of the trajectory up to some "point of no return", where the eventual fate of the system is decided. This initial stage of the reaction is too short to permit appreciable changes in macroscopic shape or fragmentation. Although complicated rearrangements of nuclear matter will in general occur during later stages of the reaction, their details have no pronounced effect on the magnitude of L_{fu}.

One rather obvious result in this type of model is that fusion is possible only for systems where the effective two-body potential for zero angular momentum has a maximum $(= B_{\text{fu}})$. This condition is sometimes expressed by stating that the potential must have a "pocket", in which the system can be trapped. As a consequence, fusion is completely forbidden for very heavy systems, like $^{238}\text{U} + ^{238}\text{U}$, at all energies.

Apart from the common features discussed above, different assumptions have been made by different authors concerning the nucleus–nucleus potential, the strength and radial dependence of the friction forces, and the coupling between relative and intrinsic angular momentum. In the following we describe in some detail the model proposed by Bass [Ba 73a,b; Ba 74 a,b]; subsequently we comment briefly on its relationship to other models.

We begin by defining three different regions in radial space: an outer region (region I) at $r > R_{\text{int}}$, an intermediate region (region II) at $R_{12} < r < R_{\text{int}}$, and an inner region (region III) at $r < R_{12}$ (see Fig. 7.1). Here as before we denote by $R_{12} = R_1 + R_2$ the sum of the half-density matter radii of the two interacting nuclei (assumed of spherical shape). As shown in Sect. 3.3, the dominant process in region I is elastic scattering. Upon entering region II, the system is eliminated from the elastic channel; in the language of the optical model, it undergoes "absorption". In a classical description, however, intrinsic quantum states are not specified, and therefore this type of absorption is irrelevant except for the accompanying changes of macroscopic variables. We assume now that the system continues to follow an undisturbed classical trajectory in region II without loss of relative energy or angular momentum. This implies that friction is completely neglected in regions I and II; throughout both regions the system remains in the "quasi-elastic channel" which is macroscopically indistinguishable from the elastic channel. Finally at $r = R_{12}$, strong radial and tangential friction forces are assumed to act with the result that the relative radial momentum drops sharply to zero and the relative angular momentum is lowered according to the rolling or sticking condition (see Table 6.4). Thus the system is effectively prevented from entering region III. At this point it may be "absorbed" from the quasi-elastic channel by collapsing to form a compound nucleus. Alternatively, the fragments may separate again and hence undergo inelastic scattering. In order to decide which of these two possibilities is realized we use a very simple criterion: fusion is assumed to take place if the effective radial force at the contact point is attractive, and scattering, if that force is repulsive.

Obviously this model represents a gross simplification of the real situation. It may be argued, however, that a detailed description of the interaction in region II is not necessary to calculate fusion cross sections, as long as the essential (geometrical) features of the problem are correctly incorporated. Crucial ingredients, in this respect, are the nucleus–nucleus potential, and the distance at which fragments moving along the limiting trajectory start to experience strong friction. The choice of the half-density distance for the latter parameter can be justified, for example, by the discussion given in Sect. 6.4, which relates friction to nucleon diffusion across the neck [see (6.39-42)]: at the contact point, nucleons can move freely from one fragment to the other and will thus rapidly dissipate relative momentum. Alternative (but not necessarily unrelated) arguments in favor of a strong increase of friction near the half-density distance may be derived from a consideration of nucleon–nucleon collisions in the overlap region [Ba 74a, Al 77], or of microscopic rearrangements due to level crossings [Gl 74a, Gl 76].

Perhaps the weakest point of the model, as outlined above, is the schematic criterion used to determine the further evolution of the system after complete relaxation of the relative motion has occurred. Clearly at this stage of the reaction comparatively "slow" degrees of freedom like fragment deformation and mass asymmetry come into play, and the stability of the system is no longer determined by the radial degree of freedom alone. A more realistic—but also more complicated—criterion for fusion would be to demand that the shape of the composite system is "inside the fission saddle point" as derived from a multidimensional calculation [Sw 70, Sw 72, Si 74, Ni 77]. As noted earlier, and shown in more detail below, this problem does not affect fusion cross sections for comparatively light systems at moderate energies, but becomes essential for heavy systems or at high energies (high angular momenta).

It is now straightforward to calculate fusion cross sections for a given nucleus–nucleus potential. The general procedure is illustrated in Fig. 7.19, which shows effective potentials for different relative angular momenta L as a function of distance r. In order to result in fusion, a trajectory must reach the shaded region which is bounded by the potential for $L = 0$, the vertical straight line given by $r = R_{12}$, and a curve connecting the maxima of the effective potentials for different angular momenta. We first consider an incident energy (in the centre-of-mass system) $E < E_1$ between the fusion barrier and the upper boundary E_1 of the shaded region. Here the limiting angular momentum for fusion, L_{fu}, is that value of L for which the maximum of the effective potential equals E. Denoting by r_{fu} the distance at which this maximum occurs, the assumption of energy conservation yields

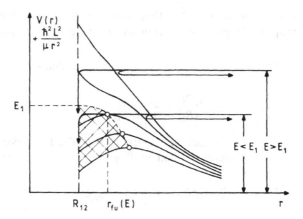

Fig. 7.19. Effective two-body potentials for different values of angular momentum L, and schematic trajectories leading to either scattering or fusion

$$E = V(r_{fu}) + \frac{\hbar^2 L_{fu}^2}{2\mu r_{fu}^2}, \tag{7.23}$$

$$L_{fu}^2 = \frac{2\mu}{\hbar^2} r_{fu}^2 [E - V(r_{fu})]. \tag{7.24}$$

By definition the distance r_{fu} is that value of r which minimizes the expression $r^2 [E - V(r)]$ for $r \geq R_{12}$, and in general depends on energy. The fusion cross section is now given by the well-known classical formula

$$E\sigma_{fu} = \frac{\pi\hbar^2}{2\mu} L_{fu}^2 = \pi r_{fu}^2 [E - V(r_{fu})]. \tag{7.25}$$

For incident energies $E > E_1$, (7.23–25) remain valid, provided the energy loss by tangential friction at $r = R_{12}$ is sufficient to reduce the total (potential + centrifugal) relative energy at that point to a value below E_1. As before, r_{fu} is the classical turning point of the limiting trajectory (for $L = L_{fu}$); however, there is no longer a maximum of the effective potential for $r > R_{12}$, and thus we have $r_{fu} = R_{12}$. We emphasize that tangential friction has important qualitative consequences for the fusion cross section in this energy region: its effect is to reduce the centrifugal forces at contact for a given asymptotic angular momentum, and hence to increase the cross section.

At still higher bombarding energies, not all fragment pairs penetrating to R_{12} can be stabilized by tangential friction, i.e. can reach the shaded region in Fig. 7.19. This situation will arise for incident energies

$$E > E_2 \approx E_1 + \frac{1 - f^2}{f^2} [E_1 - V(R_{12})], \tag{7.26}$$

where $f < 1$ is a numerical factor giving the reduction in orbital angular momentum at $r = R_{12}$. Depending on the assumed angular momentum coupling, f is given by either of the following two expressions (see Table 6.4):

$$f = \frac{5}{7} \qquad \text{(rolling)}, \qquad (7.27)$$

$$f = \left[1 + \frac{2}{5} \left(\frac{A_1 R_1^2}{A_{12} R_{12}^2} + \frac{A_2 R_2^2}{A_{12} R_{12}^2} \right) \right]^{-1} \qquad \text{(sticking)}. \qquad (7.28)$$

For $E > E_2$ the limiting angular momentum for fusion will no longer increase with increasing bombarding energy, but will saturate at

$$L_{\text{fu}}^2(E_2) = \frac{2\mu}{\hbar^2} R_{12}^2 [E_2 - V(R_{12})] \approx \frac{1}{f^2} L_{\text{fu}}^2(E_1). \qquad (7.29)$$

We can summarize this discussion of the energy dependence of the fusion cross section by first noting that three different energy regions must be distinguished: one at low energies ($B_{\text{fu}} \leq E \leq E_1$), one at intermediate energies ($E_1 \leq E \leq E_2$), and one at high energies ($E > E_2$). In the first region, the fusion cross section is given by (7.25) with an energy-dependent value of r_{fu} ($> R_{12}$), obtained by minimizing the expression $r^2 [E - V(r)]$. In the second region, we have $r_{\text{fu}} = R_{12}$, and hence

$$E\sigma_{\text{fu}} = \pi R_{12}^2 [E - V(R_{12})] \qquad (E_1 \leq E \leq E_2). \qquad (7.30)$$

Finally, in the third region, the limiting angular momentum remains constant, and the fusion cross section is given by

$$E\sigma_{\text{fu}} = \pi R_{12}^2 [E_2 - V(R_{12})] \qquad (E > E_2). \qquad (7.31)$$

It should be emphasized that the model predictions for region three are more speculative than those for regions one and two, as they depend explicitly on the stability condition discussed earlier, and on assumptions concerning angular momentum transfer by tangential friction.

It remains to specify the nucleus–nucleus potential $V(r)$. We recall that, in the present model, this potential is only required for $r \geq R_{12}$. In contrast to optical model potentials, it is real by definition; the only "absorptive process" is fusion which occurs at $r = R_{12}$.

The nuclear part of the potential has been derived from the liquid-drop model, including effects of the finite range of nuclear forces, in the following way [Ba 74a]: We consider nuclear matter in two semi-infinite regions, bounded by plane, parallel, diffuse surfaces at a distance s (measured between the planes of half-maximum density, see Fig. 7.20a). For the surface energy per unit area of this system we use the following ansatz:

$$\frac{dE_s}{dS} = 2\gamma[1 - e(s)] \qquad (s \geq 0). \qquad (7.32)$$

Fig. 7.20. Liquid-drop model surface energy with finite range of interaction for two different geometries: **a** two plane, parallel surfaces; **b** two spherical nuclei. (From Ba 74a)

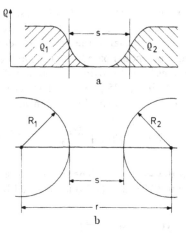

Here γ is the specific surface energy of the liquid-drop model, and the function $e(s)$ has the properties

$$e(s) = 1 \text{ for } s = 0; \quad e(s) \to 0 \quad \text{for } s \to \infty. \tag{7.33}$$

For two spheres with half-density radii R_1 and R_2 (Fig. 7.20b) one obtains from (7.32) by integration, retaining only terms of lowest order in s/R_1 and s/R_2,

$$E_S = \gamma\left(S_1 + S_2\right) - 4\pi\gamma\,\frac{R_1 R_2}{R_1 + R_2}\int_s^\infty e(s')\,ds'. \tag{7.34}$$

The nuclear potential can now be identified with the difference in surface energies between finite and infinite separation s:

$$V_N(s) = -4\pi\gamma\,\frac{R_1 R_2}{R_{12}}\int_s^\infty e(s')ds' = -4\pi\gamma\,\frac{R_1 R_2}{R_{12}}f(s) = -\frac{R_1 R_2}{R_{12}}g(s). \tag{7.35}$$

The importance of this result lies in the fact that the functions $e(s), f(s)$, and $g(s)$ are universal functions of the separation coordinate s which do not depend on properties of the fragments involved. Therefore the system dependence of the nuclear potential is completely specified by the factor $R_1 R_2/R_{12}$, which is sometimes called a reduced radius. In addition, (7.33) leads to the following restriction for the function $f(s)$:

$$\frac{df(s)}{ds} = -1 \quad \text{for} \quad s = 0. \tag{7.36}$$

Results equivalent to (7.35) and (7.36) have been obtained independently by Wilczynski [Wi 73] and, in a somewhat more general context, by Randrup, Swiatecki, and Tsang [Ra 74, Bl 77; see also My 74]. The latter authors have expressed the contents of these equations as a theorem (the "proximity theorem") which states that "the force between rigid gently curved surfaces is proportional to the potential per unit area between flat surfaces." Forces and potentials based on (7.35) and (7.36) are often referred to as "proximity forces" or "proximity potentials", respectively.

Using (7.35), the total effective nucleus–nucleus potential for angular momentum L can be written as

$$V_L(r) = V(r) + \frac{\hbar^2 L^2}{2\mu r^2} = \frac{Z_1 Z_2 e^2}{r} + \frac{\hbar^2 L^2}{2\mu r^2} - \frac{R_1 R_2}{R_{12}} g(s), \tag{7.37}$$

with $s = r - R_{12}$. The simplest possible ansatz for the function $g(s)$ consistent with (7.33) and (7.36) is obtained by writing [Ba 74a]

$$e(s) = \exp\left(-\frac{s}{d}\right), \tag{7.38a}$$

$$f(s) = \int_s^\infty e(s')ds' = d \exp\left(-\frac{s}{d}\right), \tag{7.38b}$$

$$g(s) = 4\pi\gamma f(s) = 4\pi\gamma d \exp\left(-\frac{s}{d}\right), \tag{7.38c}$$

where the parameter d is a measure of the range of nuclear forces. Setting further $R_i = r_0 A_i^{1/3}$ ($i = 1, 2$) and $4\pi R_{iy}^2 = a_s A_i^{2/3}$, we arrive at the following potential:

$$V_L(r) = \frac{Z_1 Z_2 e^2}{r} + \frac{\hbar^2 L^2}{2\mu r^2} - \frac{d a_s A_1^{1/3} A_2^{1/3}}{R_{12}} \exp\left(-\frac{s}{d}\right). \tag{7.39}$$

This formula contains three parameters d, r_0, and a_s, which—in the spirit of the liquid-drop model—should be determined empirically by fitting average nuclear properties. The following values have been used: $d = 1.35$ fm (based on interaction and fusion barriers), $r_0 = 1.07$ fm (based on electron scattering data), and $a_s = 17$ MeV (based on ground-state masses).

The influence of system properties on the potential given by (7.39) is conveniently expressed in terms of the dimensionless parameters

$$x_{12} = \frac{Z_1 Z_2 e^2}{R_{12} a_s A_1^{1/3} A_2^{1/3}} = \frac{1}{12.6} \frac{Z_1 Z_2}{A_1^{1/3} A_2^{1/3}(A_1^{1/3} + A_2^{1/3})}, \tag{7.40}$$

$$y_{12} = \frac{\hbar^2}{2\mu R_{12}^2 a_s A_1^{1/3} A_2^{1/3}} = 1.06 \frac{A_1 + A_2}{A_1^{4/3} A_2^{4/3}(A_1^{1/3} + A_2^{1/3})^2}, \tag{7.41}$$

which are defined in analogy to the parameters x, y of the rotating liquid-drop model [see (7.21), (7.22); note, however, that y_{12} does not contain the factor L^2]. The potential now assumes the form

$$V_L(r) = A_1^{1/3}A_2^{1/3}a_s\left[x_{12}\left(\frac{R_{12}}{r}\right) + y_{12}L^2\left(\frac{R_{12}}{r}\right)^2 - \frac{d}{R_{12}}\exp\left(-\frac{s}{d}\right)\right], \qquad (7.42)$$

and the radial force at the contact point $(r = R_{12})$ is given by

$$\left(\frac{dV_L}{dr}\right)_{r=R_{12}} = \frac{A_1^{1/3}A_2^{1/3}a_s}{R_{12}}(1 - x_{12} - 2y_{12}L^2). \qquad (7.43)$$

Clearly this force is attractive only for values of L which satisfy the condition

$$L^2 \le \frac{1 - x_{12}}{2y_{12}} = L_{fu}^2(E_1), \qquad (7.44)$$

where E_1 is defined as in Fig. 7.19 and (7.26) and (7.29). It follows immediately from (7.43) and (7.44) that for systems with $x_{12} > 1$, the potential has no "pocket" for $r > R_{12}$, and hence fusion is forbidden in the present model. Equation (7.21) shows, on the other hand, that a compound nucleus, if it could be formed under such circumstances, would have no fission barrier. This indicates that the distinction between limitations arising from entrance--channel dynamics and from compound nucleus properties tends to break down for systems approaching the limit $x_{12} = 1$.

It is instructive to consider the consequences of the present model in an L^2 versus E representation. This is shown for a typical system of intermediate mass and charge in Fig. 7.21. According to the model, the shaded area should

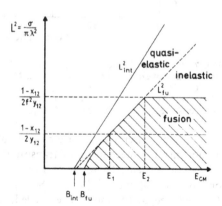

Fig. 7.21. Angular momentum limits for different processes as functions of bombarding energy. For definition of the symbols see text. (From Ba 74a)

contribute to the fusion cross section; its upper boundary represents—apart from a constant factor $\pi\hbar^2/2\mu$—the product $E\sigma_{fu}$ as a function of centre-of--mass energy E. The three energy regions discussed above are indicated. The characteristic energies E_1, E_2 can be expressed in terms of the parameter x_{12} as follows:

$$E_1 = \frac{Z_1 Z_2 e^2}{R_{12}}\left(1 + \frac{1 - x_{12}}{2x_{12}} - \frac{1}{x_{12}}\frac{d}{R_{12}}\right), \tag{7.45}$$

$$E_2 = \frac{Z_1 Z_2 e^2}{R_{12}}\left(1 + \frac{1 - x_{12}}{2f^2 x_{12}} - \frac{1}{x_{12}}\frac{d}{R_{12}}\right). \tag{7.46}$$

The model predictions have been compared extensively with experimental data (see, for example, Ba 74a, Na 75, Br 75, Br 76, Sc 76, Le 76b). An example is shown in Fig. 7.22, taken from an article of Namboodiri et al. [Na 75]. From many such comparisons (both published and unpublished), we can conclude that the model reproduces the overall dependence of the fusion cross section on system parameters and bombarding energy with remarkable accuracy (see also Figs. 7.26 and 7.27). Significant discrepancies are observed, on the other hand, for some light systems where the use of average radius parameters yields unrealistic fusion barriers (see Table 7.7), and in situations where $E \gtrsim E_2$, i.e. where L_{fu} is predicted to saturate.

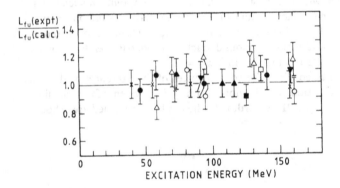

Fig. 7.22. Comparison of experimental and calculated values of L_{fu} for different systems as a function of excitation energy. (From Na 75):

● $^{12}C + ^{27}Al$;	○ $^{12}C + Ti$;	× $^{12}C + Ni$;
■ $^{14}N + ^{27}Al$;	□ $^{14}N + Cr$;	▽ $^{14}N + Ni$;
▲ $^{40}Ar + ^{77}Se$;	▼ $^{40}Ar + ^{107}Ag$;	△ $^{40}Ar + ^{121}Sb$

The latter discrepancies are of particular interest as they could provide insight into the mechanism of transition from the entrance channel to the compound nucleus. There are indications from experiments with ^{14}N beams on ^{27}Al [Na 75] and ^{12}C [St 76a, St 77] targets that the saturation limit of L_{fu} predicted by the model is exceeded, i.e. that saturation is reached at a higher bombarding energy. This is in contrast to some results obtained with ^{32}S and ^{40}Ar beams [Ko 75, Br 76, Si 68] where the deduced values of L_{fu} are either lower than or consistent with the predicted saturation values. Unfortunately, in these latter experiments the distinction between compound nucleus formation and deep-inelastic scattering (or quasi-fission) often meets with technical and conceptual problems which throw doubts on the significance of a comparison with model calculations.

An attempt has been made, nevertheless, to improve the applicability of the model to heavy systems ($x_{12} \lesssim 1$) by explicitly including axially symmetric deformations of the fragments after contact has been established [Ba 74a]. Under the influence of the Coulomb and centrifugal forces the fragments assume prolate deformations, which tend to reduce the nuclear attraction and thus lower the saturation value of L_{fu}. As a further consequence, fusion cannot take place in nearly symmetric systems for a certain range of x_{12} values below unity even at zero angular momentum, as illustrated in Fig. 7.23.

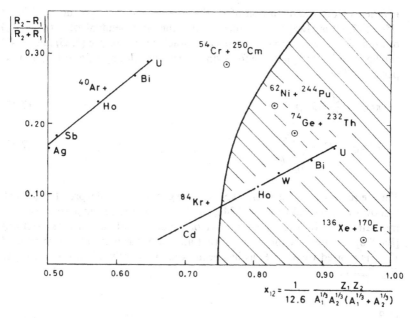

Fig. 7.23. Calculated regions of fusion (unshaded) and quasi-fission (shaded) for two originally spherical fragments in contact at zero angular momentum. The straight lines connect systems produced by either ^{40}Ar or ^{84}Kr bombardment of different target nuclei; the symbol ⊙ denotes systems of interest for the production of superheavy nuclei. (From Ba 74a)

For the lighter systems and moderate energies, the precision of the model predictions can be improved by introducing certain refinements. One obvious step in this direction is to calculate the half-density radii R_1, R_2 not with an average radius parameter r_0, but with some prescription taking into account the empirical mass dependence of these radii [see, for example, My 73 and (3.12), (7.50)]. This results, in particular, in a significantly improved reproduction of fusion barriers for light systems (see Table 7.7).

Next we can modify our simple exponential ansatz for the shape functions $e(s)$, $f(s)$, and $g(s)$ (7.38). One way to proceed would be to derive these functions from a model calculation, as in the "proximity potential" of the Berkeley group (see Fig. 7.29). Alternatively, we can try to deduce an empirical shape function from a systematic analysis of experimental fusion cross sections [Ba 77b]. Such an analysis can test the consistency of the model and—if successful—provide a basis for reliable predictions. The required input data are fusion excitation functions measured with good absolute precision ($\leq 10\%$) over a large range of bombarding energies. The analysis is based on (7.25); by differentiation with respect to energy, we obtain

$$\frac{d}{dE}(E\sigma_{\text{fu}}) = \pi r_{\text{fu}}^2 + \frac{dr_{\text{fu}}}{dE}\left(\frac{d}{dr}\left\{\pi r^2[E - V(r)]\right\}\right)_{r=r_{\text{fu}}}. \tag{7.47}$$

It follows from the definition of r_{fu} that the second term on the right-hand side of (7.47) is always zero, since r_{fu} is either independent of E ($r_{\text{fu}}= R_{12}$) or minimizes the expression in the curly brackets. Therefore (7.25) and (7.47) can be combined to yield the following equations relating r_{fu} and $V(r_{\text{fu}})$ to experimental data:

$$\pi r_{\text{fu}}^2 = \frac{d}{dE}(E\sigma_{\text{fu}}), \tag{7.48}$$

$$V(r_{\text{fu}}) = E - \frac{E\sigma_{\text{fu}}}{\dfrac{d}{dE}(E\sigma_{\text{fu}})}. \tag{7.49}$$

These equations can be given a very simple geometrical interpretation, which is shown schematically in Fig. 7.24a, and has been used to deduce corresponding pairs r_{fu}, $V(r_{\text{fu}})$ from a large set of experimental fusion cross sections [Ba 77b]. We note, without proof, that an analogous method can be applied if σ_{fu} is plotted versus $1/E$, as has been done frequently in the literature (see Fig. 7.24b).

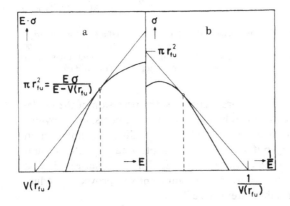

Fig. 7.24. Graphical method used in the classical analysis of fusion excitation functions: **a** plotting $E\sigma$ versus E; **b** plotting σ versus $1/E$. (From Ba 77b)

In order to deduce the function $g(s)$, the results of such an analysis must be referred to a common scale of separation s. This means that a prescription to calculate the radii R_1, R_2 must be adopted. The derivation of (7.34) and (7.35) implies that R_1 and R_2 are much larger than the range of the nuclear force; in this limit, the half-density radius is the appropriate choice for R. For finite nuclei, R should be interpreted as an "effective half-density radius" which depends, in principle, on properties of the interaction [Ra 74, Bl 77]. We adopt [Ba 77b]

$$R = 1.16 A^{1/3} - 1.39 A^{-1/3} \quad [\text{fm}] , \tag{7.50}$$

where the first term, proportional to $A^{1/3}$, reproduces half-density matter radii in the limit of large A [My 73]. The coefficient in the second term has been chosen—somewhat arbitrarily—to minimize the scatter in the resulting values for the function $g(s)$, but is not sharply defined by the data. It should be noted, however, that the deduced function $g(s)$ depends on the coefficients in (7.50).

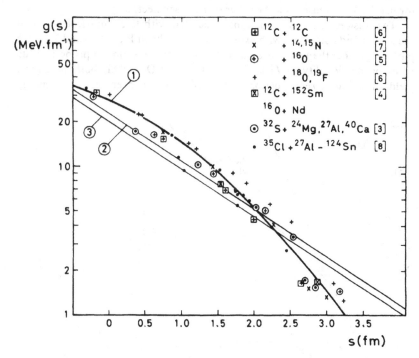

Fig. 7.25. Empirical nucleus–nucleus potential, expressed in terms of the universal function $g(s)$ as defined by (7.35) and (7.50). "Experimental" points for different systems are distinguished by different symbols (for references, see Table 7.5 and Ba 77b). Curve 1 is a fit to the data according to (7.51). The straight lines marked 2 and 3 represent the exponential potential of (7.39) with $R_{12} = r_0(A_1^{1/3} + A_2^{1/3})$ and $r_0 = 1.00$ fm (line 2) or 1.07 (line 3), respectively

The result of the analysis is shown in Fig. 7.25. The "experimental points" for the function $g(s)$ clearly follow a universal trend, which can be fitted with the expression

$$g(s) = \left[A \exp\left(\frac{s}{d_1}\right) + B \exp\left(\frac{s}{d_2}\right) \right]^{-1}, \qquad (7.51)$$

where $A = 0.0300$ MeV^{-1} fm, $B = 0.0061$ MeV^{-1} fm, $d_1 = 3.30$ fm, and $d_2 = 0.65$ fm (see heavy curve in Fig. 7.25). Moreover, the fit function approximately satisfies the condition (7.36). The analysis, therefore, lends strong support not only to the concept of a proximity potential, but also to the assumptions leading to (7.25) for the fusion cross section. The simple exponential potential of (7.38)—also shown in Fig. 7.25 for two typical values of the radius parameter r_0—reproduces the empirical fit function quite well on the average, but falls off too slowly at large distances. We note further that the empirical potential defined by (7.37) and (7.51), together with (7.50), is in excellent agreement with a semiclassical analysis of elastic scattering data performed by Christensen and Winther [Ch 76; see Sect. 3.3].

Two fusion excitation functions calculated with this empirical potential are shown and compared with experimental results in Figs. 7.26 and 7.27 (we plot $E\sigma_{fu}$ rather than σ_{fu}, since the former quantity has simpler properties due to its proportionality to L_{fu}^2). For the system ^{16}O + ^{27}Al, both potentials yield an adequate reproduction of the cross section over the entire energy

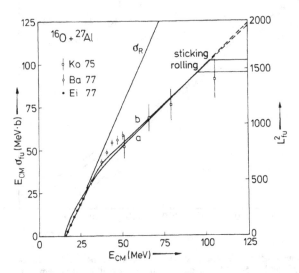

Fig. 7.26. Comparison of calculated and experimental fusion cross sections (expressed as $E\sigma_{fu}$) for the system ^{16}O + ^{27}Al as a function of energy. The curves marked a and b are calculated with the exponential ($r_0 = 1.00$ fm) and empirical potential, respectively. The total reaction cross section σ_R (calculated classically with $R_{int} = R_{12} + 3$ fm) is shown for comparison

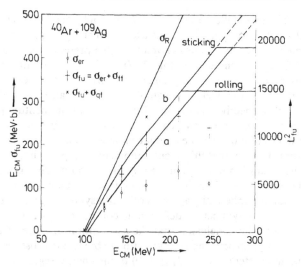

Fig. 7.27. Comparison of calculated fusion cross sections with different experimental cross sections (expressed as $E\sigma$) for the system ^{40}Ar $+$ ^{109}Ag as a function of energy. For explanation of a, b, and σ_R see caption of Fig. 7.26. The experimental data are from [Br 76]

range studied; significant differences between the calculations occur only at the lowest energies, where the calculation with the exponential potential underestimates the barrier (this could be repaired by a slight adjustment of r_0; see Table 7.7). Clearly additional measurements at the higher energies, with smaller errors and extending into the predicted saturation region, would be of interest. Fits of comparable quality to those shown in Fig. 7.26 have been obtained for most of the lighter systems where fusion cross sections have been measured.

For the system ^{40}Ar $+$ ^{109}Ag (Fig. 7.27), the near-barrier data are reproduced somewhat better with the exponential potential. At higher energies, up to about 220 MeV, both calculations provide a satisfactory fit to the data, but at the highest measured energy of 246 MeV, the cross section remains well below the model predictions. This discrepancy, if real, could be attributed to a lowering of the saturation limit to L_{fu} by deformation as discussed earlier and in the following. It should be noted, however, that the experimental data shown in Fig. 7.27 are based on a separation of the total fission yield into components arising from fusion-fission and quasi-fission, respectively, where only the former contributes to the fusion cross section. Such an analysis involves uncertainties which are difficult to judge, but could be large enough to render the present comparison between experiment and model calculation inconclusive. This situation is typical for argon bombardments of medium and heavy targets at higher energies.

The system ^{40}Ar $+$ ^{109}Ag may also be discussed from a somewhat different point of view, which emphasizes properties of the compound nucleus

^{149}Tb. This is illustrated in Fig. 7.28 which correlates the angular momentum J and the compound-nucleus excitation energy E_{ex}. The limiting angular momentum for fusion, as derived from the present model calculations with two different potentials (see Fig. 7.27), is shown as a hatched band. The curves marked E_{min} (J) and E_{sp} (J) give the predictions of the rotating liquid-drop model for the energy at equilibrium deformation and at the fission saddle point, respectively, as a function of angular momentum [Pl 74b]. The difference between these two curves represents the fission barrier and hence their intersection defines the angular momentum at which the fission barrier vanishes. The fact that this angular momentum (approximately 88 $\hbar$) is well below the saturation limits predicted by the fusion model can be understood qualitatively by recalling that the latter arise from a one-dimensional stability criterion for spherical fragments of fixed asymmetry, whereas the fission barrier refers to a symmetric, deformable compound system. Thus we are led once more to the conclusion that the assumption of frozen fragment shapes and fixed asymmetry tends to overestimate the fusion cross section for heavy compound systems ($A_1 + A_2 \gtrsim 150$) in the energy region where L_{fu} is predicted to saturate.

In addition to the limits already discussed, we have included in Fig. 7.28 a vertical line marked $B_f = 9$ MeV. Compound nuclei formed to the left of this line are expected to decay predominantly by neutron evaporation, whereas to

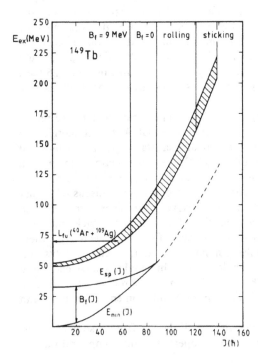

Fig. 7.28. Relationship between various angular momentum limits and excitation energy for the compound nucleus ^{149}Tb. The curves marked $E_{min}(J)$ and $E_{sp}(J)$ are based on liquid-drop-model calculations (Pl 74b)

the right of this line, fission decay should be more probable. The corresponding angular momentum can therefore be interpreted as the limiting angular momentum for evaporation residue formation (a comparison with the data given in Fig. 7.27 shows that the calculation slightly under-estimates the experimental angular momentum limit). The question of competition between neutron evaporation and fission at high angular momentum will be taken up again in Sect. 8.5.

A sensitive test of nucleus–nucleus potentials to be used in fusion calculations is provided by a systematic comparison of measured and calculated barriers and fusion radii. Such a comparison is presented in Table 7.7 for the exponential and empirical nuclear potentials. In order to exhibit the extreme sensitivity of the barrier heights to the (effective) half-density distances, two sets of calculations have been performed for the exponential potential: first the barriers and barrier radii were calculated with R_{12} values derived from (7.50), corresponding to $r_0 = 1.16 - 1.39 (A_1 A_2)^{-1/3}$ fm. Subsequently, the r_0 values were readjusted to fit exactly the experimental barrier heights (values given in parentheses). It can be seen that, except for the lightest systems, the resulting r_0 values differ only slightly from the original values, and vary smoothly with system mass. The conspicuous failure of the exponential potential for very light systems can be traced back to the fact that for these systems the barriers are located at large separation s, where the exponential shape function overestimates the nuclear potential. For heavier systems, the agreement between measured and calculated barriers is of comparable quality for the two potentials. The agreement is usually within $\pm 3\%$ except for some very heavy systems where the experimental values have larger errors. There is a systematic tendency for the calculated barrier radii to be larger than the measured ones, especially with the empirical nuclear potential. Whether this indicates a failure of the potential or of the classical method used to determine experimental barrier radii [based on (7.25), which is expected to become inaccurate close to the barrier] is not clear.

In conclusion, we can state that the classical scattering model described above accounts quantitatively for fusion barriers and fusion cross sections over large ranges of system mass and energy. When carefully applied (i.e. under conditions where fusion is limited by entrance-channel dynamics and can be experimentally identified, and with appropriately chosen radii), the predictive power of the model appears to be comparable to present experimental accuracy. The regime in which the model may be safely applied is bounded at low energies (i.e. near and below the fusion barrier) by the need to use quantum mechanics, and at high energies (or for heavy systems), by the inadequacy of the "frozen-shape approximation" in treating the stability problem of the composite system.

Several authors have proposed fusion models similar in spirit and content to the model described so far. Differences between the various approaches arise from the use of different nucleus–nucleus potentials, and from different assumptions on the radial dependence of dissipative forces. (Note that the latter may be implicit in geometrical postulates like that of a "critical distance

Table 7.7. Comparison of experimental and calculated fusion barriers (B_{fu}) and barrier radii (r_B).

System	Experimental results			Exponential potential[a]			Empirical potential[b]	
	$B_{fu}{}^c$ [MeV]	$r_B{}^d$ [fm]	References	$r_0{}^e$ [fm]	B_{fu} [MeV]	r_B [fm]	B_{fu} [MeV]	r_B [fm]
^{12}C + ^{14}N	7.00	7.0	Co 76	0.91 (0.45)	5.84	8.9	7.03	8.0
^{15}N	6.80	7.35		0.91 (0.50)	5.77	8.9	6.95	8.0
^{16}O	7.50	7.6	Sp 76a	0.92 (0.65)	6.68	8.8	7.95	8.0
^{18}O	7.45	7.8	Sp 76b	0.93 (0.65)	6.55	9.1	7.78	8.3
^{19}F	8.20	7.7		0.93 (0.73)	7.46	8.9	8.76	8.2
^{16}O + ^{27}Al	15.9	8.4	Ei 77	0.98 (0.81)	14.44	8.9	15.98	8.7
^{58}Ni	31.4		Ro 76	1.02 (0.97)	30.4	9.1	31.8	9.5
$^{60-62}$Ni	29.9			1.02 (1.03)	30.05	9.3	31.5	9.6
^{64}Ni	30.8			1.02 (0.97)	29.7	9.4	31.2	9.7
^{63,65}Cu	31.0			1.02 (1.02)	31.0	9.3	32.4	9.7
^{64}Zn	32.5			1.02 (1.01)	32.2	9.3	33.6	9.6
$^{66-68}$Zn	32.5			1.02 (1.00)	31.9	9.3	33.3	9.7
^{70}Zn	32.1			1.03 (1.00)	31.6	9.4	33.0	9.8
^{18}O + ^{58}Ni	30.9			1.02 (0.97)	29.8	9.3	31.2	9.6
$^{60-62}$Ni	30.3			1.02 (0.98)	29.4	9.5	30.9	9.7
^{64}Ni	29.9			1.03 (0.99)	29.2	9.6	30.6	9.9
^{63,65}Cu	30.9			1.03 (1.00)	30.4	9.5	31.8	9.8
^{64}Zn	32.7			1.03 (0.98)	31.6	9.5	33.0	9.8
$^{66-68}$Zn	32.6			1.03 (0.97)	31.3	9.6	32.7	9.9
^{70}Zn	33.2			1.03 (0.93)	31.0	9.6	32.4	10.0
^{12}C + ^{152}Sm	46.8	10.9	Br 75	1.05 (1.01)	45.7	10.3	46.8	10.8
^{16}O + ^{134}Ba	55.5		De 77a	1.05 (1.06)	55.9	10.0	56.4	10.7
^{148,150}Nd	57.8	9.8	Br 75	1.06 (1.08)	58.9	10.3	59.4	11.0
^{32}S + ^{24}Mg	28.3	8.8	Gu 73a,b	1.01 (0.92)	26.7	8.8	28.2	9.1
^{27}Al	29.9	8.8		1.01 (0.95)	28.7	9.0	30.1	9.3
^{40}Ca	43.6	9.35		1.03 (1.05)	44.0	9.0	44.7	9.6
^{58}Ni	59.5	8.5		1.05 (1.06)	60.2	9.3	60.1	10.0
^{35}Cl + ^{27}Al	30.7	8.4	Sc 76	1.02 (1.00)	30.3	9.0	31.6	9.4
^{48}Ti	49.2	8.7		1.04 (1.06)	49.9	9.3	50.4	10.0
^{58}Ni	61.3	9.0		1.05 (1.10)	63.5	9.3	63.2	10.1
^{60}Ni	61.0	9.2		1.05 (1.10)	63.0	9.3	62.8	10.2
^{62}Ni	60.8	9.6		1.05 (1.09)	62.5	9.5	62.4	10.3
^{64}Ni	60.3	9.7		1.05 (1.09)	62.1	9.6	62.0	10.4
^{90}Zr	84.0	9.8		1.07 (1.11)	86.9	9.8	85.3	10.8
^{116}Sn	102.3	9.8		1.07 (1.11)	105.8	10.1	103.1	11.1
^{124}Sn	104.0	10.1		1.07 (1.07)	104.1	10.3	101.8	11.3
^{40}Ar + ^{110}Pd	101		De 77a	1.08 (1.08)	101.4	10.3	99.3	11.2
^{116}Sn	111	10.5	Ga 74d	1.08 (1.07)	110.2	10.3	107.4	11.3
^{121}Sb	111		Ga 74c	1.08 (1.08)	111.6	10.4	108.8	11.4
^{197}Au	158	10.5	Ng 77	1.09 (1.13)	163.3	11.1	157.3	12.2
^{208}Pb	160		Og 74,75	1.09 (1.15)	168.1	11.2	161.8	12.3
^{52}Cr + ^{208}Pb	203			1.10 (1.21)	222.3	11.3	211.9	12.5
^{54}Cr + ^{207}Pb	201.5			1.10 (1.21)	221.0	11.3	210.9	12.5
^{74}Ge + ^{232}Th	310		Og 73	1.11 (1.11)	311.7	11.8	295.3	13.0
^{84}Kr + ^{72}Ge	147		Ga 72	1.08 (1.02)	140.1	10.4	135.1	11.4
^{116}Cd	204			1.09 (1.07)	200.7	10.9	191.6	12.1

[a] Values calculated with the nuclear potential given by (7.39).
[b] Values calculated with the nuclear potential given by (7.37), (7.50), and (7.51).
[c] Typical uncertainties of B_{fu} are about 1 % for projectiles lighter than ^{40}Ar, 2–3 % for ^{40}Ar, and 3–5 % for heavier projectiles.

of approach".) In the following we give a brief survey of the potentials and then comment on the problem of dissipation.

A "single folding potential" was used by Gross and Kalinowski in their earlier work on fusion cross sections [Gr 74a, b]. This potential is obtained by folding the nucleon density distribution of one fragment with a standard optical model potential for nucleon scattering from the other fragment. The calculation neglects the saturating properties of nuclear forces and thus produces potentials which are too deep, especially at the small separations relevant to the fusion problem. This difficulty is avoided by the following potentials which are derived from the liquid-drop model or some other model of nuclear matter. Krappe and Nix proposed a "double folding potential", which is obtained by folding a Yukawa-type interaction into two homogeneous, sharply bounded matter distributions [Kr 74a]. This model may be considered a simplified version of the "generalized liquid-drop model" of Scheid and Greiner [Sc 68, Sc 69; see (2.5)]. The range parameter of the Krappe–Nix potential ($d = 1.4$ fm) has been adjusted to fit experimental interaction barriers, and its strength is then determined by comparison with the liquid-drop model mass formula. The resulting potential is very similar to the phenomenological exponential potential discussed above (note, however, that the shape function of the Krappe–Nix potential is not a simple exponential). More ambitious approaches, based on the Thomas–Fermi method, have been followed in deriving the "proximity potential" of Randrup et al. [Ra 74, Bl 77], and in the application of the "energy density formalism" of Brueckner et al. [Br 68a, b] by the Orsay group [Ba 72, Ga 74a, Ng 75a, b].

It should be noted that all of these potentials make use of the "sudden approximation", which implies frozen density distributions, and of parameters fitted to reproduce macroscopic nuclear properties. A comparison between the Berkeley potential, the Orsay potential, and the empirical potential of (7.51) for the system ^{40}Ar + ^{121}Sb is shown in Fig. 7.29. It can be seen that only minor differences exist between these potentials in the relevant range of distances. Comparisons of experimental fusion barriers with theoretically calculated barriers have been published by Krappe [Kr 74b] and by Ngô et al. [Ng 75a].

Energy and angular momentum losses by friction are considered explicitly in some models by introducing a phenomenological friction tensor. An example is the work of Gross and Kalinowski which was discussed in detail in Sect. 6.4. In this approach, fusion cross sections are computed by numerical integration of the equations of motion for each angular momentum; the limiting angular momentum for fusion is then identified with the highest value of L for which the trajectory does not re-emerge from the interaction region. The strength and radial dependence of the friction tensor are deduced from a

---

^d Typical uncertainties of r_B are about 5%.

^e $r_0 = 1.16 - 1.39/A_1^{1/3} A^{1/3}$ fm. The numbers given in brackets are those values of r_0 which exactly reproduce the experimental fusion barriers.

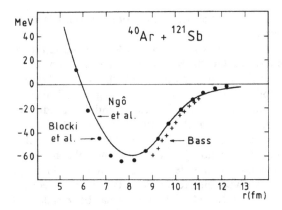

Fig. 7.29. Nuclear potentials for the system ^{40}Ar + ^{121}Sb. The empirical potential deduced from fusion cross sections (crosses) is compared with two theoretical potentials (taken from Bl 77): the "energy-density potential" of the Orsay group (full drawn curve) and the "proximity potential" of the Berkeley group (full points)

universal fit to experimental data on fusion and deep-inelastic scattering. In this way, Gross and Kalinowski were able to reproduce fusion cross sections for many systems and energies with remarkable overall accuracy, considering that no individual adjustments were made [Gr 74a, Gr 74a,b, De 76]. It was found, however, that the use of different potentials requires different assumptions on friction as pointed out in Sect. 6.4.

An alternative class of models disregards friction completely at distances larger than a "critical distance". This approach has been followed, in particular, by Bass (as described in detail above), and by the Orsay group [Ga 74a; see also Le 74b, c, Le 76a, b]. The latter group has deduced the "concept of a critical distance" empirically from an analysis of experimental fusion cross sections in terms of the "energy-density potential". We note that the Orsay work is equivalent to the model of Bass, except that the role played by friction at the critical distance is not discussed explicitly, and hence no saturation limit to L_{fu} is predicted. There can be no doubt, however, that such a limit must exist, even if it cannot be predicted accurately from one-dimensional considerations. The concept of a critical distance has also been used by Glas and Mosel to derive a formula which parametrizes the fusion cross section in terms of two characteristic distances (the barrier distance and the critical distance) and the corresponding potentials [Gl 74b, Gl 75]. This formula emphasizes the change in slope near $E = E_1$ which occurs in plots of σ_{fu} versus $1/E$ (or of $E\sigma_{fu}$ versus E), and has been used extensively to fit experimental data.

A question of some concern is why such different assumptions on friction as made in the Gross and Kalinowski treatment and in the critical-distance models can reproduce experimental fusion cross sections similarly well. Clearly there is some ambiguity with respect to potential and friction in the sense that deeper potentials require more radial friction at distances larger than the half-density distance. The empirical analyses show that the more realistic potentials, based on nuclear-matter arguments, are consistent with negligible friction beyond R_{12}, as far as fusion is concerned. However, moderate losses of angular momentum close to the turning point of the limiting trajectory, or of radial momentum for more central trajectories, would have little effect on the fusion cross section and can therefore not be excluded. In the absence of more direct evidence we conclude, that the neglect of friction beyond R_{12} is probably a reasonable first-order approximation in the sense that trajectories remaining outside R_{12} do not suffer massive losses of energy and angular momentum, at least up to the turning point. This view is qualitatively consistent with ideas which link friction with nucleon transfer (see Sect. 6.4), since the latter mechanism is inhibited at larger distances but should rapidly increase in importance as the half-density distance is approached. Trajectory calculations of Birkelund et al., based on the proximity potential and transfer-induced friction, show that indeed the fusion cross section is practically unaffected by friction at low energies ($< E_1$), but is increased appreciably due to angular momentum transfer by tangential friction at higher energies [Bi 78]. These are precisely the features predicted in the schematic model of Bass.

It remains to comment briefly on possible improvements of the simple models discussed so far. Clearly the next step should be to include explicitly the mass asymmetry and deformation degrees of freedom. An attempt has been made by Bass to estimate the effects of axially symmetric fragment deformation on fusion cross sections in an approximate way [Ba 74a; see Fig. 7.23]. More elaborate calculations were performed by Sierck and Nix, who parametrized the nuclear surface (following hard contact) in terms of smoothly joined portions of three quadratic surfaces of revolution [Si 74, Ni 77, Si 77]. In this work, fusion was defined by the requirement that the dynamical trajectory of the system must pass inside the fission saddle point, as given by the multidimensional treatment. Fusion excitation functions were calculated for a number of moderately heavy, symmetric systems, using the Krappe–Nix potential and neglecting dissipation (assuming, however, full angular momentum transfer leading to rigid rotation at contact). The results confirm that including deformation tends to inhibit fusion for heavy systems, but has little consequence for the lighter systems (and at moderate energies), where all trajectories passing over the one-dimensional fusion barrier end up in fusion. As an example of the former situation, we show in Fig. 7.30 fusion cross sections for the system $^{124}\text{Sn} + ^{124}\text{Sn}$ as deduced both from a one-dimensional and a multidimensional calculation [Ni 77]. The different energy regions discussed earlier are evident and their limits are marked by arrows in Fig. 7.30. In the one-dimensional case, the two arrows correspond to E_1 and E_2, respectively [see Fig. 7.21 and (7.45), (7.46)], while in the multidimensional case, the lowest-energy region is absent and the remaining arrow marks the energy where the limiting angular momentum is determined by the disappearance of the fission saddle point. It should be noted that multidimensional calculations of this type are not only mathematically more complicated than the one-dimensional models, but in addition involve model assumptions on potential-energy surfaces, mass surfaces, and dissipation functions which are not easily tested by direct comparison with experiment. Such calculations can

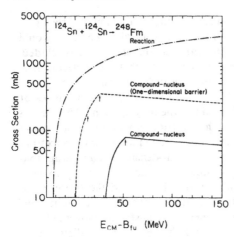

Fig. 7.30. Theoretical fusion and reaction cross sections for the system $^{124}Sn + ^{124}Sn$. The full curve represents a multidimensional calculation of the fusion cross section (without friction) and the broken curve a one-dimensional calculation. (From Ni 77)

therefore, at the present stage, not be expected to reproduce or predict fusion cross sections quantitatively under conditions where the multidimensional effects are important. The significance of such calculations lies rather in the fact that they can provide a qualitative understanding of these effects.

7.5 Superheavy Elements

A major part of the work on fusion cross sections, both experimental and theoretical, was motivated by predictions concerning the possible existence of an "island of stability" beyond the presently known mass region of particle-stable nuclei. Extrapolations of the nuclear shell model towards larger masses indicated that the next major shell closures should occur at $Z = 114$, $N = 184$. In a limited mass region around these magic numbers, the stability of nuclear ground states against spontaneous fission should be strongly enhanced; this was shown by applying shell corrections to liquid-drop-model estimates of ground-state masses and fission barriers. Fission barriers of the order of 10 MeV were deduced for the near-magic superheavy nuclei. If these numbers are correct, then one should not only expect long-lived ground states, but also a fair chance of survival against fission after formation of such a superheavy compound nucleus at an excitation of several tens of MeV.

Many theorists have made important contributions to the exciting field of superheavy nuclei. It is impossible to give proper credit to each of them here; instead we refer the reader to several comprehensive review articles and conference proceedings [Se 68, He 69, Le 72, Ni 72, RO 74]. In Fig. 7.31, we show contour plots of predicted half-lives with respect to different decay modes for nuclei in the superheavy mass region, as calculated by Fiset and Nix [Fi 72; see also Ts 70]. It should be noted that, mainly due to uncertainties in the calculated fission barriers, these lifetimes are uncertain by many orders of magni-

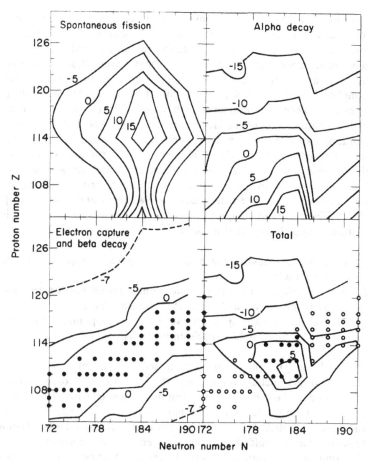

Fig. 7.31. Contour plots of predicted half-lives for even super-heavy nuclei with respect to different decay modes, and of the corresponding total half-lives, versus neutron number N and proton number Z. The contours are labelled by the logarithms (to base 10) of the half-lives in years. The points indicate beta-stable nuclei. (From Fi 72, Ni 72)

tude. There is no unambiguous evidence for the occurrence of superheavy nuclei in nature so far.

Although the formation of superheavy nuclei by successive absorption of neutrons or light charged particles has been discussed in the literature, collisions between two heavy nuclei, associated with substantial mass transfer, are now believed to be the most promising method of production. Three different mechanisms have been discussed: complete fusion, followed by emission of a few light particles (mainly neutrons and α particles); complete fusion, followed by fission ("fusion-fission"); and incomplete fusion, followed by multi-nucleon transfer (deep-inelastic transfer). The latter two mechanisms involve large composite systems decaying by emission of at least two heavy fragments. It is clear from our discussion of the systematics of fusion cross sections that such a reaction cannot proceed through a classical compound nucleus, and therefore superheavy production by fusion-fission seems an unrealistic concept. We are thus left with two basic mechanisms: complete fusion of two fragments with $A_1 + A_2 \approx 300$, and multi-nucleon transfer between two fragments with $A_1 + A_2$ much larger than 300. The physical background of the latter approach is discussed in Chap. 6; first exploratory experiments with the systems Xe + U and U + U have been performed recently at the UNILAC accelerator in Darmstadt [Hi 77c, Sc 78]. In the following, we consider the more conventional approach to superheavy production based on fusion reactions. The main problems encountered in this approach have been discussed by Swiatecki and Björnholm [Sw 70, Sw 72]. There are three important conditions which must be satisfied in order to form a superheavy final nucleus:

(i) The combined mass-to-charge ratio of target and projectile must be of the correct magnitude.

(ii) After passing over the (one-dimensional) fusion barrier, the composite system must evolve towards a nearly spherical shape.

(iii) The excited (nearly spherical) superheavy compound nucleus must survive de-excitation by light-particle emission without undergoing fission.

Condition (i) implies that neutron-rich isotopes must be used as projectiles and target nuclei; even so, only the proton-rich ("north-western") part of the predicted island of stability can be reached (see Fig. 7.32). This should not prevent the production of comparatively long-lived superheavy nuclei, if the island is predicted correctly. A slight shift of the island towards more neutron-rich nuclei could, however, have serious consequences.

Condition (ii) imposes probably the most important restriction. The saddle-point shapes of superheavy nuclei are much less deformed than those of the known fissionable nuclei, in the actinide region. This means that a very compact, almost spherical shape must be reached, before the superheavy compound nucleus is stabilized by shell effects. As emphasized by Swiatecki, these nuclei are very "brittle" in the sense that they can only withstand small distortions without disintegrating [Sw 70]. When target and projectile come into solid contact, their combined shape is much less compact than that of the saddle-point configuration. Consequently the dependence of the potential energy on deformation (dominated by Coulomb forces) tends to drive the composite system into a stretched configuration and eventually to scission. For a given compound system this tendency is more pronounced with nearly symmetric fragmentations than with highly asymmetric fragmentations. These qualitative arguments are supported by the empirical evidence on fusion with projectiles heavier than argon, and by the theoretical results given in Fig. 7.23 which shows allowed and forbidden regions for fusion (at zero angular momentum) as a function of asymmetry and the generalized fissility parameter x_{12}. These results have been derived from a very crude model where the fragmentation is assumed "frozen" and the fragments are allowed to undergo axially symmetric deformations. Nevertheless they show clearly the increasing tendency of contraction towards more spherical configurations with increasing initial asymmetry, and the role played by dynamic deformation in reducing the stability of the composite system.

So far we have disregarded the influence of angular momentum and of friction. It is qualitatively clear that both these features tend to reduce the fusion probability. Although some friction is needed in principle in order to trap the system in the compound state, its main effect will be to prevent the system from reaching a relatively "cold" spherical configuration at bombarding energies close to the barrier. Even in the absence of friction, the calculations of Sierck and Nix suggest that bombarding energies significantly above the barrier are needed to drive symmetrical systems over the barrier to form a compound nucleus (see Fig. 7.30). With strong friction, a compound nucleus may not be formed at all, or only at high excitation.

However, it follows from condition (iii) that superheavy compound nuclei must be produced at the lowest possible excitation energy. This is so not only because fission competes with light-particle emission at each step of de-excitation, but also because the stabilizing effects of shell structure, leading to increased fission barriers, tend to disappear at higher excitation. Consequently, the chance of survival of a superheavy compound nucleus formed initially at 50 MeV excitation may well be many orders of magnitude smaller, than if the same compound nucleus was formed at only 20 MeV excitation.

In Fig. 7.32 and Table 7.8, we present some pertinent data, based on cal-

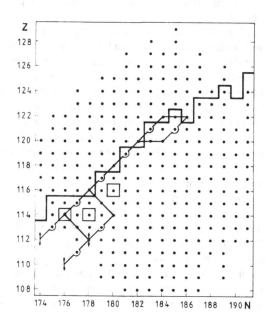

Fig. 7.32. Schematic diagram illustrating the production of superheavy nuclei near $Z = 114$, $N = 184$ by fusion reactions, according to theoretical predictions by Fiset and Nix [Fi 72]

culations by Fiset and Nix [Fi 72]. The solid points in Fig. 7.32 mark the location of nuclei in the relevant part of the Z-N plane for which the fission barriers are predicted to be larger than the neutron separation energy. These nuclei, if formed by fusion, might have a fair chance of survival. However, only points which lie above the heavy line (running zig-zag across the figure) are accessible with projectiles and target nuclei considered practicable at present. The three "islands" in the inaccessible part of the Z-N plane are obtained by combining ^{48}Ca projectiles with the target nuclei ^{242,244}Pu and ^{248}Cm; further such islands would result if more exotic target nuclei, like ^{250}Cm and ^{252}Cf, were included.

The thin lines extending from the point $Z = 122$, $N = 186$ towards the lower left illustrate probable decay patterns of the appropriate compound nucleus (formed, for example, in the reaction ^{76}Ge + ^{232}Th → ^{308}122; see Table 7.8). It has been assumed here that the primary compound nucleus evaporates two neutrons and an α particle prior to de-excitation by electromagnetic transitions. The subsequent radioactive decay is expected to occur by sequential α emission and electron capture. Finally the decay chain is terminated by spontaneous fission at $Z = 112$ or 110. According to the theoretical predictions, the signature of such a reaction would be the emission of several high-energy α particles ($E_\alpha = 7.5$–12 MeV) with half-lives from milliseconds to many days, and of two fission fragments accompanied by an unusually large number of neutrons (about eight to ten). These conspicuous features should be very helpful in recognizing the production of a superheavy nucleus experimentally. We note that the centre of the predicted island of stability cannot be reached by this mechanism; nevertheless, rather long-lived products in the neutron-deficient part of the island may be expected.

Table 7.8 gives a list of interesting target–projectile combinations, together with calculated Q values and fission and fusion barriers. The (one-dimensional) fusion barriers have been calculated with the exponential nuclear potential of (7.36), using $r_0 = 1.14$ fm (see Table 7.7) and with the empirical nuclear potential of (7.37), (7.50), and (7.51). The results obtained with these two potentials are seen to agree within about 3%. The most interesting numbers in Table 7.9 are the fission barriers on one hand, and the excess energies at the one-dimensional fusion barrier relative to the ground state of the compound nucleus on the other hand. The latter values ($= B_{fu} + Q_{fu}$) can be taken as a lower limit for the excitation energy of the compound nucleus, and range from around zero for nearly symmetric systems to about 50 MeV for highly asymmetric systems.

Clearly the fusion of nearly symmetric fragments—sometimes also referred to as "inverse fission"—implies that the fragments must overcome the fission barrier. Therefore, in symmetric systems, the fission barrier defines a more relevant lower limit of compound-nucleus excitation. In order to reach the fission barrier, considerable interpenetration of the fragments is necessary which may be inhibited by friction, or, in more general terms, by the coupling between radial motion and intrinsic or deformation degrees of freedom.

Table 7.8. Fission and fusion barriers for superheavy compound systems

Compound nucleus				Projectile	Target	$-Q_{fu}$[a] [MeV]	Exponential Potential[b]		Empirical Potential[c]	
Z	N	A	B_f[a] [MeV]				B_{fu} [MeV]	$B_{fu}+Q_{fu}$ [MeV]	B_{fu} [MeV]	$B_{fu}+Q_{fu}$ [MeV]
116	180	296	11.0	^{48}Ca	^{248}Cm	164.0	202.8	38.8	201.1	37.1
118	180	298	11.4	^{50}Ti	^{248}Cm	181.9	224.3	42.4	221.4	39.5
				^{54}Cr	^{244}Pu	195.0	239.6	44.6	235.8	40.8
119	180	299	11.5	^{51}V	^{248}Cm	182.2	235.1	52.9	231.6	49.4
				^{55}Mn	^{244}Pu	202.3	250.1	47.8	245.7	43.4
120	182	302	12.5	^{54}Cr	^{248}Cm	201.6	244.4	42.8	240.4	38.8
				^{64}Ni	^{238}U	231.9	272.1	40.2	266.4	43.5
				^{76}Ge	^{226}Ra	261.6	295.7	34.6	288.5	26.9
121	182	303	11.8	^{55}Mn	^{248}Cm	209.4	255.1	45.7	250.5	41.1
				^{65}Cu	^{238}U	239.0	282.4	43.4	276.1	37.1
				^{71}Ga	^{232}Th	253.8	294.3	40.5	287.2	33.4
122	184	306	11.7	^{64}Ni	^{242}Pu	239.7	277.7	38.0	271.7	32.0
				^{68}Zn	^{238}U	250.0	291.1	41.1	284.3	34.3
				^{74}Ge	^{232}Th	265.2	302.8	37.6	295.3	30.1
				^{130}Te	^{176}Yb	368.1	377.8	9.7	365.5	−2.6
				^{136}Xe	^{170}Er	373.8	380.9	7.1	368.4	−5.4
122	186	308	9.8	^{64}Ni	^{244}Pu	239.3	277.2	37.9	271.2	31.9
				^{70}Zn	^{238}U	254.2	289.7	35.5	283.0	28.8
				^{76}Ge	^{232}Th	270.7	301.5	30.8	294.0	23.3
				^{82}Se	^{226}Ra	285.9	312.4	26.5	304.3	18.4
124	188	312	7.5	^{64}Ni	^{248}Cm	249.8	282.7	32.9	276.5	26.7
				^{74}Ge	^{238}U	276.2	308.6	32.4	300.7	24.5
				^{80}Se	^{232}Th	292.4	319.8	27.4	311.3	18.9
				^{86}Kr	^{226}Ra	309.7	330.4	20.7	321.1	11.4
				^{136}Xe	^{176}Yb	390.0	390.6	0.6	377.6	−12.4
124	190	314	5.8	^{70}Zn	^{244}Pu	264.3	295.1	30.8	288.1	23.8
				^{76}Ge	^{238}U	280.5	307.2	26.7	299.4	18.9
				^{82}Se	^{232}Th	297.7	318.5	20.8	310.0	12.3

[a] Based on calculations by Fiset and Nix (Fi 72).
[b] Based on (7.39) with $R_{12} = 1.14 (A_1^{1/3} + A_2^{1/3})$ fm.
[c] Based on (7.37), (7.50), and (7.51).

Asymmetric fragmentations should be more favourable in this respect, since the potential energy at the one-dimensional fusion barrier is above that at the fission saddle point, and dynamical paths leading to a spherical configuration could conceivably exist even with moderately strong friction.

These arguments thus support our previous conclusion that fusion is more likely to occur with more asymmetric fragmentations. The latter provide, in addition, better possibilities of producing very neutron-rich compound nuclei. Large asymmetries, on the other hand, imply moderately large excitation energies and hence a reduced chance of survival. Considerations of this nature suggest that the overall success of superheavy production may depend rather critically on a proper choice of entrance channel and bombarding energy.

Finally, we consider possible effects of a static deformation of one of the fragments which have been neglected so far. The results of Stokstad et al. [St 78], discussed in Sect. 7.2 (see Fig. 7.10), show that nuclear deformation tends to enhance the sub-barrier fusion cross section. This may have important consequences for the production of superheavy elements by fusion reactions involving target nuclei or projectiles in the actinide region. Using the empirical nuclear potential derived in Sect. 7.4 [(7.37) and (7.51)], and the liquid-drop model prescription for its dependence on local surface separation and curvature (the proximity theorem; see Fig. 7.20), we obtain the following estimate for the nucleus–nucleus potential acting between a spherical projectile and a deformed target nucleus with rotational symmetry and quadrupole deformation β:

$$V(r, \Theta) = V_C(r) + V_N(r) \tag{7.52}$$

$$= \frac{Z_1 Z_2 e^2}{r} \left[1 + \frac{3}{5} \frac{\beta R_2^2}{r^2} Y_{20}(\Theta) \right]$$

$$- \frac{R_1 R_2}{R_1 + R_2} \left[1 - \frac{5}{2} \frac{\beta R_1}{R_1 + R_2} Y_{20}(\Theta) \right] g[r - (R_1 + R_2) - \beta R_2 Y_{20}(\Theta)].$$

Here Θ is the angle between the symmetry axis of the deformed nucleus and the centre-to-centre line, while the radii R_1, R_2 and the function g are defined by (7.50) and (7.51), respectively. All terms of higher than first order in the deformation β are omitted in (7.52).

Figure 7.33 shows potentials for the system ^{65}Cu + ^{238}U ($\beta \approx 0.3$) and different angles Θ, which were calculated with (7.52). We note that the most favourable orientation of the uranium nucleus ($\Theta = 0$) leads to a substantially lower fusion barrier than expected for a spherical nucleus of equal mass [$\Theta = 55°$, $Y_{20}(\Theta) = 0$]. Thus fusion should become possible at lower bombarding energies and the prospects for producing "cold" compound nuclei should be enhanced. On the other hand, the overall configuration of the system at the one-dimensional barrier would be more elongated, and dissipative effects might prevent a contraction to nearly spherical shapes.

Many attempts have been made, in practically all major heavy-ion laboratories, to synthesize superheavy nuclei by fusion reactions. None of these has been successful, however. Table 7.9 gives a summary of relevant work. The experimental methods may be classified according to the accessible range of half-lives. Products with half-lives above about 10^4 s are most sensitively detected by radiochemical techniques. Here the cross section limits are typically in the range 10^{-34} to 10^{-35} cm^2, or more than ten orders of magnitude

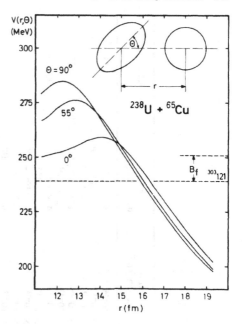

Fig. 7.33. Influence of static deformation on the nucleus–nucleus potential for the system ^{65}Cu + ^{238}U, according to (7.52). The potential is shown for three different orientations of the symmetry axis of the ^{238}U nucleus; note that for $\Theta = 55°$ the potential is the same as for spherical nuclei with the same average radii. Predictions for the ground state energy and fission barrier of the superheavy compound nucleus ^{303}121 (Fi 72) are incluced for comparison

below the geometrical cross sections. Shorter half-lives, down to about 10^{-3}s, can be measured by using some means of fast mechanical transport from the target position to the detector, like a rotating wheel or a gas-jet system. Finally Armbruster et al. at Darmstadt have used a specially designed electromagnetic velocity filter [Ew 76] to detect prompt evaporation residues in delayed coincidence with decay products, following separation from the primary beam [Ar 78, Mü 78]. With this technique it was possible to extend the range of half-lives down to 10^{-6}s. In general, the sensitivity of the different methods decreases somewhat with increasing speed (see Table 7.9). Nevertheless, the failure to detect superheavy nuclei with half-lives above 10^{-6} s suggests that regions of still smaller half-lives should be explored.

So far we do not know whether the negative outcome of these experimental searches is primarily caused by a failure of the theoretical predictions —yielding too optimistic estimates of nuclear stability in the superheavy mass region—or whether it merely reflects the difficulties associated with the production mechanism. In any case, it seems clear that the cross sections are exceedingly low; to understand why this is so, and to develop improved theoretical and experimental methods, must be the next step. Undoubtedly the search for superheavy nuclei will continue to provide valuable information on the static and dynamic properties of heavy nuclear systems, even if the actual goal of the search should not be reached.

Table 7.9. Search for superheavy elements produced in fusion reactions

System	Method	Detected decay[a]	Upper limit of cross section [cm²]	Range of half-lives [s]	References
^{74,76}Ge + ^{232}Th	rotating wheel chemical sep.	SF	1×10^{-34}	$5 \times 10^{-3} - 10^{5}$	Og 73
			4×10^{-35}	$10^{4} - 10^{7}$	
^{76}Ge + ^{238}U	rotating wheel chemical sep.		1×10^{-34}	$5 \times 10^{-3} - 10^{5}$	Hu 77
^{48}Ca + ^{248}Cm		α, SF	2×10^{-34}	$10^{4} - 10^{5}, 10^{7} - 10^{8}$	
			2×10^{-35}	$10^{5} - 10^{7}$	
^{48}Ca + ^{242}Pu, ^{243}Am	chemical sep.	α	2×10^{-34}	$4 \times 10^{4} - 10^{6}$	Og 78
		SF	1×10^{-35}	$10^{4} - 10^{7}$	
^{48}Ca + ^{246}Cm		α	5×10^{-34}	$4 \times 10^{4} - 10^{6}$	
		SF	2×10^{-35}	$10^{4} - 10^{7}$	
^{48}Ca + ^{248}Cm		α	5×10^{-33}	$4 \times 10^{4} - 10^{6}$	
		SF	2×10^{-34}	$10^{0} - 10^{7}$	
^{48}Ca + ^{248}Cm	He-jet	α, SF	1×10^{-34}	$10^{0} - 10^{7}$	Il 78
^{48}Ca + ^{248}Cm	chemical sep.	SF	5×10^{-35}	$4 \times 10^{4} - 10^{7}$	Ot 78
^{65}Cu + ^{238}U	velocity filter	α,SF	4×10^{-34}	$1 \times 10^{-6} - 10^{5}$	Ar 78, Mü 78
^{136}Xe + ^{170}Er		α,SF	2×10^{-33}	$1 \times 10^{-6} - 10^{5}$	Ar 77

[a] α: α decay; SF: spontaneous fission

References Chapter 7

(For AS 63, CA 76, HE 69, HE 74, MU 73, NA 74, RO 73 see Appendix D)

Ag 76   Agarwal, S. et al.: Z. Phys. *A278*, 265 (1976)
Al 77   Albrecht, K.; Stocker, W.: Nucl. Phys. *A278*, 95 (1977)
Ar 77   Armbruster, P. et al.: GSI-Report J-1-77 (1977), p. 74
Ar 78   Armbruster, P.; Reisdorf, W.: GSI-Report M-2-78 (1978)
Ba 71   Barnes, C.A.: *Adv. Nucl. Phys.* Vol. 4, ed. by M. Baranger; M. Vogt, p. 133. New York: Plenum Press 1971
Ba 72   Basile, R. et al.: J. Phys. (Paris) *33*, 9 (1972)
Ba 73a Bass, R.: MU 73 Vol.1 p. 614
Ba 73b Bass, R.: Phys. Lett. *47B*, 139 (1973)
Ba 74a Bass, R.: Nucl. Phys. *A231*, 45 (1974)
Ba 74b Bass, R.: NA 74 Vol.1 p. 117
Ba 77a Back, B.B. et al.: Nucl. Phys. *A285*, 317 (1977)
Ba 77b Bass, R.: Phys. Rev. Lett. *39*, 265 (1977)
Bi 78   Birkelund, J.R. et al.: Phys. Rev. Lett. *40*, 1123 (1978)
Bl 77   Blocki, J. et al.: Ann. Phys. (N.Y.) *105*, 427 (1977)
Br 68a Brueckner, K.A. et al.: Phys. Rev. *171*, 113 (1968)
Br 68b Brueckner, K.A.; Buchler, J.R.; Kelly, M.M.: Phys. Rev. *173*, 944 (1968)
Br 74   Brink, D.M.; Rowley, N.: Nucl. Phys. *A219*, 79 (1974)
Br 75   Broda, R. et al.: Nucl. Phys. *A248*, 356 (1975)
Br 76   Britt, H.C. et al.: Phys. Rev. *C13*, 1483 (1976)
Br 77   Britt, H.C. et al.: Phys. Rev. Lett. *39*, 1458 (1977)
Bu 57   Burbridge, E.M. et al.: Rev. Mod. Phys. *29*, 547 (1957)
Ca 75   Cabot, C. et al.: J. Phys. (Paris) *L36*, 289 (1975)
Ch 73   Chulick, E.T.; Natowitz, J.B.; Schnatterly, C.: Nucl. Instrum. Methods *109*, 171 (1973)
Ch 76   Christensen, R.; Winther, A.: Phys. Lett. *65B*, 19 (1976)
Ch 77   Christensen, P.R.; Switkowski, Z.E.; Dayras, R.A.: Nucl. Phys. *A280*, 189 (1977)
Co 63   Cohen, S.; Plasil, F.; Swiatecki, W.J.: AS 63 p. 325
Co 74   Cohen, S.; Plasil, F.; Swiatecki, W.J.: Ann. Phys. (N.Y.) *82*, 557 (1974)
Co 76   Conjeaud, M. et al.: CA 76 p. 116
Cu 76   Čujec, B.; Barnes, C.A.: Nucl. Phys. *A266*, 461 (1976)
Da 75   Dauk, J.; Lieb, K.P.; Kleinfeld, A.M.: Nucl. Phys. *A241*, 170 (1975)
Da 76a Dayras, R.A. et al.: Nucl. Phys. *A261*, 478 (1976)
Da 76b Dayras, R.A. et al.: Nucl. Phys. *A265*, 153 (1976)
De 76   De, J.N.; Gross, D. H. E.; Kalinowski, H.: Z. Phys. *A227*, 385 (1976)
De 77a Della Negra, S. et al.: Z. Phys. *A282*, 65 (1977)
De 77b Della Negra, S. et al.: *A282*, 75 (1977)
Ei 77   Eisen, Y. et al.: Nucl. Phys. *A291*, 459 (1977)
Ew 76   Ewald, H. et al.: Nucl. Instrum. Methods *139*, 223 (1976)
Ey 76   Eyal, Y. et al.: Phys. Rev. *C13*, 1527 (1976)
Fi 72   Fiset, E.; Nix, J.R.: Nucl. Phys. *A193*, 647 (1972)
Fo 75a Fowler, W.A.; Caughlan, G.R.; Zimmerman, B.A.: Ann. Rev. Astron. Astrophys. *13*, 6 (1975)
Fo 75b Fowler, M.M.; Jared, R.C.: Nucl. Instrum. Methods *124*, 341 (1975)
Ga 72   Gauvin, H. et al.: Phys. Rev. Lett. *28*, 697 (1972)
Ga 74a Galin, J. et al.: Phys. Rev. *C9*, 1018 (1974)
Ga 74b Galin, J. et al.: Phys. Rev. *C9*, 1113, 1126 (1974)
Ga 74c Gauvin, H.; Le Beyec, Y.; Porile, N.T.: Nucl. Phys. *A223*, 103 (1974)
Ga 74d Gauvin, H. et al.: Phys. Rev. *C10*, 722 (1974)
Ga 75   Gauvin, H. et al.: Phys. Lett. *58B*, 163 (1975)
Gl 74a Glas, D.; Mosel, U.: Phys. Lett. *B49*, 301 (1974)
Gl 74b Glas, D.; Mosel, U.: Phys. Rev. *C10*, 2620 (1974)
Gl 75   Glas, D.; Mosel, U.: Nucl. Phys. *A237*, 429 (1975)
Gl 76   Glas, D.; Mosel, U.: Nucl. Phys. *A264*, 268 (1976)
Gr 67a Grover, J.R.; Gilat, J.: Phys. Rev. *157*, 802, 814, 823 (1967)
Gr 67b Grover, J.R.: Phys. Rev. *157*, 832 (1967)
Gr 74a Gross, D.H.E.; Kalinowski, H.: Phys. Lett. *48B*, 302 (1974)

Gr   74b Gross, D.H.E.; Kalinowski, H.; De, J.N.: HE 74 p. 194
Gu   73a Gutbrod, H.H.; Winn, W.G.; Blann, M.: Phys. Rev. Lett. *30*, 1259 (1973)
Gu   73b Gutbrod, H.H.; Winn, W.G.; Blann, M.: Nucl. Phys. *A213*, 267 (1973)
Gu   74   Gutbrod, H.H. et al.: RO 73 Vol.II p. 309
Ha   75   Hagemann, G.B. et al.: Nucl. Phys. *A245*, 166 (1975)
Ha   78   Halbert, M.L. et al.: Phys. Rev. *C17*, 155 (1978)
He   69   Herrmann, G.; Seyb, K.E.: Naturwissenschaften *56*, 590 (1969)
Hi   53   Hill, D.L.; Wheeler, J.A.: Phys. Rev. *89*, 1102 (1953)
Hi   65   Hiebert, J.C.; McIntyre, J.A.; Couch, J.G.: Phys. Rev. 138, B346 (1965)
Hi   69   Hillman, M.; Grover, J.R.: Phys. Rev. *185*, 1303 (1969)
Hi   76a High, M.D.; Čujec, B.: Nucl. Phys. *A259*, 513 (1976)
Hi   76b Hille, P. et al.: Nucl. Phys. *A266*, 253 (1976)
Hi   77a High, M.D.; Čujec, B.: Nucl. Phys. *A278*, 149 (1977)
Hi   77b High, M.D.; Čujec, B.: Nucl. Phys. *A282*, 181 (1977)
Hi   77c Hildenbrand, K.D. et al.: Phys. Rev. Lett. *39*, 1065 (1977)
Hu   77   Hulet, E.K. et al.: Phys. Rev. Lett. *39*, 385 (1977)
Il   78   Illige, J.D. et al.: Phys. Lett. *78B*, 209 (1978)
Jo   67   Johansson, S.A.E.; Wene, C.O.: Ark. Fys. *36*, 353 (1967)
Ka   64   Kalinkin, B.N.; Petkov, I.Z.: Acta. Phys. Polon. *25*, 265 (1964); Univ. of Calif.
          Lawrence Radiation Lab. Rep. UCRL Trans-1151 (1964)
Ke   77   Kettner, K.U. et al.: Phys. Rev. Lett. *38*, 337 (1977)
Ko   68   Kowalski, L.; Jodogne, J.C.; Miller, J.M.: Phys. Rev. *169*, 894 (1968)
Ko   74   Kozub, R.L. et al.: Phys. Rev. *C10*, 214 (1974)
Ko   75   Kozub, R.L. et al.: Phys. Rev. *C11*, 1497 (1975)
Ko   76   Kolata, J.J. et al.: Phys. Lett. *65B*, 333 (1976)
Kr   74a Krappe, H.J.; Nix, R.: RO 73 Vol. I p. 159
Kr   74b Krappe, H.J.: HE 74 p. 24
Ku   64   Kuehner, J.A.; Almqvist, E.: Phys. Rev. *134B*, 1229 (1964)
Le   72   Lefort, M.: J. Phys. (Paris) *33*, C5-73 (1972)
Le   74a Lefort, M.: J. Phys. *A7*, 107 (1974)
Le   74b Lefort, M.: Phys. Scr. *10A*, 101 (1974)
Le   74c Lefort, M.; Le Beyec, Y.; Péter, J.: NA 74 Vol. 2 p. 81
Le   75   Lefort, M.: Orsay Report IPNO 75–05 (1975)
Le   76a Lefort, M.: Rep. Prog. Phys. *39*, 129 (1976)
Le   76b Lefort, M.: J. Phys. (Paris) *37*, C5-57 (1976)
Le   77   Lefort, M.: Orsay Report IPNO-RC-7709 (1977)
Lu   76   Lu, N.H. et al.: Phys. Rev. *C13*, 1496 (1976)
Ma   73   Mazarakis, M.G.; Stephens, W.E.: Phys. Rev. *C7*, 1280 (1973)
Mi   70   Michaud, G.J.; Scherk, L.; Vogt, E.: Phys. Rev. *C1*, 864 (1970)
Mi   72   Michaud, G.J.; Vogt, E.W.: Phys. Rev. *C5*, 350 (1972)
Mi   73   Michaud, G.J.: Phys. Rev. *C8*, 525 (1973)
Mü   78   Münzenberg, G. et al.: GSI-Report J-1-78, 75 (1978)
My   73   Myers, W.D.: Nucl. Phys. *A204*, 465 (1973)
My   74   Myers, W.D.: NA 74 Vol.2 p.1
Na   70   Natowitz, J.B.: Phys. Rev. *C1*, 623 (1970)
Na   72   Natowitz, J.B.; Chulik, E.T.; Namboodiri, M.N.: Phys. Rev. *C6*, 2133 (1972)
Na   73   Natowitz, J.B.; Chulik, E.T.; Namboodiri, N.M.: Phys. Rev. Lett. *31*, 643 (1973)
Na   75   Namboodiri, M.N. et al.: Phys. Rev. *C11*, 401 (1975)
Ng   75a Ngô, C. et al.: Nucl. Phys. *A240*, 353 (1975)
Ng   75b Ngô, C. et al.: Nucl. Phys. *A252*, 237 (1975)
Ng   77   Ngô, C. et al.: Z. Phys. *A283*, 161 (1977)
Ni   72   Nix, J.R.: Ann. Rev. Nucl. Sci. *22*, 65 (1972)
Ni   77   Nix, J.R.; Sierk, A.J.: Phys. Rev. *C15*, 2072 (1977)
Og   73   Oganessian, Yu. Ts.: MU 73 Vol. 2 p. 351; Flerov, G. N. et al.: JINR-Report
          P7-7409 (1973)
Og   74   Oganessian, Y.T.: HE 74 p. 222
Og   75   Oganessian, Y.T. et al.: Yad. Fiz. *21*, 239 (1975); Sov. J. Nucl. Phys. *21*, 126 (1975)
Og   78   Oganessian, Yu. Ts. et al.: Nuc.. Phys. *A294*, 213 (1978)
Ol   74   Olmer, C. et al.: Phys. Rev. *C10*, 1722 (1974)
Ot   78   Otto, R.J. et al.: J. Inorg. Nucl. Chem. *40*, 589 (1978)
Pa   69   Patterson, J.R.; Winkler, H.; Zaidins, C.S.: Astrophys. J. *157*, 367 (1969)

Pa 71   Patterson, J.R. et al.: Nucl. Phys. *A165*, 545 (1971)
Pe 75   Péter, J.; Ngô, C.; Tamain, B.: Nucl. Phys., *A250*, 351 (1975)
Pl 74a Plasil, F.: Report ORNL-TM-4599 (1974)
Pl 74b Plasil, F.: NA 74 Vol. 2 p. 107
Pü 75   Pühlhofer, F. et al.: Nucl. Phys. *A244*, 329 (1975)
Ra 74   Randrup, J.; Swiatecki, W.J.; Tsang, C.F.: Lawrence Berkeley Laboratory Report
        No. 3603 (1974)
Re 66   Reeves, H.: Astrophys. J. *146*, 447 (1966)
Ro 76   Robinson, R.L. et al.: Phys. Rev. *C14*, 2126 (1976)
Sa 76   Sarantites, D.G. et al.: Phys. Rev. *C14*, 2138 (1976)
Sc 68   Scheid, W.; Ligensa, R.; Greiner, W.: Phys. Rev. Lett. *21*, 1479 (1968)
Sc 69   Scheid, W.; Greiner, W.: Z. Phys. *226*, 364 (1969)
Sc 75   Scobel, W. et al.: Phys. Rev. *C11*, 1701 (1975)
Sc 76   Scobel, W. et al.: Phys. Rev. *C14*, 1808 (1976)
Sc 78   Schädel, M. et al.: Phys. Rev. Lett. *41*, 469 (1978)
Se 68   Seaborg, G.T.: Ann. Rev. Nucl. Sci. *18*, 53 (1968)
Se 70   Seeger, P.A.: In *Proc. of the Intern. Conf. on the Properties of Nuclei far from the
        Region of β-Stability*, Leysin, Aug./Sept. 1970; CERN-Report 70–30, p. 217
Si 62   Sikkeland, T.; Haines, E.; Viola, V.E.: Phys. Rev. *125*, 1350 (1962)
Si 63   Sikkeland, T.; Viola, V.E.: AS 63 p. 232
Si 64   Sikkeland, T.: Phys. Rev. *135*, B 669 (1964)
Si 68   Sikkeland, T.: Phys. Lett. *27B*, 277 (1968)
Si 74   Sierk, A.J.; Nix, J.R.: RO 73 Vol.II p. 273
Si 77   Sierk, A.J.; Nix, J.R. : Phys. Rev. *C16*, 1048 (1977)
Sp 74   Spinka, H.; Winkler, H.: Nucl. Phys. *A233*, 456 (1974)
Sp 76a Sperr, P. et al.: Phys. Rev. Lett. *36*, 405 (1976)
Sp 76b Sperr, P. et al.: Phys. Rev. Lett. *37*, 321 (1976)
St 76a Stokstad, R.G. et al.: Phys. Rev. Lett. *36*, 1529 (1976)
St 76b Stokstad, R.G. et al.: Phys. Rev. Lett. *37*, 888 (1976)
St 77   Stokstad, R.G. et al.: Phys. Lett. *70B*, 289 (1977)
St 78   Stokstad, R.G. et al.: Phys. Rev. Lett. *41*, 465 (1978)
Sw 70   Swiatecki, W.J.: HE 69 p. 729
Sw 72   Swiatecki, W.J.; Björnholm, S.: Phys. Rep. *4*, 325 (1972)
Sw 76   Switkowski, Z.E.; Stokstad, R.G.; Wieland, R.M.: Nucl. Phys. *A274*, 202 (1976)
Sw 77   Switkowski, Z.E.; Stokstad, R.G.; Wieland, R.M.: Nucl. Phys. *A279*, 502 (1977)
Ta 75   Tamain, B. et al.: Nucl. Phys. *A252*, 187 (1975)
Ts 70   Tsang, C.F.; Nilsson, S.G.: Nucl. Phys. *A140*, 289 (1970)
Vi 62   Viola, V.E.; Sikkeland, T.: Phys. Rev. *128*, 767 (1962)
Vi 76   Viola, V.E. et al.: Nucl. Phys. *A261*, 174 (1976)
Vo 64   Vogt, E.W. et al.: Phys. Rev. *136*, B 99 (1964)
Wa 71   Wapstra, A.H.; Gove, N.B.: Nucl. Data *9A*, 265 (1971)
We 76   Weidinger, A. et al.: Nucl. Phys. *A263*, 511 (1976)
Wi 73   Wilczynski, J. Nucl. Phys. *A216*, 386 (1973)
Ze 73   Zebelman, A.M.; Miller, J.M.: Phys. Rev. Lett. *30*, 27 (1973)
Ze 74   Zebelman, A.M. et al.: Phys. Rev. *C10*, 200 (1974)

8. Compound-Nucleus Decay

8.1 General Aspects

We start this chapter with a survey of qualitative features of the reaction mechanism. Comprehensive review articles on compound-nucleus reactions induced by heavy ions have been published by Thomas [Th 68] and by Lefort [Le 76], and further details may be found there. In addition, there are several instructive articles summarizing certain parts of the field, such as compound-nucleus reactions in light systems [St 74], (heavy-ion,xn) reactions [Ne 70] and heavy-ion-induced fission [Hu 62, Hy 64b, Pl 74].

Fusion reactions with heavy projectiles usually produce compound nuclei with high excitation energies (up to several hundred MeV) and large angular momenta (up to about 100 $\hbar$). These features offer unique possibilities of studying nuclear properties under conditions which are not accessible in other types of reactions. Clearly under these circumstances one does not deal with individual quantum states (at least not in the primary compound nucleus) but with statistical distributions of overlapping levels. At first sight, the nature of such a compound state is quite different from the states observed as isolated resonances near the nucleon binding energy. In particular, as discussed in Sect. 6.1, the lifetime is much shorter (typically 10^{-20} to 10^{-21}s) and therefore the question arises whether the traditional concept of a compound nucleus in interal thermal equilibrium is applicable. Related to this question is the problem of how to distinguish pre-equilibrium emission of particles from compound-nucleus decay. While these questions are still open to some extent, we shall see that the main features of heavy-ion-induced fusion reactions can be understood in complete analogy to conventional compound-nucleus reactions.

In discussing experimental results on these reactions, one must realize that there are a number of complicating features, which are direct consequences of the high excitation energies and broad angular momentum distributions of the primary compound nuclei. High excitation energies, for example, in general lead to sequential emission of several particles and gamma quanta. Experimentally one may hope to identify the final products or certain transitions between low-lying excited states, but not the origin of intermediate radiations which emerge from different nuclei with different excitation energies and angular momenta. Under these circumstances, detailed studies of the reaction products are not very meaningful, and the experimental information is best expressed in terms of average quantities like the total energy or multiplicity associated with a given type of decay.

Another point to be noted is that the decay pattern of a given compound nucleus depends in general on angular momentum. The observed yield of reaction products represents an average over the broad angular momentum spectrum of the compound nucleus (see Fig. 7.2) and therefore, as discussed in Sect. 7.1, the independence of compound-nucleus formation and decay does not apply. More specifically, the branching ratios for different decay modes at a given excitation energy depend on the angular momentum distribution and hence on the entrance channel. Thus a measurement of relative intensities for different decay products may give information on the angular momentum distribution of the compound nucleus. Similarly the excitation function for a given product will be affected by the angular momentum distribution, as discussed in more detail below.

In order to illustrate the correlation between angular momentum and decay of the compound nucleus, we show schematically in Fig. 8.1 possible paths of de-excitation for a medium heavy compound nucleus ($A \approx 150$) in the energy versus angular momentum plane. The accessible part of that plane is bounded by the y-rast line which connects the states of lowest energy for each angular momentum J [Gr 67b; see also Sect. 7.4]. For simplicity it is assumed in Fig. 8.1 that neutron emission is the dominating mode of particle decay at all excitation energies and angular momenta; the effects of competing decay modes are considered later. Each step of neutron emission removes a

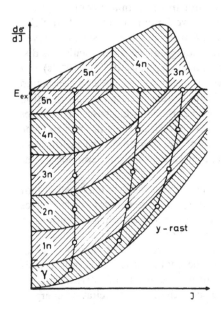

Fig. 8.1. Decay of a highly excited compound nucleus ($A \approx 150$) by neutron and gamma emission. Top: angular momentum distribution of the primary compound nucleus; bottom: typical decay paths in the E_{ex}–J plane

certain amount of excitation energy, equal to the sum of the neutron separation energy and the kinetic energy of the emitted neutron. The associated change in angular momentum is relatively small due to the small mass and average energy of the evaporated neutrons, which result typically in orbital angular momenta of about one or two units. In addition, the neutron angular momenta are randomly oriented with respect to the angular momentum of the emitting nucleus in the region of small J, where neutron emission is not affected by restrictions in the availability of high angular momentum levels in the daughter nucleus. These restrictions become increasingly important as the y-rast line is approached, and then produce a preferential alignment of the neutron angular momentum parallel to the initial angular momentum. As a result, a limited amount of angular momentum is removed by neutron emission in the vicinity of the y-rast line. Finally, when the excitation energy above the y-rast line is about equal to or less than the neutron separation energy, neutron emission is strongly inhibited and gamma emission takes over. The remaining excitation energy and angular momentum are then dissipated by a series of electromagnetic transitions which proceed initially as a "statistical cascade" towards the y-rast line, and eventually through a sequence of y-rast states to the ground state. This decay pattern results in a characteristic enhancement of collective transitions within the ground-state rotational band of deformed even–even final nuclei, as was first demonstrated for α-induced reactions by Morinaga and Gugelot [Mo 63] and with heavier projectiles by Stephens, Lark, and Diamond [St 65]. First measurements of gamma multiplicities and of the total energy emitted as gamma radiation in heavy-ion reactions were performed by Mollenauer [Mo 62]. The results of this work provided clear evidence that the number and total energy of the gamma rays increases with angular momentum for a given system and excitation energy, and that with heavy projectiles (^{12}C) the total gamma energy exceeds the neutron binding energy.

The qualitative arguments given above can be used to estimate the most probable number of neutrons emitted in the decay of a compound nucleus with a given excitation energy and angular momentum. This is shown schematically in Fig. 8.1 where the allowed region of the E_{ex}–J plane is subdivided into parts according to the (dominant) emission of 1, 2, 3, 4, or 5 neutrons. The angular momentum distribution of the primary compound nucleus is given at the top of the figure, and it can be seen that the number of emitted neutrons decreases with increasing angular momentum. The decay paths originating from different parts of the angular momentum spectrum are indicated by the (nearly) vertical lines. It should be noted that each decay passes through a sequence of neighbouring isotopes which are not distinguished in Fig. 8.1.

It is clear from Fig. 8.1 that the effective threshold energy for emission of a certain number of neutrons depends on angular momentum. Experimental excitation functions for (heavy-ion, xn) reactions, corresponding to a specified value of x, represent an average over the angular momentum spectrum of the primary compound nucleus. Since higher angular momenta produce partial excitation functions which are shifted towards higher bombarding energies, the resulting total excitation function is broadened compared to that expected for a narrow angular momentum distribution. Excitation functions for (p, xn) and (α,xn) reactions are typically 10–15 MeV wide, while with heavier projectiles in the mass range $A_1 = 12$–40 widths of 20–50 MeV have been observed. An example which illustrates the correlation between projectile mass and excitation function is shown in Fig. 8.2, where excitation functions

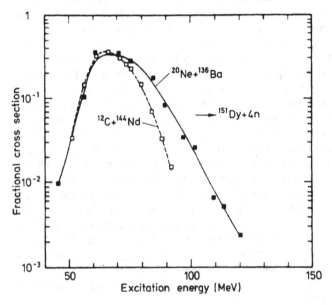

Fig. 8.2. Excitation functions for (HI, 5n) reactions involving the same compound and final nuclei, but different entrance channels. (From Al 64)

for two reactions leading to the same compound and final nucleus are compared [Al 64]. Near threshold the reactions are dominated by low-angular--momentum components, and therefore the initial rise of the excitation function is practically independent of projectile mass. On the high-energy side, however, the reaction yield arises from high-angular-momentum compound states, and therefore the excitation function for the ^{20}Ne projectile extends to higher energies than that for the ^{12}C projectile. Exceptions to this trend occur in cases where the high-angular-momentum components are eliminated by competing reactions, so that only comparatively low angular momenta contribute to the measured yield. Examples of this kind are the observation of a low-spin isomer in the final nucleus [Al 63; see Fig. 8.4] or measurements in the heavy-element region where compound nuclei with large angular momenta decay preferentially by fission [Th 63, Si 67, Si 68a,b, Ga 73; see Fig. 8.12].

So far we have concentrated on neutron and gamma-ray emission, neglecting other decay channels like charged-particle evaporation or fission. This may be justified to some extent for compound nuclei of intermediate mass ($A = 100$–180) and moderate angular momentum (i.e. not too close to the y-rast line), by noting that under these circumstances neutron emission from particle-unstable states is strongly favoured by the absence of a Coulomb barrier. Charged particles, like protons or α particles, must be emitted with higher kinetic energies to overcome the respective Coulomb barriers; consequently the final nuclei are produced at lower excitation where the density of levels—and hence the number of open decay channels—is strongly reduced. These arguments are not valid in general for lighter compound nuclei ($A < 100$), where the Coulomb barriers are lower, especially since the separation energies for charged particles are often smaller than those for neutrons. Similarly for heavy compound nuclei ($A \gtrsim 200$) the fission barrier (at zero angular momentum) becomes comparable with typical nucleon separation energies, and α emission is favoured by positive Q values. Therefore, in general, charged-particle emission (including fission) is expected to compete significantly with neutron emission both for light and for very heavy compound nuclei. The extent of this competition in individual cases will depend on a number of factors, such as shell and odd–even effects, charge-to-mass ratio, and, in particular, on angular momentum.

The influence of angular momentum on the relative decay probabilities is a direct consequence of the different capability of various decay modes to remove angular momentum. As a crude indication of this capability, we can define the ratio of the angular momentum (in units of $\hbar$) carried away by the emitted radiation to the associated change in excitation energy. This ratio is typically 0.1 to 0.2 MeV^{-1} for nucleon emission (n or p), 0.25 to 0.4 MeV^{-1} for α emission, and larger than 1 MeV^{-1} for fission and gamma emission. Thus we expect that the latter decay modes should be strongly enhanced compared to nucleon emission for compound nuclei with large angular momenta, especially near the y-rast line. These qualitative considerations are confirmed both by extensive statistical model calculations and by a fairly large body of experimental data; some examples will be discussed in Sects. 8.3–5.

8.2 Statistical Theory

In this section we summarize the formalism needed for a quantitative discussion of compound-nucleus cross sections. It was originally introduced by Wolfenstein [Wo 51], and by Hauser and Feshbach [Ha 52], and was afterwards extended and generalized by many authors. It implies that, once a compound nucleus has been formed with given excitation energy, angular momentum, and parity, its decay is completely determined by the statistical weights of the various possible final states, irrespective of nuclear structure. The angle-integrated cross section for producing an exit channel β from an entrance channel α can then be written as [see (7.8), (7.13)]

$$\sigma_{\alpha\beta} = \sum_J \sigma_J(\alpha) \frac{\Gamma_J(\beta)}{\Gamma_J} = \frac{\pi \lambdabar_\alpha^2}{(2I_1 + 1)(2I_2 + 1)} \sum_J (2J + 1) \frac{T_J(\alpha)T_J(\beta)}{\sum_\lambda T_J(\lambda)}. \quad (8.1)$$

Here the index λ runs over all possible two-body exit channels (defined by discrete quantum states of the fragments), and $T_J(\lambda)$ denotes a generalized transmission coefficient which is related to the transmission coefficients introduced in Sect. 7.1 by the definition

$$T_J(\lambda) = \sum_{lS} T_{lSJ}(\lambda) = \sum_{lj} T_{ljJ}(\lambda). \quad (8.2)$$

Depending on the angular momentum coupling scheme used to describe the various channels, the sums on the right-hand side of (8.2) include all combinations of orbital angular momentum l and channel spin S, or of l and total angular momentum j of the emitted radiation, which are allowed by the selection rules. The partial width for decay of the compound nucleus with excitation energy E and angular momentum J into channel λ is related to the transmission coefficient $T_J(\lambda)$ by

$$\Gamma_J(\lambda) = \frac{T_J(\lambda)}{2\pi\rho(E, J)}, \quad (8.3)$$

where $\rho(E,J)$ is the spin-dependent level density of the compound nucleus. The theoretical derivation of, and experimental evidence on, the function $\rho(E,J)$ is described extensively in the literature (see, for example, Hu 72) and will not be discussed here. Simple analytical expressions are given in (8.4) and (8.27).

In principle, the decay widths $\Gamma_J(\lambda)$ depend on the change of intrinsic parities associated with the decay into channel λ, and therefore (8.1) should include an explicit summation over both parities of the compound nucleus. For simplicity we include here in the definition of $T_J(\lambda)$ a summation over both parities of the compound nucleus (or, equivalently, of channel λ). It is easily verified that this approximation does not affect the calculated cross sections as long as final-state parities are not observed and levels of both parities are equally available. Moreover, it should be noted that explicitly parity-dependent calculations are generally insensitive to final-state parity, especially in cases where many partial waves are involved.

In many practical applications, individual final states are not energetically resolved, and instead continuous energy spectra of emitted particles are measured. Under these circumstances we can re-interpret the channel index λ as denoting a given type of emitted particle and the corresponding residual nucleus, rather than a two-body channel composed of discrete states. For the spin-dependent level density of the residual nucleus in channel λ we write (see, for example, Er 60, Hu 72)

$$\rho_\lambda(E, I) = \bar\rho_\lambda(E) \frac{(2I + 1)}{2\sigma_\lambda^2} \exp\left(-\frac{I(I + 1)}{2\sigma_\lambda^2}\right), \quad (8.4)$$

where $\bar{\rho}_\lambda(E)$ is the total level density at energy E (summed over all spin values I), and σ_λ is called the "spin cut-off parameter". The quantity σ_λ^2 controls the decrease in level density at high angular momentum, and hence has important consequences for the angular momentum dependence of relative decay widths. It is related to the moment of inertia $\mathscr{I}_\lambda$ and the nuclear temperature t by

$$\sigma_\lambda^2 = \frac{\mathscr{I}_\lambda t}{\hbar^2}. \tag{8.5}$$

Numerical values of σ_λ^2 range from about 10 for light nuclei to about 30 for the heaviest nuclei at excitation energies near 10 MeV, and are expected to increase roughly as the square root of the excitation energy. The differential cross section for producing a final state in channel β with kinetic energy ε_β and residual spin I is now given by

$$\left(\frac{d\sigma_{\alpha\beta}}{d\varepsilon_\beta}\right)_I = \sum_J \frac{\sigma_J(\alpha)\rho_\beta(E - S_\beta - \varepsilon_\beta, I)T_J(\varepsilon_\beta, I)}{\sum_\lambda \sum_{I'} \int_0^{E-S_\lambda} \rho_\lambda(E - S_\lambda - \varepsilon_\lambda, I')T_J(\varepsilon_\lambda, I')\, d\varepsilon_\lambda}, \tag{8.6}$$

where S_β is the separation energy in channel β.

A further simplification arises when the final state spin is not observed (or need not be specified for the calculation of additional decay steps). In order to obtain compact and transparent expressions for the cross sections, we now define spin-averaged transmission coefficients $\bar{T}_J(\varepsilon_\lambda)$:

$$\bar{T}_J(\varepsilon_\lambda) = \sum_I \frac{(2I + 1)}{\sigma_\lambda^2} \exp\left(-\frac{I(I + 1)}{2\sigma_\lambda^2}\right) T_J(\varepsilon_\lambda, I) \tag{8.7a}$$

$$= \sum_{l=0}^{\infty} g_\lambda f_\lambda(l, J)T_l(\varepsilon_\lambda) = \sum_{l,j} f_\lambda(j, J)T_{lj}(\varepsilon_\lambda). \tag{8.7b}$$

Here $g_\lambda = 2i_\lambda + 1$ ($i_\lambda =$ spin of the light particle emitted in channel λ) is a statistical weight factor for channel λ, and the auxiliary functions $f_\lambda(l,J)$ and $f_\lambda(j,J)$ are defined by

$$f_\lambda(l, J) = \sum_{I=|J-l|}^{J+l} \frac{(2I + 1)}{2\sigma_\lambda^2} \exp\left(-\frac{I(I + 1)}{2\sigma_\lambda^2}\right) \tag{8.8a}$$

$$\approx 2\sinh\left(\frac{1}{2}\frac{(2J + 1)(2l + 1)}{2\sigma_\lambda^2}\right) \exp\left(-\frac{J(J + 1) + l(l + 1)}{2\sigma_\lambda^2}\right). \tag{8.8b}$$

We further introduce an energy-integrated transmission function $F_J(\lambda)$ for channel λ by writing

$$F_J(\lambda) = \sum_I \int_{E_\lambda=0}^{E-S_\lambda} \rho_\lambda(E_\lambda, I)T_J(\varepsilon_\lambda, I)\, dE_\lambda = \int_{\varepsilon_\lambda=0}^{E-S_\lambda} \bar{\rho}_\lambda(E-S_\lambda-\varepsilon_\lambda)\bar{T}_J(\varepsilon_\lambda)\, d\varepsilon_\lambda \tag{8.9}$$

This yields for the differential cross section with respect to the kinetic energy ε_β in channel β (summed over final state spins and integrated over all angles)

$$\frac{d\sigma_{\alpha\beta}}{d\varepsilon_\beta} = \bar{\rho}_\beta(E - S_\beta - \varepsilon_\beta) \sum_J \frac{\sigma_J(\alpha)\bar{T}_J(\varepsilon_\beta)}{\sum_\lambda F_J(\lambda)}, \tag{8.10}$$

and, for the total cross section in channel β (integrated, in addition, over all energies ε_β),

$$\sigma_{\alpha\beta} = \sum_J \frac{\sigma_J(\alpha)F_J(\beta)}{\sum_\lambda F_J(\lambda)}. \tag{8.11}$$

The dominant decay modes of the compound nucleus are often those involving the emission of light particles, in particular nucleons and α particles. For these channels the transmission coefficients $\bar{T}_J(\lambda)$ are generally calculated with the optical model, since surface reactions are not important in the relevant range of energies. Other decay modes, like fission and gamma emission, can be included in (8.10) and (8.11) provided appropriate expressions are used to calculate the corresponding transmission coefficients. Gamma emission is usually restricted to dipole ($j = 1$) and quadrupole ($j = 2$) radiation, with transmission coefficients proportional to $\varepsilon_\gamma^{2j+1}$ [Gr 67a]:

$$T_j(\varepsilon_\gamma) = 2\pi\xi_j\, \varepsilon_\gamma^{2j+1}, \tag{8.12}$$

$$\bar{T}_j(\varepsilon_\gamma) = 2\pi \sum_J f_\gamma(j, J)\xi_j\, \varepsilon_\gamma^{2j+1}, \tag{8.13}$$

$$F_J(\gamma) = 2\pi\rho(E, J)\Gamma_\gamma(E, J) = 2\pi \sum_j f_\gamma(j, J)\xi_j \int_{\varepsilon_\gamma=0}^E \bar{\rho}_\gamma(E-\varepsilon_\gamma)\varepsilon_\gamma^{2j+1}\, d\varepsilon_\gamma. \tag{8.14}$$

Here j denotes the total angular momentum (or multipole order) of the emitted radiation, and the quantities ξ_j are empirical parameters adjusted to fit experimental data on radiative widths. The coefficients T_j and ξ_j are defined here to include summation over both parities, i.e. they include electric and magnetic transitions of multipole order j (in many practical applications, however, only electric transitions are considered). We note that the functions f_γ, $\bar{\rho}_\gamma$ refer to the initial nucleus, since the nuclear species is not changed by an electromagnetic transition.

The total fission width (integrated over all fragment masses and kinetic energies) can be calculated according to Bohr and Wheeler by treating the levels available at the saddle-point deformation ("fission channels") as final states, which must be reached for fission to take place [Bo 39]. The energy--integrated transmission function for fission is then given by

$$F_J(f) = \int_{\varepsilon_f=-\infty}^{E-B_f} \sum_{K=-J}^{+J} \rho_f((E - B_f - \varepsilon_f),J,K)T_{JK}(f)\, d\varepsilon_f \tag{8.15}$$

where ε_f is the kinetic energy in the fission mode and B_f is the fission barrier at zero angular momentum. The function ρ_f describes the density of levels at the saddle-point deformation as a function of total angular momentum J

and angular momentum projection K on the intrinsic nuclear symmetry axis. This function may be approximated by [Ha 58, St 61]

$$\rho_f((E - B_f - \varepsilon)J, K) = \bar{\rho}_f(E - B_f - \varepsilon) \frac{(2J + 1)}{2\sigma_f^2}$$

$$\exp\left(-\frac{J(J + 1)}{2\sigma_f^2}\right) P(K), \tag{8.16a}$$

$$P(K) = \frac{\exp\left(-\frac{K^2}{2K_0^2}\right)}{\sum_{K'=0,1/2}^{J} \exp\left(-\frac{(K')^2}{2K_0^2}\right)} \approx \left(\frac{2}{\pi K_0^2}\right)^{1/2} \exp\left(-\frac{K^2}{2K_0^2}\right). \tag{8.16b}$$

Here the quantities σ_f^2 and K_0^2 are related to the nuclear temperature t and the moments of inertia of the saddle point configuration parallel ($\mathscr{I}_{\parallel}$) and perpendicular ($\mathscr{I}_{\perp}$) to the nuclear symmetry axis by the following equations:

$$\sigma_f^2 = \frac{\mathscr{I}_{\perp} t}{\hbar^2}; \ K_0^2 = \frac{\mathscr{I}_{eff} t}{\hbar^2}, \text{ with } \mathscr{I}_{eff} = \frac{\mathscr{I}_{\parallel} \mathscr{I}_{\perp}}{\mathscr{I}_{\perp} - \mathscr{I}_{\parallel}}. \tag{8.17}$$

The coefficients $T_{JK}(f)$ in (8.15) give the transmission probability through the fission barrier. In statistical model calculations, these coefficients are usually assumed to be independent of J and K and given by the Hill–Wheeler formula for a parabolic barrier [Hi 53, see (7.18), (7.19)]

$$T_{JK}(f) \equiv T(\varepsilon_f) = \left[1 + \exp\left(-\frac{2\pi\varepsilon_f}{\hbar\omega_f}\right)\right]^{-1}, \tag{8.18}$$

where the parameter $\hbar\omega_f(\approx 0.5 - 1.0 \text{ MeV})$ measures the curvature of the barrier [see (7.19)]. With (8.16) and (8.18), the transmission function $F_J(f)$ can be rewritten as

$$F_J(f) = \frac{(2J + 1)}{2\sigma_f^2} \exp\left[-\frac{J(J + 1)}{2\sigma_f^2}\right] \int_{\varepsilon_f = -\infty}^{E - B_f} \bar{\rho}_f(E - B_f - \varepsilon_f) T(\varepsilon_f) \, d\varepsilon_f. \tag{8.19}$$

An alternative form of (8.19) is obtained by using instead of (8.16) a level density which explicitly depends on the energy of the saddle-point configuration at angular momentum J, $E_J(f)$, as derived, for example, from the rotating liquid-drop model:

$$F_J(f) = \frac{(2J + 1)}{2\sigma_f^2} \int_{\varepsilon_f = -\infty}^{E - E_J(f)} \bar{\rho}_f(E - E_J(f) - \varepsilon_f) T(\varepsilon_f) \, d\varepsilon_f. \tag{8.20}$$

This equation may be simplified further by noting that the barrier transmission coefficient $T(\varepsilon_f)$ rises rather steeply from zero to unity in the vicinity of $\varepsilon_f = 0$. Since the width of this rise is comparable with typical uncertainties in the quantity $E_J(f)$, one may use, without significant loss of accuracy, a sharp cut-off approximation for $T(\varepsilon_f)$, which finally yields

$$F_J(f) = \frac{(2J + 1)}{2\sigma_f^2} \int_{\varepsilon_f=0}^{E-E_J(f)} \bar{\rho}_f(E - E_J(f) - \varepsilon_f) \, d\varepsilon_f . \qquad (8.21)$$

Clearly the formalism summarized in (8.1–21) requires rather lengthy and complicated computations when applied to actual cases, where high excitation energies and angular momenta are involved, and the compound nucleus decays by a multistep cascade. A number of sophisticated codes have been developed which will treat this problem numerically with the help of large, high-speed computers. Some of these calculate the various distributions step by step using analytical techniques and numerical integrations (GROGI [Gr 67a, Gr 70]; ALICE [Bl 66, Bl 73, Bl 76, Pl 77]; CASCADE [Pü 77]), while others follow individual decays using the Monte Carlo method (JULIAN [Hi 76]; LILITA [Go 77]). The merits of the different approaches cannot be discussed here, and we refer the reader to the original literature for further details. The motivation to perform such calculations comes mainly from two sources. Firstly, one should like to gain some understanding of the general mechanism and selectivity of compound nucleus reactions induced by heavy projectiles. A major part of the early work in this direction has been performed by Grover and collaborators [Gr 61, Gr 62, Gr 67a] and by Thomas and collaborators [Th 64, Wi 67, Th 68]. Secondly, one wishes to interpret specific experimental data by relating observable quantities to properties of the original compound nucleus or of unobserved intermediate products. The experimental data considered in the latter context may be excitation functions, mass and charge distributions of residual nuclei, or multiplicities, energy spectra, and angular distributions of emitted light particles or gamma quanta. The information to be deduced may concern the excitation energy or angular momentum distribution of the primary system, or level densities, y-rast energies, or angular-momentum-dependent fission barriers of nuclei along the decay chain. In any case the results will be somewhat indirect due to the complexity of the reaction mechanism.

Next we discuss briefly some approximations in the formalism which have been widely used in the literature and can serve to illustrate the qualitative trends associated with different parameters. We first consider a situation where

$$\frac{1}{4}\left(\frac{(2J + 1)(2l + 1)}{2\sigma_\lambda^2}\right)_\lambda^2 \ll 1 \qquad (8.22)$$

for all channels λ and the significant ranges of angular momenta J,l. Equation (8.22) implies that either J or l is confined to small values compared to σ_λ. Important limiting cases of condition (8.22) are (i) that particle emission occurs only with $l = 0$ ("s-wave approximation") or (ii) that the spin cut-off parameter σ_λ goes to infinity [in the latter case, the angular momentum restriction in channel λ is removed and the decay proceeds as if the level density in that channel was simply proportional to $(2l + 1)_\lambda$]. The hyperbolic sine function in (8.8b) can now be replaced by its argument and, as a consequence, the quantities $\bar{T}_J(\varepsilon_\lambda)$ and $F_J(\lambda)$ defined by (8.7) and (8.9) factor into parts depending respectively only on energy and only on angular momentum:

$$\bar{T}_J(\varepsilon_\lambda) = \frac{2J+1}{2\sigma_\lambda^2} \exp\left(-\frac{J(J+1)}{2\sigma_\lambda^2}\right) g_\lambda \sum_l (2l+1) T_l(\varepsilon_\lambda) \tag{8.23a}$$

$$= \frac{2J+1}{2\sigma_\lambda^2} \exp\left(-\frac{J(J+1)}{2\sigma_\lambda^2}\right) \frac{2g_\lambda\mu_\lambda}{\pi\hbar^2} \varepsilon_\lambda\sigma(\lambda) \tag{8.23b}$$

$$F_J(\lambda) = \frac{2J+1}{2\sigma_\lambda^2} \exp\left(-\frac{J(J+1)}{2\sigma_\lambda^2}\right) \frac{2g_\lambda\mu_\lambda}{\pi\hbar^2}$$

$$\int_{\varepsilon_\lambda=0}^{E-S_\lambda} \varepsilon_\lambda\sigma(\lambda)\bar{\rho}_\lambda(E - S_\lambda - \varepsilon_\lambda)\, d\varepsilon_\lambda. \tag{8.24}$$

Here $\sigma(\lambda)$ denotes the "inverse cross section" for compound-nucleus formation in channel λ and all other symbols have the same significance as in (8.7–9). Assuming further that σ_λ^2 has the same value for all channels, and that gamma or fission decay can be neglected, one arrives at the conclusion that the branching ratios for particle emission are independent of angular momentum, and hence independence of compound nucleus formation and decay is effectively restored. Equation (8.10) for the cross section then assumes the form

$$\frac{d\sigma_{\alpha\beta}}{d\varepsilon_\beta} = \sigma(\alpha)\frac{g_\beta\mu_\beta\varepsilon_\beta\sigma(\beta)\bar{\rho}_\beta(E - S_\beta - \varepsilon_\beta)}{\sum_\lambda g_\lambda\mu_\lambda \int_{\varepsilon_\lambda=0}^{E-S_\lambda} \varepsilon_\lambda\sigma(\lambda)\bar{\rho}_\lambda(E - S_\lambda - \varepsilon_\lambda)\, d\varepsilon_\lambda}. \tag{8.25}$$

It should be noted that the inequality (8.22) and the resulting equations (8.23–25) are often not well satisfied in heavy-ion-induced reactions and must therefore be used with caution. In particular, these equations ignore differences in the availability of high-spin final states in the different exit channels, and hence cease to be applicable as the decay approaches the y-rast line. A basically similar, but somewhat more flexible, approximation consists of subtracting from the total excitation energy E the rotational energy (or y-rast energy) $E_J(\lambda)$ appropriate to angular momentum J and channel λ before calculating the level density for that channel. Including now also the competing fission channel [using (8.21)] we can write the cross section for particle emission in channel β as

$$\frac{d\sigma_{\alpha\beta}}{d\varepsilon_\beta} = \sum_J \sigma_J(\alpha)\left(\frac{g_\beta\mu_\beta\varepsilon_\beta\sigma(\beta)\bar{\rho}_\beta(E - E_J(\beta) - S_\beta - \varepsilon_\beta)}{G_J(f) + \sum G_J(\lambda)}\right), \tag{8.26a}$$

$$\text{with}\quad G_J(f) = \frac{\pi\hbar^2}{2} \int_{\varepsilon_f=0}^{\varepsilon_f(\text{max})} \bar{\rho}_f(E - E_J(f)) - \varepsilon_f)\, d\varepsilon_f, \tag{8.26b}$$

$$G_J(\lambda) = g_\lambda\mu_\lambda \int_{\varepsilon_\lambda=0}^{\varepsilon_\lambda(\text{max})} \varepsilon_\lambda\sigma(\lambda)\bar{\rho}_\lambda(E - E_J(\lambda) - S_\lambda - \varepsilon_\lambda)\, d\varepsilon_\lambda. \tag{8.26c}$$

This approach implies that the rotational component of the excitation energy is not available for particle evaporation. Consequently the excited nucleus initially emits particles without loss of angular momentum, until the excess energy above the y-rast line is too small to emit a further particle. The remaining excitation energy and angular momentum are then dissipated by gamma--ray emission (or fission). This procedure tends to over-estimate the enhancement of gamma emission and the corresponding overall reduction of particle emission due to angular momentum, since in reality some angular momentum is also removed by particle evaporation (see Fig. 8.1). At the same time, the effect of angular momentum on the relative decay widths into different particle channels, resulting from the different capabilities to remove angular momentum, is neglected.

In the method of analysis described in the preceding paragraph, the rotational energy plays the role of an effective barrier for particle emission. This feature is demonstrated particularly clearly in a simple version of the method which might be called the "constant-temperature effective-barrier model" and has been applied successfully by Sikkeland [Si 67] and others. Here one uses for the level density of the residual nucleus in channel λ at excitation energy E_λ and angular momentum J the "constant-temperature formula"

$$\rho_\lambda(E_\lambda, J) = \frac{2J+1}{2\sigma_\lambda^2} \bar{\rho}_\lambda(E_\lambda - E_J(\lambda)) = C_\lambda \frac{(2J+1)}{2\sigma_\lambda^2} \exp\left(\frac{E_\lambda - E_J(\lambda) - \delta_\lambda}{t_\lambda}\right) \tag{8.27}$$

where C_λ, t_λ, and σ_λ are adjustable constants (often assumed equal for all channels), and the quantity δ_λ corrects for pairing in the ground state of the residual nucleus.

We note that the second part of (8.27) is formally equivalent to (8.4) in combination with (8.5) if the spin I of the residual nucleus is identified with the compound nucleus spin J.

For the inverse cross sections $\sigma(\lambda)$, we write

$$\sigma(\lambda) = \pi R_\lambda^2 \left(1 - \frac{E_C(\lambda)}{\varepsilon_\lambda}\right), \tag{8.28}$$

denoting by $E_C(\lambda)$ the Coulomb barrier in channel λ. Equations (8.27) and (8.28) can be inserted into (8.26) to calculate cross sections, branching ratios, and energy spectra. The energy-integrated transmission functions are now given by

$$F_J(\lambda) = (2J+1) \frac{C_\lambda t_\lambda g_\lambda \mu_\lambda R_\lambda^2}{\mathscr{I}_\lambda} \left[\exp\left(\frac{E - B_J(\lambda)}{t_\lambda}\right) - \left(1 + \frac{E - B_J(\lambda)}{t_\lambda}\right)\right], \tag{8.29}$$

$$F_J(f) = (2J+1) \frac{C_f \hbar^2}{2\mathscr{I}_f} \left[\exp\left(\frac{E - B_J(f)}{t_f}\right) - 1\right], \tag{8.30}$$

where the symbols λ and f, as before, denote light particle emission of type λ and fission, respectively, and $\mathscr{I}_\lambda, \mathscr{I}_f$ are appropriately defined moments of inertia [$\mathscr{I}_f = \mathscr{I}_\perp$, see (8.16) and (8.17)]. In addition, we have introduced angular-momentum-dependent effective barriers $B_J(\lambda)$, $B_J(f)$ by writing

$$B_J(\lambda) = E_C(\lambda) + E_J(\lambda) + S_\lambda + \delta_\lambda, \tag{8.31}$$

$$B_J(f) = E_J(f) + \delta_f. \tag{8.32}$$

At excitation energies below the appropriate effective barrier, no final states are available for a given channel, and hence the decay is inhibited. Well above the barriers, on the other hand, the expressions on the right-hand sides of (8.29) and (8.30) are dominated by the exponential functions; ignoring the channel dependence of C, R, and $\mathcal{I}$ in the pre-exponential factor and assuming a uniform temperature t for all channels, the branching ratios can then be written approximately as

$$\left(\frac{\Gamma_\beta}{\Gamma_\lambda}\right)_J = \frac{F_J(\beta)}{F_J(\lambda)} \approx \frac{g_\beta \mu_\beta}{g_\lambda \mu_\lambda} \exp\left(\frac{B_J(\lambda) - B_J(\beta)}{t}\right), \tag{8.33}$$

$$\left(\frac{\Gamma_\beta}{\Gamma_f}\right)_J = \frac{F_J(\beta)}{F_J(f)} \approx \frac{2g_\beta \mu_\beta R_\beta^2 t}{\hbar^2} \exp\left(\frac{B_J(f) - B_J(\beta)}{t}\right). \tag{8.34}$$

According to (8.33) and (8.34), the branching ratio for a given pair of channels should be mainly determined by the difference in effective barriers, and should depend only weakly on excitation energy for fixed angular momentum. A case of particular interest is the ratio of the widths for neutron emission and fission; with $g_n = 2$ and $R_n = r_0 A^{1/3}$ we obtain from (8.34)

$$\left(\frac{\Gamma_n}{\Gamma_f}\right)_J \approx \frac{4mr_0^2}{\hbar^2} t A^{2/3} \exp\left(\frac{B_J(f) - B_J(n)}{t}\right). \tag{8.35}$$

This equation, and similar ones derived by using the Fermi-gas model for the level densities, have been discussed extensively by Huizenga and Vandenbosch [Hu 62]. The systematic trends of Γ_n/Γ_f, as deduced from light-projectile--induced fission, appear to be well reproduced by this simple approach. For heavy-projectile-induced fission, where the angular momentum dependence of Γ_n/Γ_f has a decisive influence, the evidence is much less direct due to a number of complicating factors, like unkown angular momentum distributions and unknown contributions from multiple chance fission. Qualitatively, (8.35) predicts a strong decrease of Γ_n/Γ_f with increasing angular momentum in agreement with observation. This is a consequence of the reduced rotational energy of the strongly deformed saddle-point shape (as compared to the less deformed equilibrium shape) for a given angular momentum, which results in a decrease of the difference $B_J(f) - B_J(n)$ with increasing angular momentum (see Fig. 7.28).

We conclude our theoretical survey of compound-nucleus decay with some remarks on angular distributions. The theory of angular distributions of evaporated particles is well known in principle and has been worked out in detail [see, for example, Sa 56, Do 59, Bi 59, Th 68]. We shall not present the results of an exact treatment of the problem, however, since the relevant formulae are rather complicated and most of the experimental data are not sufficiently detailed to make a direct comparison meaningful. Instead we quote some results of a semiclassical treatment of angular distributions, originally due to Ericson and Strutinsky [Er 58], which are physically more transparent and should be adequate for a semiquantitative discussion of the data. This approach implies that the relevant total and orbital angular momenta are large compared to $\hbar$ and to the intrinsic spins of the fragments, assumptions which seem not unreasonable for reactions induced by heavy projectiles.

The semiclassical formula for the doubly differential cross section for light-particle emission in channel β can be written as

$$\frac{\partial^2 \sigma_{\alpha\beta}}{\partial \varepsilon_\beta \partial \Omega_\beta} = \bar{p}_\beta(E - S_\beta - \varepsilon_\beta) \sum_J \frac{\sigma_J(\alpha) W_J(\varepsilon_\beta, \Theta_\beta)}{\sum_\lambda F_J(\lambda)}. \tag{8.36}$$

This equation differs from the corresponding one for the angle-integrated cross section (8.10) by the replacement of the factor $\bar{T}_J(\beta)$ with the function $W_J(\varepsilon_\beta, \Theta_\beta)$ which is given by [Er 60]

$$W_J(\varepsilon_\beta, \Theta_\beta) = \frac{1}{4\pi} \sum_{l=0}^\infty g_\beta \frac{(2J + 1)(2l + 1)}{2\sigma_\beta^2} \exp\left(-\frac{J(J + 1) + l(l + 1)}{2\sigma_\beta^2}\right) T_l(\varepsilon_\beta)$$
$$\times \sum_{\text{even} k} (-1)^{k/2}(2k + 1) \left\{\frac{k!}{(k!!)^2}\right\}^2 j_k \left\{\frac{i}{2} \frac{(2J + 1)(2l + 1)}{2\sigma_\beta^2}\right\}$$
$$P_k(\cos\Theta_\beta). \tag{8.37}$$

Here the second sum on the right-hand side is taken to include only even values of k, and hence only Legendre polynomials $P_k(\cos\Theta)$ of even order are present. The angular distribution is therefore symmetric with respect to $\theta = 90°$, as expected for the decay of a quasi-stationary compound system. The functions j_k are the spherical Bessel functions as defined, for example, by Blatt and Weisskopf [Bl 52]. Noting that $xj_0(ix) = \sinh x$, it is readily shown with the help of (8.7 a,b) and (8.8 a,b) that

$$\int_{4\pi} W_J(\varepsilon_\beta, \Theta_\beta)\, d\Omega_\beta = \bar{T}_J(\varepsilon_\beta) \tag{8.38}$$

as required.

It is interesting to consider certain limiting cases of (8.36) and (8.37) which arise when the argument of the spherical Bessel functions assumes either very small ($\ll 1$) or very large ($\gg 1$) values. In the former case, corresponding to the inequality (8.22), the orbital angular momenta of the emitted particles are not strongly aligned with respect to the direction of the total angular momentum, a situation usually referred to as "weak coupling". In the extreme weak-coupling limit, all spherical Bessel functions of order $k > 0$ vanish and the angular distribution becomes isotropic. We recall that in the same limit, the branching ratios for different decay channels become independent of angular momentum as discussed previously. Going one step further in an expansion with respect to the coupling parameter, we obtain the following approximate result for the angular distribution:

$$\frac{\partial^2 \sigma_{\alpha\beta}}{\partial \varepsilon_\beta \partial \Omega_\beta} \approx \frac{d\sigma_{\alpha\beta}}{d\varepsilon_\beta} \frac{1}{4\pi} \left(1 + \frac{\langle J(J+1)\rangle \langle l(l+1)\rangle}{12\sigma_\beta^4} P_2(\cos\Theta_\beta) \right). \tag{8.39}$$

Here $\langle J(J+1)\rangle$ and $\langle l(l+1)\rangle$ denote averages of the corresponding distributions, weighted with the appropriate contributions to the angle-integrated cross section as given by (8.7a,b) and (8.10). Equation (8.39) should be adequate in most cases where light-particle emission (n,p,α) is considered because of the limited angular momenta involved. It has been used in practice to deduce either compound-nucleus angular momenta or spin–cut-off parameters (and hence moments of inertia) for residual nuclei from measured angular anisotropies. It should be noted that the results of such an analysis are often associated with rather large uncertainties, both due to experimental problems and because not all parameters entering the calculation are well known.

The other extreme case of interest is the "strong-coupling limit", where the orbital angular momentum of the emitted particles is completely aligned with the compound nucleus angular momentum as a consequence of restrictions in the residual fragment spins. This limit applies if

$$\frac{1}{4} \left(\frac{(2J+1)(2l+1)}{2\sigma_\beta^2} \right)_\beta \gg 1 \tag{8.40}$$

for the significant range of angular momenta. The angular distribution is then given by the classical formula

$$\frac{\partial^2 \sigma_{\alpha\beta}}{\partial \varepsilon_\beta \partial \Omega_\beta} = \frac{d\sigma_{\alpha\beta}}{d\varepsilon_\beta} \frac{1}{2\pi^2 \sin\Theta_\beta}. \tag{8.41}$$

It should be noted that most of our discussion so far refers (either explicitly or implicitly) to the emission of particles from the primary compound nucleus. Experimentally, however, one normally deals with sequential evaporation from highly excited compound nuclei, and particles emitted from different nuclei along the evaporation chain cannot be distinguished. It is therefore necessary to compute averages for the entire evaporation cascade for comparison with the experimental distributions. In addition, one may be interested in the energy spectra or angular distributions of a specific type of radiation (for example, γ radiation) emitted after a sequence of preceding decays. In principle, such calculations can be performed by applying the formalism discussed above step by step; in practice, however, a more approximate statistical treatment of the decay may be much simpler and yet adequate to extract the essential features of the process. In this spirit, Halpern et al. have presented a classical statistical model for the evolution of the angular momentum distribution along the evaporation cascade, which allows one to calculate relatively simply angular distributions and other quantities of interest appropriate to various stages of the reaction [Ha 68].

The angular distribution of fission fragments emitted in compound nucleus decay has been studied theoretically by a number of authors [Ha 58, Gr 59, Hu 69; see also Hu 62, Hy 64b, Th 68]. We quote here an approximate semi--classical formula derived by Huizenga et al. [Hu 69] for cases where projectile and target have zero spin:

$$\frac{d\sigma_{af}}{d\Omega_f} = \sum_J \frac{\sigma_J(\alpha) \, \bar{W}_J(\Theta_f)}{\sum_\lambda F_J(\lambda)}, \tag{8.42}$$

$$\bar{W}_J(\Theta_f) = \frac{F_J(f)}{4\pi} \frac{2}{\sqrt{\pi}} \frac{x \exp\left(-\frac{x^2 \sin^2\Theta}{2}\right) J_0\left(i\frac{x^2 \sin^2\Theta}{2}\right)}{\mathrm{erf}(x)}. \tag{8.43}$$

In (8.43), $F_J(f)$ is the transmission function for the fission channel as introduced previously [see (8.19–21)],

$$\mathrm{erf}(x) = \left(\frac{2}{\sqrt{\pi}}\right)\int_0^x \exp(-t^2)\, dt$$

is the error function, J_0 denotes the Bessel function of order zero (and imaginary argument), and the parameter x is defined by

$$x^2 = \frac{J(J+1)}{2K_0^2}. \tag{8.44}$$

The quantity K_0^2 is the mean-square projection of the intrinsic nuclear spin on the symmetry axis of the deformed saddle-point shape [see (8.16) and (8.17)] and determines the degree of coupling between the directions of the total and orbital angular momentum. We note that K_0^2 increases with increasing nuclear temperature (roughly proportional to the square root of the excitation energy) and with decreasing deformation at the fission saddle point (i.e. with increasing mass and charge of the fissioning compound nucleus). In the weak-coupling approximation ($x^2 \ll 1$), the angular distribution can be written as

$$\frac{d\sigma_{af}}{d\Omega_f} \approx \frac{\sigma_{af}}{4\pi}\left[1 + \frac{1}{3}\frac{\langle J(J+1)\rangle}{2K_0^2} P_2(\cos\Theta_f)\right], \tag{8.45}$$

whereas in the strong-coupling limt ($x^2 \gg 1$), it is given by

$$\frac{d\sigma_{af}}{d\Omega_f} = \frac{\sigma_{af}}{2\pi^2 \sin\Theta_f}. \tag{8.46}$$

There is a considerable body of experimental results on fragment angular distributions in heavy-ion-induced fission (see, for example, Go 60, Vi 60, Ko

71, Ze 74). The data are generally consistent with (8.42–46); in particular, the expected dependence on the coupling parameter x^2 is qualitatively confirmed. The observed anisotropies increase with increasing projectile mass (increasing $\langle J(J + 1)\rangle$), but decrease with increasing target mass (increasing K_0^2) at comparable bombarding energies. As the bombarding energy is raised, there is an initial increase in anisotropy (increasing $\langle J(J + 1)\rangle$) followed by a more or less constant behaviour (increasing K_0^2 tends to counter-balance increasing $\langle J(J + 1)\rangle$). Whereas nucleon-induced fission typically corresponds to a weak-coupling situation, fission induced by projectiles from ^{12}C to ^{40}Ar is closer to the strong-coupling limit. The angular distributions are usually well approximated by $(\sin\Theta)^{-1}$ in a fairly large angular region around $\Theta = 90°$ with significant deviations only at small forward ($\Theta \lesssim 20°$) and large backward ($\Theta \gtrsim 160°$) angles. This feature, together with uncertainties in the primary angular momentum distribution and in the nature of the fissioning nuclei, makes it difficult, however, to extract quantitative information on K_0^2 and hence on the saddle-point deformation.

8.3 Competition Between Neutron and Gamma-Ray Emission

In the following sections we give a brief survey of representative theoretical and experimental results on specific aspects of compound-nucleus decay. As before, we concentrate on the reaction mechanism; applications to problems of nuclear structure or to the production of exotic nuclear species are not included.

The competition between neutron and gamma emission was discussed in great detail from a theoretical point of view by Grover and collaborators [Gr 61, Gr 62, Gr 67a]. In this work the importance of angular momentum effects and the crucial role played by the y-rast levels was clearly recognized. It was pointed out, in particular, that gamma emission should occur with high probability from levels excited to energies well above the neutron binding energy, and predictions were made for the number, multipole order, and energy spectrum of emitted photons as a function of excitation energy (E_{ex}) and angular momentum (J).

The dependence of the principal decay mode on E_{ex} and J is shown qualitatively in Fig. 8.3 for a nucleus in the rare-earth region ($A \approx 150$). The region of predominant gamma decay ("gamma-cascade band") has a width (along the energy axis) approximately equal to the neutron binding energy, and is cut off towards large values of E_{ex} and J by the onset of α emission or fission. A detailed knowledge of the boundaries of this region (i.e. the y-rast line and the entry line) would enable one to deduce information on such interesting nuclear properties as distortion, moment of inertia, and average electromagnetic matrix elements under conditions of high angular momentum and excitation energy. Consequently considerable efforts have been devoted to an experimental study of this region.

Relevant evidence has been derived so far either from measurements of

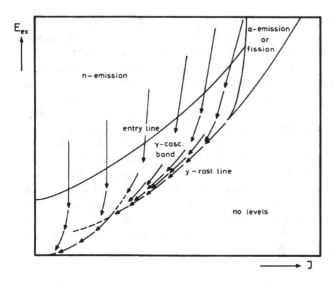

Fig. 8.3.   Competition between various decay modes in the E_{ex}–J plane. Gamma emission is expected to dominate within the gamma-cascade band, bounded towards lower E_{ex} by the y-rast line and towards higher E_{ex} by the entry line

(HI, xn) excitation functions, or from observations of gamma emission following (HI, xn) reactions. We comment first on the former approach. In most of the measurements the residual nuclei have been detected by activation techniques (especially α counting), but in-beam gamma spectroscopy has also been used. Both methods allow the reaction products to be identified unambiguously; however, for quantitative studies an accurate knowledge of level and decay schemes is necessary. The principal quantity of interest is the probability $P_x(E_{ex},J)$ for emission of x neutrons from a given primary compound nucleus with excitation energy E_{ex} and angular momentum J. In terms of this quantity, the cross section can be expressed as

$$\sigma(\alpha, xn) = \sum_J \sigma_J(\alpha)P_x(E_{ex}, J) = \sigma(\alpha) \langle P_x(E_{ex}, J)\rangle_{E_{ex}} \qquad (8.47)$$

where α denotes the entrance channel, and the average on the right-hand side is taken with respect to angular momentum at constant excitation energy. The problem is, of course, that only this average is directly observable [provided the fusion cross section in channel α, $\sigma(\alpha)$, is known], but not the detailed dependence of $P_x(E_{ex},J)$ on J. As discussed in Sect. 8.1, the angular momentum dependence of $P_x(E_{ex},J)$ gives rise to a broadening of (HI, xn)-excitation functions where broad angular momentum distributions are involved. At the same time, the peak of the excitation function shifts to higher excitation energy compared to a reaction involving only low angular momenta (see Fig. 8.4). As shown by Alexander and Simonoff [Al 63, Si 63, Al 64, Si 64b] these effects may be discussed quantitatively by plotting $\langle P_x(E_{ex},J)\rangle \approx \sigma(\alpha,xn)/\sigma_R$ as a function of the "available energy per emitted neutron" $(E_{ex} - \sum_x S_n)/x$. The latter quantity is equal to the average kinetic energy of the emitted neutrons plus the total gamma energy divided by x. In Fig. 8.4 such plots are shown for a "normal" (HI, xn) reaction (where presumably the total yield of the final nucleus has been measured) and a reaction populating only a low--spin isomer of the final nucleus (^{149g}Tb, $I = 5/2$). The difference is quite striking: in the latter case the peak cross section is strongly reduced, the position of the peak is shifted to lower excitation and the excitation function is considerably narrower. These effects are readily explained by assuming that only a restricted range of angular momenta ($J_{max} = 7.5 \pm 1.5\ \hbar$) contributes to the excitation function for production of ^{149g}Tb [Al 63].

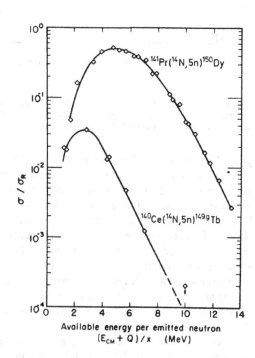

Fig. 8.4. Reduced cross sections σ/σ_R for (^{14}N, 5n) reactions populating ^{150}Dy and ^{149g}Tb, versus available energy per emitted neutron. (From Al 63)

Figure 8.5 compares in the same manner several "normal" (HI, xn)-excitation functions ($x = 5,6,7$), where the same compound and final nuclei have been produced with either ^{12}C or ^{16}O projectiles. The observation of very similar reduced excitation functions for the two entrance channels is indicative of a classical compound nucleus mechanism, in connection with similar angular momentum distributions. It is important to note that all the maxima in Fig. 8.5 occur at "available energies" between 5 and 6 MeV per neutron, whereas the corresponding plots for low-spin reactions usually peak around 3 MeV per neutron. This difference is related to the energy dissi-

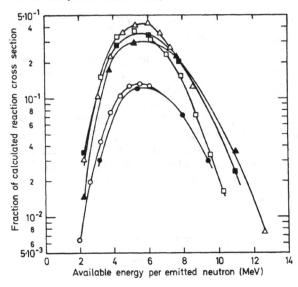

Fig. 8.5. Reduced cross sections σ/σ_R for (HI,xn) reactions populating $^{149-151}$Dy from different entrance channels (open symbols: ^{12}C + ^{144}Nd, closed symbols: ^{16}O + ^{140}Ce), versus available energy per emitted neutron. (From Si 63)

pated by photon emission and may be used to estimate the latter quantity.

It is also possible to estimate the average kinetic energy of the *first* neutron by comparing effective average excitation energies corresponding to the emission of x and x-1 neutrons [Al 64], or to derive average *total* neutron kinetic energies from the angular distribution of the evaporation residues (as deduced from measurements of projected recoil ranges, see Si 64b). The total energy dissipated by photon emission can then be calculated by subtracting from the initial excitation energy of the compound nucleus the neutron binding energies and the total neutron kinetic energy. In addition, the total gamma energy may be related to the total angular momentum carried away by gamma rays (as deduced from indirect arguments) and hence an estimate of the entry line may be obtained [Al 64].

Alexander and Simonoff analysed a considerable body of data on (HI, xn) reactions (A_1 = 10–22, x = 3–11), leading either to various Dy isotopes or to the isomer ^{149g}Tb, in the manner indicated above. Their conclusions may be summarized as follows:

(i) the reactions proceed by compound-nucleus formation;

(ii) the (xn) channels account for about 90% of the total reaction cross section at E_{ex} = 45 MeV, and for about 40% at E_{ex} = 120 MeV;

(iii) the average kinetic energy of the neutrons increases approximately as the square root of the excitation energy;

(iv) in the "normal" (HI, xn) reactions, corresponding to a broad angular momentum distribution of the compound nucleus, approximately one-half of the total "available" energy is dissipated by photon emission (i.e. about 30 MeV for available energies of 50–60 MeV);

(v) for these reactions, the relationship between total energy and total angular momentum of the photons indicates a large proportion of quadrupole emission;

(vi) in contrast, the "low-spin reactions" (i.e. those populating the isomer ^{149g}Tb) are characterized by relatively small total photon energies (less than 12 MeV).

These results are qualitatively consistent with the calculations of Grover and collaborators.

A relatively simple formalism for the quantitative description of (HI, xn)-excitation functions was developed by Sikkeland [Si 67]. It is based on the Jackson model of multiple neutron emission as originally applied to (p, xn) reactions [Ja 56]. Angular momentum effects were incorporated by Sikkeland by assuming that the rotational part of the excitation energy is not available for neutron evaporation. This leads to the constant-temperature effective--barrier model outlined in Sect. 8.2 [see (8.27–35)]. The principal parameters of this model are the nuclear temperature t (assumed independent of mass number and excitation energy) and the ratio of the moment of inertia $\mathscr{I}$ to that of a rigid sphere, $\mathscr{I}_{rig}^{0}$. The former of these parameters reflects the level densities of the daughter nuclei as a function of excitation, whereas the latter controls the rotational energy and hence the influence of angular momentum. In addition, an overall normalization factor is applied to each excitation function in order to account for competing decay channels other than neutron emission. The angular momentum distribution in the entrance channel is calculated with a standard optical model potential applying, if necessary, an upper cut-off l_{fu}.

Using this simple approach, excellent fits were obtained to experimental excitation functions (see Fig. 8.6). The best-fit parameters for nuclei in the rare-earth region are $t = 2.0$ MeV and $\mathscr{I}/\mathscr{I}_{rig}^{0} = 1.5$, while $t = 1.3$ MeV was found for heavy nuclei in the actinide region ($\mathscr{I}$ is not determined by the data in the latter case, as the angular momentum distribution is limited by fission decay; see Si 67). Clearly in view of the approximations of the model, the interpretation of these parameter values is not obvious. In particular, the deduced moment of inertia in excess of the rigid-sphere value may be partly due to nuclear deformation [Si 67], but probably also reflects the approximate treatment of angular momentum in the model (which neglects, for example, angular momentum removal by neutron emission). Nevertheless the fact that the experimental data are consistently reproduced with a small number of relatively stable parameters has made this model extremely useful for predicting (HI, xn)-excitation functions, especially in the heavy-element region.

Much more elaborate calculations of (HI, xn)-excitation functions, including explicitly the angular momentum dependence of the level densities and the competition between neutron and gamma emission, were performed more recently by Gilat, Jones, and Alexander [Gi 73]. These authors concluded that agreement with experimental excitation functions could be ob-

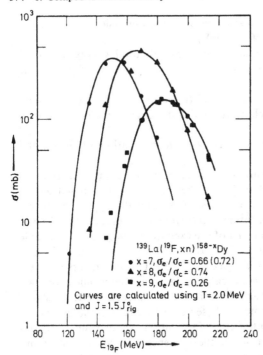

Fig. 8.6. Theoretical fits by Sikkeland to experimental excitation functions for the reactions ^{139}La(^{19}F,xn) $^{158-x}$Dy ($x = 7,8,9$; from Si 67, experimental results from Al 64).

tained only by allowing for a strong enhancement of the gamma transition strengths [ξ_J, see (8.12–14)] compared with values obtained by fitting slow-neutron-resonance data. A reasonable reproduction of the (HI, xn) data was achieved either by applying a constant factor (approximately 100) to the empirical value of $\xi_1(10^{-7})$, or by assuming ξ_1 to increase in proportion to angular momentum as predicted theoretically by Sperber [Sp 65].

Next we consider experiments where the gamma rays from fusion reactions are observed. Pioneering work in the field was performed by Mollenauer, who studied the yields and energy spectra of (unresolved) gamma rays emitted when the same compound nuclei (^{63}Cu, Ce) were produced with either α particles or ^{12}C nuclei as projectiles [Mo 62]. The deduced total gamma energies were between 9 and 17 MeV for the ^{12}C bombardments and less than 7 MeV for the α bombardments, i.e. above the neutron binding energy in the former, and below the neutron binding energy in the latter case. At the same excitation energy in ^{63}Cu more photons were emitted with ^{12}C than with α projectiles; the absolute gamma multiplicities ($M_\gamma \approx 0.5\langle J\rangle$) and the observed anisotropies suggested preferential emission of quadrupole radiation. These findings provided the first direct evidence for angular momentum effects in the competition between particle and gamma emission.

Additional information has been derived from studies of resolved gamma transitions between low-lying states of the final nuclei. In general the transitions along the lower part of the y-rast line collect intensities from many decay branches originating in regions of higher excitation energy and angular momentum (see Fig. 8.3). Hence these transitions stand out strongly in the spectra and can easily be studied. Two questions are of particular interest with respect to the reaction mechanism: what is the highest spin that can be identified, and how are the relative intensities of successive transitions along the y-rast line affected by "side-feeding" from regions of higher excitation energy.

Following earlier work by Morinaga and Gugelot with α particles [Mo 63], Stephens et al. observed transitions from states up to spin 16 within the ground-state rotational bands of nuclei in the rare-earth region using heavier projectiles ($A_1 \leq$ 19; St 65).

Subsequently similar experiments were carried out in many laboratories with a variety of target nuclei and projectiles up to ^{40}Ar (see, for example, Ne 70). An important result of these studies was that the highest spins seen did not increase as expected with increasing projectile mass (and hence angular momentum). For a long time, no spins significantly above 20 $\hbar$ could be observed no matter what projectiles were used. More recently, transitions from states with spins up to 30 $\hbar$ could be identified with improved techniques using ^{40}Ar projectiles [Le 77]; however, this is still far below the maximum compound-nucleus spin values of about 70 $\hbar$, as deduced from cross sections for evaporation residue formation.

The reasons for the apparent cut-off of the y-rast band appear to be partly physical and partly technical: in the high-spin region, there seems to be no longer a unique y-rast band, but a variety of more or less equivalent decay paths close to the y-rast line, leading to considerable fractionation of the total gamma intensity. Consequently decreasing intensities, together with poorer peak-to-background ratios and Doppler broadening, prevent an analysis of individual transitions.

Thus the missing angular momentum must be dissipated by emission of gamma rays with an essentially continuous energy distribution. Studies of this radiation can yield information on the entry line in the E_{ex}-J diagram (see Fig. 8.3) and have met with increasing interest in recent years.

The relevant experimental methods have been refined considerably since the early work of Mollenauer quoted above. One important technique is the measurement of gamma-multiplicity distributions with multiple detector arrangements operated in coincidence [Tj 74, Ma 74, Ha 75, Sa 76, An 78]. In this type of work, one usually selects specific transitions between low-lying states with a high-resolution detector [Ge (Li)] and registers n-fold coincidences with additional detectors of lower resolution but higher efficiency (NaI). The number of n-fold coincidences as a function of n can then be used to calculate certain moments (of low order) of the multiplicity distribution for all decay cascades proceeding through the selected level. Finally the multiplicity distribution is converted to an angular momentum distribution by assuming a relationship between the total number of transitions in a cascade and the total angular momentum removed by the cascade.

It follows from a fairly large body of experimental evidence that an approximately linear relationship exists between gamma multiplicity and angular momentum change. This leads to the following expression for the first moment of the multiplicity distribution [see, for example, Sa 76, An 78]:

$$\langle M_\gamma \rangle = \frac{1}{2} \langle I - I_0 \rangle + k . \tag{8.48}$$

Here M_γ is the gamma multiplicity; I, the angular momentum of the entry state; I_0, the angular momentum where the observed cascade terminates; and k, an empirical number (usually taken in the range 3–4). Equation (8.48) may be interpreted as stating that the average cascade consists of k ($\approx$ 3–4) "statistical" transitions, which do not produce a resulting spin change, followed by a sequence of "stretched" E2 transitions, each associated with a spin change of two units. Equations based on (8.48) have also been used to relate higher moments of the multiplicity distribution (i.e. the variance and skewness) to the corresponding moments of the spin distribution.

The total gamma energy associated with a given low-lying transition may be deduced by multiplying the average multiplicity with the average gamma energy as derived from the measured spectrum. Alternatively, the total energy can be determined by summing the individual transitions in a single, large--volume detector. The latter method has received considerable attention re-recently, as a very simple and effective way to select cascades originating from regions of high spin [see, for example, Tj 78]. By combining total energies and total angular momenta [as derived by application of (8.48)], information on the shape and position of the entry line may be obtained.

In Figs. 8.7 and 8.8 we show results of gamma-multiplicity measurements taken from the work of Sarantites et al. on the system ^{20}Ne + ^{150}Nd [Sa 76]. It is evident from Fig. 8.7 that the multiplicities increase with increasing

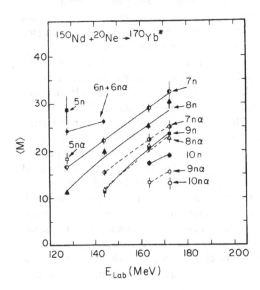

Fig. 8.7. Average gamma multiplicities corresponding to different types of particle decay, as observed for the system ^{20}Ne + ^{150}Nd, as functions of bombarding energy. (From Sa 76)

bombarding energy (average angular momentum) and with decreasing number of evaporated neutrons. These trends are consistent with the qualitative arguments presented earlier. Figure 8.8 gives the average angular momenta deduced from the same data [taking $k = 4$ in (8.48)]. The average angular

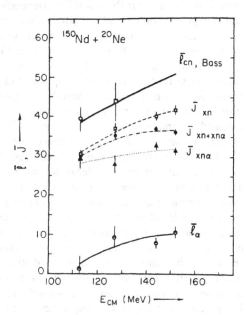

Fig. 8.8. Average angular momenta deduced from the data shown in Fig. 8.7. as functions of centre-of-mass energy. J denotes the angular momentum of entry states for a specific type of decay, and l_α the orbital angular momentum of evaporated α particles. The solid line marked $\bar{l}_{cn,Bass}$ refers to the compound nucleus and has been calculated with the potential given by (7.39); the points shown for comparison with that line are based on experimental fusion cross sections. (From Sa 76)

momenta of the entry states, as derived from this analysis, are seen to be lower than the average angular momentum of the compound nucleus. The difference, of the order of 10 ℏ, is attributed to the orbital angular momenta of emitted particles. From the difference between the (xn) and the $(xn\alpha)$ channels, it is possible to estimate the average angular momentum carried away by an α particle, and this is also shown in Fig. 8.8.

Further information on the spin distribution associated with the various channels has been derived from the second and higher moments of the measured multiplicity distributions. Several studies of this type, all performed on compound nuclei in the rare-earth region, led to the conclusion that there is considerable overlap in angular momentum between different (xn) channels for a given system and bombarding energy [Ha 75, Sa 76, An 78]. No evidence could be obtained for a lower cut-off in angular momentum associated with fusion [Sa 76, Si 77, Br 77], as suggested by excitation function measurements of the Orsay group [Ga 74b].

8.4 Competition Between Different Particle Channels

Next we consider the question of competition between different types of light-particle evaporation, and in particular, between nucleon and α emission.

The first extensive statistical model calculations concerning this point (including angular momentum effects) were performed by Grover and Gilat [Gr 67a], Jägare [Jä 67], and Williams and Thomas [Wi 67, see also Th 68]. All the calculations predict a strong enhancement of α emission from high angular momentum states near the y-rast line, depending in magnitude on the system and energy considered and the parameters entering the calculation.

Williams and Thomas presented approximate analytical formulae for the average energy of the first emitted particle, the resulting average change in angular momentum, and absolute and relative widths for emission of different particles. These formulae are applicable in cases where the classical rotational energy is much smaller than the excitation energy, i.e. far from the y--rast line. Jägare discussed the sequential emission of several particles and assumed gamma emission to take over whenever the difference between excitation energy and rotational energy becomes less than some effective threshold for particle emission. He concluded that α emission becomes the dominant mode of particle decay for states with angular momenta above about 20 $\hbar$ in the rare-earth region. The calculations of Grover and Gilat, on the other hand, explicitly include the competition between gamma and particle emission in all stages of compound-nucleus decay, and should therefore yield a more realistic description of the decay pattern, especially in the vicinity of the entry line. These authors give detailed consideration to the energy spectra of the emitted α particles and distinguish three different subspectra peaking at different energies: the type I subspectrum, produced by the decay of nuclei well above the entry line, should be concentrated near the Coulomb barrier energy; the type II subspectrum, fed from the neighbourhood of the entry line (where the abrupt disappearance of the neutron width causes a maximum in the relative α decay probability) should be associated with α energies below the Coulomb barrier; finally, at still lower energies, discrete α lines should be observed from the decay of special states at (or close to) the y-rast line, which cannot decay by gamma transitions of low multipole order ("y-rast traps"). These predictions have stimulated considerable experimental activity in recent years, but have not yet been verified in detail.

Further calculations on α emission were presented by Gilat, Jones, and Alexander [Gi 73]. Their results confirm the existence of a cut-off of the γ--cascade band towards higher angular momenta due to predominant α decay; at the same time they show that the location of this cut-off depends sensitively on assumptions about the gamma transition strength and the effects of nuclear distortion (produced by high angular momenta) on the Coulomb barrier for α emission.

In bombardments of medium and heavy target nuclei with light heavy ions ($A_1 = 12$–16) well above the Coulomb barrier, appreciable fractions of precompound α particles were identified on the basis of angular distributions [Br 61, Ga 74a]. Little or no evidence for precompound α emission was obtained, on the other hand, in low-energy work with light systems [Ha 63; see also chapter 2] or in argon-induced reactions [Ga 74a].

The effect of high angular momenta on the angular distribution of evaporated α particles was studied by Halbert and Durham [Ha 63, Du 65] and by Galin et al. [Ga 74a]. These authors measured continuous energy spectra and analysed their angular distributions in the framework of the semiclassical theory of Ericson and Strutinsky [Er 58, Er 60; see (8.36–41)].

As discussed in Sect. 8.2 the distributions are symmetric with respect to 90°, with anisotropies depending on the angular momentum spectrum of the compound nucleus and the spin dependence of the level densities in the various final nuclei. It is important to realize that both of these factors enter the theoretical expressions and cannot be determined separately from the data. More detailed considerations show, however, that lower-energy data may be used with reasonable confidence to extract spin cut-off parameters for the dominating exit channels [Ha 63, Du 65], whereas higher-energy data can yield information on the angular momentum limit for compound nucleus formation [Ga 74a].

More detailed and accurate information than from measurements of continuous energy spectra can be deduced from experiments where individual final states are resolved. For technical reasons, however, such studies have so far been limited to light systems (A_1, $A_2 < 20$). Extensive work was reported especially for systems made up of ^{12}C and ^{16}O fragments with the aim of investigating structure in the excitation functions. Resonance spins were determined from α-particle angular distributions, and average lifetimes of the compound nuclei were deduced from statistical analyses of fluctuating excitation functions. These aspects are discussed in Chap. 2 and will not be taken up here again. In addition, however, a considerable body of information on limiting angular momenta for fusion, spin cut-off parameters, and final state spins is available from Hauser–Feshbach analyses of nonfluctuating cross sections (including angular distributions) for the population of discrete final states. An excellent summary of results available by 1973 was presented by Stokstad at the Nashville conference [St 74].

Among the systems most widely discussed in this context is ^{12}C + ^{14}N. Following earlier speculations concerning the importance of direct multi--nucleon transfer contributions to some exit channels, it was shown by several groups that the bulk of the available data could be reproduced very well by statistical Hauser–Feshbach calculations [St 74, Ha 74, Kl 74b, Vo 75]. An important result of these analyses was that the relative yields of different exit channels depend sensitively on the angular momentum of the compound nucleus; hence different regions of the angular momentum spectrum can be examined by selecting appropriate exit channels. This is illustrated in Fig. 8.9, which shows calculated cross sections for emission of different products as functions of the angular momentum of the compound nucleus ^{26}Al at $E_{CM} = 36$ MeV. It is evident from this figure that the emission of ^{6}Li or ^{7}Be is concentrated near the upper end of the angular momentum spectrum and can thus serve as a sensitive indicator for the presence and location of an angular momentum limit for fusion.

The kinematic factors which govern the "high-spin selectivity" of certain reaction channels were discussed extensively by Klapdor and collaborators

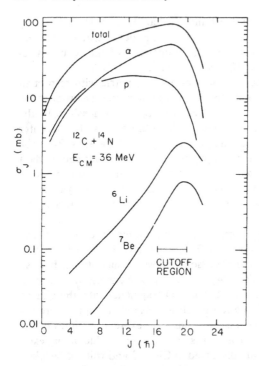

Fig. 8.9. Predicted cross sections for emission of different particles in the reaction $^{12}C + ^{14}N$ as functions of compound nucleus spin I. (From St 74)

[Kl 74a,b, Kl 75a, Kl 76]. The main point is that the angular momentum balance between the entrance and exit channel defines an accessible region in the $E_{ex} - I$ plane of the final nucleus, which depends on the bombarding energy, Q value, and mass of the emitted particle. This is shown in Fig. 8.10 for the reactions $^{10}B(^{14}N, \alpha)^{20}Ne$ and $^{10}B(^{12}C, d)^{20}Ne$. In each reaction only those residual levels (shown as horizontal bars) located within the appropriate "allowed region" should be populated with an appreciable cross section. It follows that the $(^{14}N, \alpha)$ reaction has a much larger overall cross section (corresponding to a wide angular momentum distribution) than the $(^{12}C, d)$ reaction. In addition, the latter reaction should selectively populate levels in a restricted range of excitation energies near the y-rast line in ^{20}Ne. These qualitative predictions are supported by more elaborate calculations and by experimental results [Kl 74a]. It is interesting to note that compound nucleus reactions (under conditions of "high-spin selectivity", i.e. where mainly grazing trajectories contribute to the cross section) are associated with a Q-value window similar to that encountered in direct transfer reactions (see Sect. 4.3).

The existence of a limiting angular momentum for fusion in the entrance channel (L_{fu}) will affect absolute and relative cross sections for discrete final states, as well as their dependence on angle and bombarding energy. In principle, all these dependences could be used to determine L_{fu}, provided the other parameters entering the problem are known. For example, Volant et al. deduced values of L_{fu} for the system $^{12}C + ^{14}N$ from an analysis of excitation

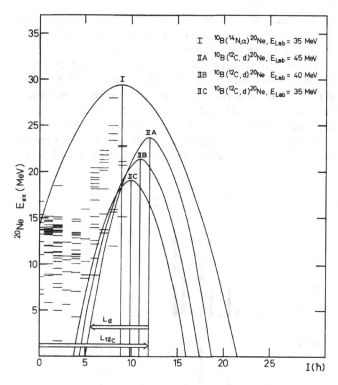

Fig. 8.10. Selectivity of the reactions $^{10}B(^{14}N, \alpha)^{20}Ne$ and $^{10}B(^{12}C, d)^{20}Ne$ as deduced from the requirement of angular momentum matching in the incident and outgoing channels. The parabolas define "allowed" regions in the E_{ex}–I diagram of the final nucleus for population by the corresponding reactions. (From Kl 75)

functions [Vo75]. Klapdor et al., on the other hand, argued that this procedure is sensitive to assumptions about level density parameters for competing decay channels, and based their analysis of L_{fu} for the same system on relative cross sections for different final states (i.e. the dependence of the cross section on final state excitation energy and spin at a fixed bombarding energy; Kl 74b, Kl 75b). It seems clear from these and other, similar studies that a reliable determination of L_{fu} must include a comprehensive analysis of all important decay channels, in order to eliminate errors due to arbitrary parameter choices.

A different method of studying the competition between different types of particle emission consists of measuring the mass and charge distribution of the evaporation residues. This can be done either with radiochemical techniques or, more efficiently, with counter techniques, involving the simultaneous measurement of energy loss, residual energy, and time of flight of the heavy recoils [Pü75, Ga75, We76, Co77, Ko77]. In this case, the limiting angular momentum for fusion is determined by the measured total fusion cross section;

the detailed mass and charge distribution of evaporation residues depends, in addition, on properties of the various nuclei produced along the decay chain, and on the mechanism of de-excitation.

The statistical model analysis of such data and its sensitivity to various parameters have been discussed by Pühlhofer [Pü77]. As an example, we show in Fig. 8.11 a comparison of measured and calculated mass distributions for the system ^{19}F + ^{27}Al, and calculated distributions for various (sharp) values of the compound-nucleus spin. The experimental overall distribution is well reproduced by the calculation, although its structure is less pronounced than the calculated one. The calculations show that there is a strong correlation

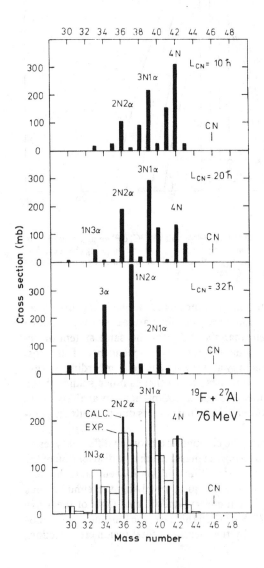

Fig. 8.11. Bottom: Experimental and calculated mass distributions of evaporation residues from the system ^{19}F + ^{27}Al. The number of nucleons and α particles emitted in the formation of the prominent product masses is indicated. Top: Mass distributions calculated for definite values of the compound nucleus angular momentum. (From Pü 77)

between angular momentum and mass distribution, and that larger angular momenta tend to enhance α emission.

The conclusion from a number of studies of this type is that the assumption of statistical compound nucleus decay correctly reproduces the main features of the experimental A- and Z-distributions. The question to what extent reliable information on details of the de-excitation process (such as the shape and position of the y-rast line) can be deduced remains open, however.

8.5 Competition Between Fission and Light-Particle Emission

We start this last section with some remarks about the theoretical situation. Although dynamical, multidimensional theories of fission have been discussed in the literature (see, for example, Ha 71, Br 72, Fi 74), they have not, to our knowledge, been developed to a point where meaningful comparisons with experimental data could be made. Thus practically all quantitative analyses of experimental results on fission competition make use of the statistical treatment of Bohr and Wheeler [Bo39], which is based on one-dimensional, quasi-static arguments. In this approach, the fission width (summed over all final fragmentations) is controlled by the number of levels available at the fission saddle point, which plays the role of a bottle-neck for the transition to fission. The branching ratios for fission and particle emission are then essentially given by ratios of level densities appropriate to the fissioning nucleus at saddle-point deformation on one hand, and to the evaporation residues on the other [see (8.9), (8.19)]. Furthermore, in most practical applications, the same analytical level-density expressions (based on the Fermi-gas or constant--temperature models, see Hu 72) were used for nuclei at ground state and saddle-point deformation—allowing, however, for a possible dependence of the level-density parameters on deformation. A number of authors also presented more sophisticated treatments, where saddle-point level densities were deduced from microscopic calculations of single-particle levels for deformed nuclear shapes [see, for example, Mo 72a, Va 72]. The latter approach was followed, in particular, in attempts to extrapolate into the superheavy mass region [Mo 72b].

The experimental evidence on fission–evaporation competition may be divided roughly into two categories. The first of these comprises studies of excitation functions for specific evaporation channels, especially (HI, xn) reactions. These channels are depleted by fission decay and, hence, represent a sensitive indicator for the latter under conditions where fission competition is strong. The second category of experimental data consists of total cross sections for fission or evaporation residue formation. Preferably, both cross sections should be measured simultaneously for the same system and bombarding energy in order to obtain reliable information [see, for example, Ze 74, Pl 75]. From this type of experiment, limiting angular momenta for evaporation residue formation can be deduced, above which the compound nuclei have only a minor chance to survive de-excitation without undergoing fission.

We note that the interpretation of experiments of either category may be complicated by a number of factors: the possibility of fission decay at various stages of de-excitation, the possible overlap of various decay modes in angular momentum space (i.e. the inadequacy of the sharp cut-off approximation), and the difficulty of distinguishing between compound-nucleus fission and quasi-fission (or deep-inelastic transfer). With increasing awareness of these problems, the methods of anlysis have been refined considerably in recent years. It is clear, on the other hand, that the quantitative conclusions drawn in earlier work must be taken with caution, and a re-analysis of some older data would be desirable. In the following we give a brief summary of experimental results and their interpretation.

The influence of the fission channel on (HI, xn)-excitation functions has been studied in many experiments with projectiles up to ^{40}Ar and target mass numbers ≥ 150 [Th 63, Si 67, Si 68a,b, Ga 73,Og 74; for references to earlier work, see Hy 74a, Hy 64b]. There are two obvious qualitative effects of fission competition: the width of the excitation function for a particular value of x is reduced as a consequence of the cut-off in angular momentum by fission, and the maximum cross section decreases strongly with increasing x. The latter observation may be explained partly by the fact that fission competes with neutron emission at each step of the evaporation cascade, and partly by the increase in primary angular momentum with bombarding energy (and hence x). These effects are illustrated in Fig. 8.12, which shows (HI, xn)-excitation functions for the system ^{16}O + ^{238}U [Si 67].

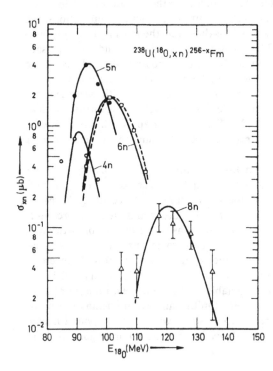

Fig. 8.12. Excitation functions for reactions ^{238}U(^{16}O, xn)$^{254-x}$Fm, with $x = 4,5,6,8$. The curves are theoretical fits to the data. (From Si 67)

Sikkeland has included fission competition in his analysis of (HI,xn)-excitation functions for heavy target nuclei by writing

$$\sigma(\alpha, xn) \approx \bar{\sigma}(\alpha,xn) \prod_{i=1}^{x} \left(\frac{\Gamma_n}{\Gamma_n + \Gamma_f}\right)_i \approx \left\langle\frac{\Gamma_n}{\Gamma_n + \Gamma_f}\right\rangle^x \bar{\sigma}(\alpha, xn), \qquad (8.49)$$

where $\bar{\sigma}(\alpha,xn)$ denotes the cross section in the absence of fission. The latter quantity was calculated originally ignoring the angular momentum dependence of $P_x(E_{ex}, J)$ [i.e. using $P_x(E_{ex}, O)$, see (8.47)], in order to allow for effects of a restricted angular momentum distribution on the shape of the excitation functions [Si67]. In subsequent work, however, improved fits to experimental data were obtained with angular-momentum-dependent values of P_x and readjusted parameters t (temperature) and $\mathscr{I}$ (moment of inertia) [Si 68a, b]. It should be noted that both approaches imply that Γ_n/Γ_f is independent of angular momentum and excitation energy, in contrast to experimental evidence and more refined calculations. The parameters Γ_n/Γ_f, t, and $\mathscr{I}$ which can be deduced from analyses of this type represent complicated averages for the relevant ranges of J and E, and are not easily interpreted in simple physical terms. The results of Sikkeland and collaboratos have, nevertheless, provided a useful parametrization of (HI, xn) cross sections in terms of an effective branching ratio Γ_n/Γ_f, and have been applied extensively to predict excitation functions, and to assign product mass numbers. Empirical systematics of Γ_n/Γ_f have been obtained in this manner as a function of neutron and proton number [Si 67, Si 68b, Og 74, Te 75, Og 78; see Fig. 8.13].

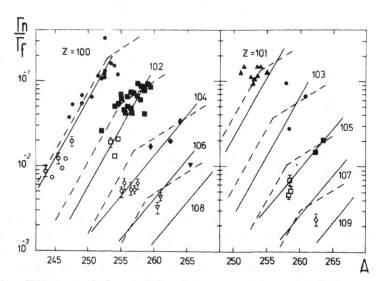

Fig. 8.13. Effective branching ratios Γ_n/Γ_f as deduced from (HI, xn) cross sections for compound nuclei with $Z = 100$–107. Closed symbols represent data obtained with light projectiles ($A \leq 20$; Si 67, Si 68b); open symbols, data obtained with heavy projectiles ($A \geq 40$; Og 74, Og 75a,b, Te 75, Og 76). The full and broken curves are empirical fit functions. (From Og 78)

An alternative source of information on Γ_n/Γ_f is provided by measurements of fission excitation functions for systems where fission is not the dominant decay channel of the compound nucleus ($A \leq 200$). An important condition in this type of study is that reactions interpreted as fission correspond to the decay of a compound nucleus, and not to deep-inelastic transfer or quasi-fission. This condition is probably satisfied with comparatively light projectiles ($A_1 < 20$) and heavy target nuclei ($A_2 = 150$–200). Such measurements have been reported, for example, by Gilmore et al. [Gi 62] and by Sikkeland et al. [Si 64a, Si 71; see Fig. 8.14]. A complete analysis of these

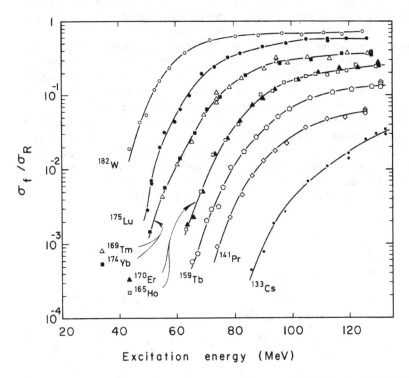

Fig. 8.14. Fraction of the total reaction cross section resulting in fission (σ_f/σ_R) for ^{16}O incident on different target nuclei, as a function of compound-nucleus excitation energy. (From Si 64a)

measurements would require a knowledge of the fusion cross sections and of the contributions from first- and higher-chance fission to the measured fission cross sections, and has not yet been performed. The qualitative influence of angular momentum on Γ_n/Γ_f was clearly established, however, by experiments where the same compound nucleus was produced from different entrance channels [Gi 62, Si 64a]. This is shown in Fig. 8.15 for the compound nucleus ^{181}Re: at a given excitation energy of the primary compound nucleus,

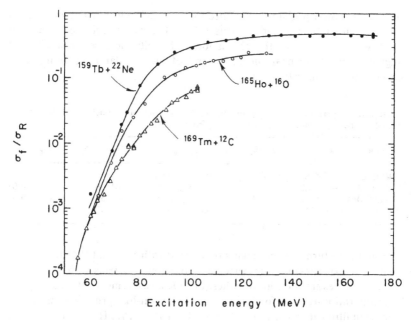

Fig. 8.15. Fraction of the total reaction cross section resulting in fission (σ_f/σ_R) for different target–projectile combinations producing the same compound nucleus ^{181}Re. (From Si 64a)

the ratio of the fission cross section to the total reaction cross section increases with projectile mass, and hence with angular momentum. A more meaningful comparison should, of course, be based on the fusion cross sections rather than the total reaction cross sections, but this would not affect the qualitative conclusions from Fig. 8.15. (Note also that the fission cross sections given in Fig. 8.15 may include higher-chance fission.)

Assuming that only first-chance fission contributes to the measured cross sections, Sikkeland compared Γ_n/Γ_f, as implied by his measurements, with statistical model calculations; he concluded that the level density parameter appropriate to the fission saddle point (a_f) must be increased over that appropriate to neutron emission (a_n) in order to reproduce the data ($a_f/a_n \approx 1.2$; Si 64a). This result, if verified by more rigorous calculations, could be explained as due to deformation-dependent shell corrections to the Fermi-gas level density [see, for example, Va 72].

In an interesting series of experiments performed at Yale by Kowalski, Zebelman, Miller et al., the compound nucleus ^{170}Yb was studied at a fixed excitation energy [107 MeV] via different entrance channels [Ko 71, Ze 74]. These authors measured cross sections for evaporation-residue formation and for fission, and deduced limiting angular momenta for complete fusion from the sum of these cross sections for each entrance channel (assuming a sharp cut-off model). From an analysis of cross-section differences, they were able to deduce fission probabilities appropriate to different angular momentum windows in the compound nucleus ^{170}Yb. The results are summarized in Table 8.1; it can be seen that the fission probability increases gradually with angular momentum near the upper end of the angular momentum range covered in this investigation.

Table 8.1. Experimental fission probabilities for the compound nucleus ^{170}Yb at $E_{ex} =$ 107 MeV (including multiple chance fission; from Ze 74)

Entrance channel	$L_{fu}[\hbar]$	σ_f/σ_{fu}	L-range [$\hbar$]	$\delta\sigma_f/\delta\sigma_{fu}$
^{11}B(115 MeV) + ^{159}Tb	40 ± 3	0.006 ± 0.001		
^{12}C(126 MeV) + ^{158}Gd	46 ± 4	0.015 ± 0.003	41–46	0.04 ± 0.03
^{16}O(137 MeV) + ^{154}Sm	58 ± 4	0.032 ± 0.006	47–58	0.06 ± 0.03
^{20}Ne(144 MeV) + ^{150}Nd	70 ± 6	0.060 ± 0.010	59–70	0.12 ± 0.07

Finally we turn to experiments performed with ^{40}Ar and ^{84}Kr ions. Comprehensive experiments were performed for several systems with compound-nucleus masses around 150–160; these include measurements of cross sections for evaporation-residue formation and for fission-like processes, as well as gamma multiplicity measurements [Gu 74, Ta 75, Ga 75, Hi 75, Br 76, Br 77]. The angular momentum distributions produced with these heavier projectiles at sufficiently high bombarding energies extend to values where fission becomes the dominant decay channel of the compound nucleus. The cross sections for evaporation-residue formation should then be limited by fission competition regardless of the question of what fraction of the observed fission-like yield might arise from noncompound processes. Figure 8.16 shows maximum angular momenta for evaporation residue formation (L_{max}), as deduced from different experiments involving compound nuclei with $A =$ 149–164 [Br 77]. The observed tendency of L_{max} to saturate near 70 is consistent with estimates based on fission competition.

Blann, Plasil, and collaborators carried out extensive statistical decay calculations for systems like ^{40}Ar + ^{109}Ag, including fission [Bl 72, Pl 74, Pl 75, Be 77a,b, Be 78]. These calculations imply that the initial angular momentum distribution of the compound nucleus extends beyond the region of fission–evaporation competition, as discussed above. They are based on fission barriers derived from the rotating liquid-drop model of Cohen, Plasil, and Swiatecki [Co 63, Co 74], and on Fermi-gas level densities. Angular momentum removal by particle evaporation was neglected in earlier versions, but later found to be important [Pl 75, Be 77a,b]. The results indicate that a simultaneous fit of the fission and evaporation-residue cross sections is necessary to remove ambiguities of the analysis, and that a significant fraction of the fission cross sections arises from higher-chance fission. From calculations with explicitly angular-momentum-dependent level densities (i.e. taking proper account of angular momentum removal by particle emission), it was concluded that the fission barriers of the rotating liquid-drop model must be

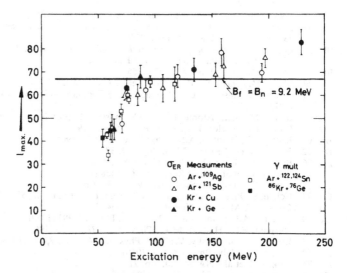

Fig. 8.16. Experimental values of the maximum angular momentum for light-particle evaporation from compound nuclei with $A = 149$–164, produced by Ar- or Kr-bombardments. The different entrance channels and methods of measurement are denoted by different symbols as explained in the figure. (From Br 77)

reduced by about 40% in order to reproduce the experimental results; no increase of the level density parameter with deformation was found necessary, however $(1.00 \leq a_f/a_n \leq 1.04$, Be 78).

An overestimation of the fission barriers by as much as 40% in the rotating liquid-drop model seems surprising at first sight; calculations of multidimensional potential-energy surfaces by Zohni et al. indicate, however, that such an effect may be explained as a consequence of the restricted shape parametrization in that model [Zo 78].

Alternatively, shortcomings of the level-density expressions or transmission coefficients used in the statistical decay calculations could necessitate the use of artificially reduced "effective fission barriers" [Be 78].

We have to conclude that a quantitative understanding of the formation and decay of these heavy compound systems requires the development of a dynamic, multidimensional theory, as well as more systematic experimental data. In this respect much work remains to be done.

References Chapter 8

(For AS 63, CA 76, HE 74, NA 74, RO 73 see Appendix D)

Al 63   Alexander, J.M.; Simonoff, G.N.: Phys. Rev. *130*, 2383 (1963)
Al 64   Alexander, J.M.; Simonoff, G.N.: Phys. Rev. *133*, B 93 (1964)
An 78   Andersen, O. et al.: Nucl. Phys. *A295*, 163 (1978)
Be 77a Beckerman, M.; Blann, M.: Phys. Rev. Lett. *38*, 272 (1977)
Be 77b Beckerman, M.; Blann, M.: Phys. Lett. *68B*, 31 (1977)

Be 78 Beckerman, M.; Blann, M.: Phys. Rev. *C17*, 1615 (1978)
Bi 59 Biedenharn, L.C.: In *Nuclear Spectroscopy*, ed. by F. Ajzenberg-Selove. Part B, p. 732. New York: Academic Press 1959
Bl 52 Blatt, J.M.; Weisskopf, V.: *Theoretical Nuclear Physics*, p. 784. Chichester: J. Wiley 1952; London: Chapman and Hall 1952
Bl 66 Blann, M.: Nucl. Phys. *80*, 223 (1966)
Bl 72 Blann, M.; Plasil, F.; Phys. Rev. Lett. *29*, 303 (1972)
Bl 73 Blann, M.; Plasil, F.: U.S. Atomic Energy Commission Report No. C00-3494-10 (1973)
Bl 76 Blann, M.: U.S. Atomic Energy Commission Report No. C00-3494-29 (1976)
Bo 39 Bohr, N.; Wheeler, J.A.: Phys. Rev. *56*, 426 (1939)
Br 61 Britt, H.C.; Quinton, A.R.: Phys. Rev. *124*, 877 (1961)
Br 72 Brack, M. et al.: Revs. Mod. Phys. *44*, 320 (1972)
Br 76 Britt, H.C. et al.: Phys. Rev. *C13*, 1483 (1976)
Br 77 Britt, H.C. et al.: Phys. Rev. Lett. *39*, 1458 (1977)
Co 63 Cohen, S.; Plasil, F.; Swiatecki, W.J.: AS 63 p. 325
Co 74 Cohen, S.; Plasil, F.; Swiatecki, W.J.: Ann. Phys. (N.Y.) *82*, 557 (1974)
Co 77 Cormier, T.M. et al.: Phys. Rev. *C15*, 654 (1977)
Do 59 Douglas, A.C.; MacDonald, M.: Nucl. Phys. *13*, 382 (1959)
Du 65 Durham, F.E.; Halbert, M.L.: Phys. Rev. *137*, B 850 (1965)
Er 58 Ericson, T.; Strutinsky, V.: Nucl. Phys. *8*, 284 (1958) *9*, 689 (1958/59)
Er 60 Ericson, T.: Adv. Phys. *63*, 479 (1960)
Fi 74 Fink, H.J. et al.: NA 74 Vol. 2 p. 21
Ga 73 Gauvin, H. et al.: Nucl. Phys. *A208*, 360 (1973)
Ga 74a Galin, J. et al.: Phys. Rev. *C9*, 1113, 1126 (1974)
Ga 74b Gauvin, H. et al.: Phys. Rev. *C10*, 722 (1974)
Ga 75 Gauvin, H. et al.: Phys. Lett. *58B*, 163 (1975)
Gi 62 Gilmore, J.; Thompson, S.G.; Perlman, I.: Phys. Rev. *128*, 2276 (1962)
Gi 73 Gilat, J.; Jones, III, E.R.; Alexander, J.M.: Phys. Rev. *C7*, 1973 (1973)
Go 60 Gordon, G.E. et al.: Phys. Rev. *120*, 1341 (1960)
Go 77 Gomez del Campo, J.: Phys. Rev. Lett. *36*, 1529 (1976)
Gr 59 Griffin, J.J.: Phys. Rev. *116*, 107 (1959)
Gr 61 Grover, J.R.: Phys. Rev. *123*, 267 (1961)
Gr 62 Grover, J.R.: Phys. Rev. *127*, 2142 (1962)
Gr 67a Grover, J.R.; Gilat, J.: Phys. Rev. *157*, 802, 815, 823 (1967)
Gr 67b Grover, J.R.; Gilat, J.: Phys. Rev. *157*, 832 (1967)
Gr 70 Grover, J.R.; Gilat, J.: Report BNL 50 246 (1970)
Gu 74 Gutbrod, H.H. et al.: RO 73 Vol. II, p. 309
Ha 52 Hauser, W.; Feshbach, H.: Phys. Rev. *87*, 366 (1952)
Ha 58 Halpern, I.; Strutinsky, V.: *Proc. 2nd Intern. Conf. on Peaceful Uses of Atomic Energy*, Geneva 1958, (United Nations, New York) Vol. 15 p. 408
Ha 63 Halbert, M.L.; Durham, F.E.: AS 63 p. 223
Ha 68 Halpern, I.; Shepherd, B.J.; Williamson, C.F.: Phys. Rev. *169*, 805 (1968)
Ha 71 Hasse, R.W.: Phys. Rev. *C4*, 572 (1971)
Ha 74 Hanson, D.L. et al.: Phys. Rev. *C9*, 929 (1974)
Ha 75 Hagemann, G.B. et al.: Nucl. Phys. *A245*, 166 (1975)
Hi 53 Hill, D.L.; Wheeler, J.A.: Phys. Rev. *89*, 1102 (1953)
Hi 75 Hille, M. et al.: Nucl. Phys. *A252*, 496 (1975)
Hi 76 Hillman, M.; Eyal, Y.: CA 76 p. 109
Hu 62 Huizenga, J.R.; Vandenbosch, R.: In *Nuclear Reactions*, Vol. II, ed. by P.M. Endt, J.R. Smith, p. 42. Amsterdam: North-Holland 1962
Hu 69 Huizenga, J.R.; Bekhami, A.N.; Moretto, L.G.: Phys. Rev. *177*, 1826 (1969)
Hu 72 Huizenga, J.R.; Moretto, L.G.: Ann. Rev. Nucl. Sci. *22*, 427 (1972)
Hy 64a Hyde, E.K.; Perlman, I.; Seaborg, G.T.: *Systematics of Nuclear Structure and Radioactivity*, Vol. I of *The Nuclear Properties of the Heavy Elements*, pp. 327-359. London: Prentice Hall 1964
Hy 64b Hyde, E.K.: *Fission Phenomena*, Vol. III of *The Nuclear Properties of the Heavy Elements*, pp. 380-399. London: Prentice Hall 1964
Ja 56 Jackson, J.D.: Can. J. Phys. *34*, 767 (1956)
Jä 67 Jägare, S.: Nucl. Phys. *A95*, 491 (1967)
Kl 74a Klapdor, H.V. et al.: Phys. Lett. *49B*, 431 (1974)

Kl 74b Klapdor, H.V.; Reiss, H.; Rosner, G.: Phys. Lett. *53B*, 147 (1974)
Kl 75a Klapdor, H.V. et al.: Nucl. Phys. *A244*, 157 (1975)
Kl 75b Klapdor, H.V.; Reiss, H.; Rosner, G.: Phys. Lett. *58B*, 279 (1975)
Kl 76 Klapdor, H.V.; Reiss, H.; Rosner, G.: Nucl. Phys. *A262*, 157 (1976)
Ko 71 Kowalski, L. et al.: Phys. Rev. *C3*, 1370 (1971)
Ko 77 Kohlmeyer, B.; Pfeffer, W.; Pühlhofer, F.: Nucl. Phys. *A292*, 288 (1977)
Le 76 Lefort, M.: In *Nuclear Spectroscopy and Nuclear Reactions with Heavy Ions*, Proc. of the Internat. School of Physics "Enrico Fermi", Course LXII, Varenna 1974, ed. by H. Faraggi, R.A. Ricci; p. 139. Amsterdam: North-Holland1976
Le 77 Lee, I.Y. et al.: Phys. Rev. Lett. *38*, 1454 (1977)
Ma 74 der Mateosian, E.; Kistner, O.C.; Sunyar, A.W.: Phys. Rev. Lett. *33*, 596 (1974)
Mo 62 Mollenauer, J.F.: Phys. Rev. *127*, 867 (1962)
Mo 63 Morinaga, H.; Gugelot, P.C.: Nucl. Phys. *46*, 210 (1963)
Mo 72a Moretto, L.G. et al.: Phys. Lett. *38B*, 471 (1972)
Mo 72b Moretto, L.G.: Nucl. Phys. *A180*, 337 (1972)
Ne 70 Newton, J.O.: Prog. Nucl. Phys. *11*, 53 (1970)
Og 74 Oganessian, Yu. Ts.: HE 74, p. 222
Og 75a Oganessian, Yu. Ts. et al.: Nucl. Phys. *A239*, 157 (1975)
Og 75b Oganessian, Yu. Ts. et al.: Nucl. Phys. *A239*, 353 (1975)
Og 76 Oganessian, Yu. Ts. et al.: Nucl. Phys. *A273*, 505 (1976)
Og 78 Oganessian, Yu. Ts.: Communication INR An SSR II-0090 (1978)
Pl 74 Plasil, F.: NA 74 Vol. 2 p. 107
Pl 75 Plasil, F.; Blann, M.: Phys. Rev. *C11*, 508 (1975)
Pl 77 Plasil, F.: Report ORNL/TM-6054 (1977)
Pü 75 Pühlhofer, F. et al.: Nucl. Phys. *A244*, 329 (1975)
Pü 77 Pühlhofer, F.: Nucl. Phys. *A280*, 267 (1977)
Sa 56 Satchler, G.R.; Phys. Rev. *104*, 1198 (1956)
Sa 76 Sarantites, D.G. et al.: Phys. Rev. *C6*, 2138 (1976)
Si 63 Simonoff, G.N.; Alexander, J.M.: AS 63 p. 345
Si 64a Sikkeland, T.: Phys. Rev. *135*, B 669 (1964)
Si 64b Simonoff, G.N.; Alexander, J.M.: Phys. Rev. *133*, B 104 (1964)
Si 67 Sikkeland, T.: Ark. Fys. *36*, 539 (1967)
Si 68a Sikkeland, T.; Maly, J.; Lebeck, D.F.: Phys. Rev. *169*, 1000 (1968)
Si 68b Sikkeland, T.; Ghiorso, A.; Nurmia, M.J.: Phys. Rev. *172*, 1232 (1968)
Si 71 Sikkeland, T. et al.: Phys. Rev. *C3*, 329 (1971)
Si 77 Simon, R.S. et al.: Nucl. Phys. *A290*, 253 (1977)
Sp 65 Sperber, D.: Nuovo Cimento *36*, 1164 (1965)
St 61 Strutinsky, V.M.: Sov. Phys. JETP *12*, 546 (1961)
St 65 Stephens, F.S.; Lark, N.L.; Diamond, R.M.: Nucl. Phys. *63*, 82 (1965)
St 74 Stokstad, R.G.: NA 74 Vol. 2 p. 327
Ta 75 Tamain, B. et al.: Nucl. Phys. *A252*, 187 (1975)
Te 75 Ter-Akopyan, G.M. et al.: Nucl. Phys. *A255*, 509 (1975)
Th 63 Thomas, T.D. et al.: Phys. Rev. *126*, 1805 (1963)
Th 64 Thomas, T.D.: Nucl. Phys. *53*, 577 (1964)
Th 68 Thomas, T.D.: Ann. Rev. Nucl. Sci. *18*, 343 (1968)
Tj 74 Tjöm, P.O. et al.: Phys. Rev. Lett. *33*, 593 (1974)
Tj 78 Tjöm, P.O. et al.: Phys. Lett. *72B*, 439 (1978)
Va 72 Vandenbosch, R.; Mosel, U.: Phys. Rev. Lett. *28*, 1726 (1972)
Vi 60 Viola, V.E.; Thomas, T.D.; Seaborg, G.T.: Phys. Rev. *120*, 2120 (1960)
Vo 75 Volant, C. et al.: Nucl. Phys. *A238*, 120 (1975)
We 76 Weidinger, A. et al.: Nucl. Phys. *A263*, 511 (1976)
Wi 67 Williams, D.C.; Thomas, T.D.: Nucl. Phys. *A92*, 1 (1967)
Wo 51 Wolfenstein, L.: Phys. Rev. *82*, 690 (1951)
Ze 74 Zebelman, A.M. et al.: Phys. Rev. *C10*, 200 (1974)
Zo 78 Zohni, O.; Blann, M.: Nucl. Phys. *A297* 163, (1978)

Appendices

A: Glossary of Symbols

The numbers in parentheses given in the right-hand column refer to the chapters where the symbols occur. Note that some symbols may have a different significance in different chapters.

Symbol	Significance
0	index denoting property of undeformed nucleus (3)
1	index denoting projectile (1–3, 6–8)
	index denoting initial core or bound state in a transfer reaction (2,4,5)
2	index denoting target nucleus (1–3, 6–8)
	index denoting final core or bound state in a transfer reaction (2,4,5)
3,4	indices denoting products of two-body reaction (1,6)
∞	asymptotic value at large separation (1,3)
a	one-half of the distance of closest approach in 180° Coulomb scattering (1,3)
	level density parameter (2,8)
a_S	coefficient of surface term in liquid-drop-model mass formula (7)
a_V	diffuseness of real part of Woods–Saxon potential (2,3,6,7)
a_W	diffuseness of imaginary part of Woods–Saxon potential (2,3,7)
A	mass number (1–3, 6, 7)
A	initial bound system in a transfer reaction (4,5)
A_{12}	reduced mass number for system $1+2$ (1,3)
A_l	scattering parameter for partial wave l (1,2,4)
b	impact parameter (1,3)
B	final bound system in a transfer reaction (4,5)
$B(E\lambda)$	reduced electromagnetic transition probability (3)
B_f	fission barrier (8)

Symbol	Significance
B_{fu}	fusion barrier (3,7)
B_{int}	interaction barrier (3,6,7)
$B_J(\lambda)$	effective barrier for channel λ and total angular momentum J (8)
B_l	effective barrier for partial wave l (7)
C	index denoting Coulomb interaction (1–8)
C_1	initial core in transfer reaction (2,4,5)
C_2	final core in transfer reaction (2,4,5)
C_{AB}	spectroscopic coefficient in sub-Coulomb transfer (5)
C_{lk}	Cartesian components of friction tensor (6)
C_L, C_r	spherical components of friction tensor (6)
$C_{\alpha\beta}$	transition amplitude in inelastic scattering (3)
d	range parameter of macroscopic nuclear potentials (7)
D	classical distance of closest approach (1,3,4,5)
D_C	distance of closest approach assuming pure Coulomb interaction (1,3)
$D(x,t)$	diffusion coefficient (6)
D_A	mass diffusion coefficient (6)
D_Z	charge diffusion coefficient (6)
e	elementary charge (1–8)
E	relative kinetic energy in the centre-of-mass system (2,4)
	excitation energy of compound nucleus (8)
E_A	initial binding energy in transfer reaction (5)
E_B	final binding energy in transfer reaction (5)
E_C	Coulomb barrier (1,4)
E_{CM}	relative kinetic energy in the centre-of-mass system (1–8)
E_{ex}	excitation energy (1,7)
$E_J(\lambda)$	rotational energy of residual nucleus in channel λ at total angular momentum J (8)
E_{Lab}	relative kinetic energy in the laboratory system (1–8)
E_{LD}	liquid-drop-model potential energy (6)
E_{rot}	rotational energy of binary system (6,7)
E_S	surface energy (7)
E_λ	excitation energy of residual nucleus in channel λ (8)
f	index denoting final state or partition (3–5)
	index denoting fission channel (7.8)
f	coefficient characterizing angular momentum transfer by tangential friction (7)
	argument denoting fission decay (8)
$f(\Theta)$	scattering amplitude for elastic scattering (1,2)
$f_{tr}(\Theta)$	transfer amplitude in configuration space (2,4)
$f_{tr}(l-l_0)$	transfer amplitude in angular momentum space (4)
$f_{\alpha\beta}(\Theta)$	scattering amplitude for inelastic scattering (3)
$f_\lambda(\Theta, \xi)$	function describing Coulomb excitation (3)

Symbol	Significance
$f_\lambda(l, J)$	coefficient in statistical decay formula (8)
$F(\mathbf{r}_{12})$	form factor in DWBA analysis of transfer reactions (5)
$F_J(\lambda)$	energy-integrated transmission function in channel λ(8)
F_L	tangential friction force (6)
F_r	radial friction force (6)
$F_\lambda(\Theta, \xi)$	function describing Coulomb excitation (3)
$F_\lambda(r)$	form factor in DWBA analysis of inelastic scattering (3)
$\mathscr{F}(q_i, \dot{q}_i)$	Rayleigh's dissipation function (6)
g	size parameter in sub-Coulomb fusion (7)
$g(l)$	function describing angular momentum dependence of scattering parameters
g_λ	statistical weight in channel λ (8)
$G(\mathbf{r}_i, \mathbf{r}_f)$	transfer function (5)
$\hbar$	Planck's constant, unit of angular momentum (1–8)
h_l	spherical Hankel function (5)
H	Hamiltonian (3)
H_β	energy stored in β-vibrational mode (3)
i	index denoting initial state or partition (3–5)
i_λ	spin of light fragment in channel λ (8)
I	nuclear spin (1,5,8)
$\mathscr{I}$	moment of inertia (2,6,8)
$\mathscr{I}_\lambda$	moment of inertia of residual nucleus in channel λ (8)
j	total (orbital plus intrinsic) angular momentum of transferred (4,5) or emitted (8) particle or radiation
j_l	spherical Bessel function (8)
J	angular momentum of compound nucleus or system (7,8)
J_0	cylindrical Bessel function of order zero (8)
k	asymptotic wave number (momentum in units of $\hbar$) of relative motion (1, 3–5) friction coefficient (6)
K	classical momentum in units of $\hbar$ (4) angular momentum projection on symmetry axis (8)
$K_L(r)$	local momentum in units of $\hbar$ for angular momentum L (3,5,6)
l	orbital angular momentum (quantal description) (1–5, 7, 8)
L	classical angular momentum in units of $\hbar$ (1,2,4,6,7) index denoting centrifugal pseudo-potential or tangential component of vector or tensor (3,6,7)
L_{fu}	limiting angular momentum for fusion (7)
L_{gr}	angular momentum of grazing trajectory (1,3,6)
L_{int}	limiting angular momentum for nuclear interactions (6,7)
$\mathscr{L}(q_i, \dot{q}_i)$	Lagrangian (6)
m	atomic mass unit, mass of nucleon (1,3,6) angular momentum projection (5)

Symbol	Significance
m_x	mass of transferred particle (4,5)
M	atomic mass, nuclear mass (4–7)
M_γ	multiplicity of gamma emission (8)
n	Sommerfeld parameter (1, 3–7)
N	normalization factor of bound-state wave function (5)
	number of open channels (6)
N	index denoting nuclear interaction (3,6,7)
p	momentum (1,2)
p_F	Fermi momentum (6)
P, P_{if}	transition probability (3–5)
$P(x,t)$	probability distribution (6)
P_A	attenuation factor (4)
P_{fu}	fusion probability (7)
P_l	Legendre polynomial (1,2,4,8)
$P_x(E_{ex}, J)$	probability for emission of x neutrons (8)
q	coordinate, degree of freedom (6)
Q	reaction Q value (1, 4–6)
$Q(2^+)$	static quadrupole moment of 2^+ state (3)
Q_{eff}	effective optimum Q value for transfer reaction (4)
Q_{fu}	Q value for fusion (1,7)
Q_{gg}	Q value for transition to final ground states (6)
Q_{opt}	optimum Q value for transfer reaction (4)
r	distance of mass centers of colliding nuclei (1–7)
	index denoting radial degree of freedom (6)
r_0	radius parameter (1–7)
r_{fu}	energy-dependent, effective fusion distance (7)
r_{int}	energy-dependent, effective interaction distance (3)
r_L	classical turning point (distance of closest approach) for angular momentum L (3,6)
R	(half-density) matter radius of individual nucleus (3,6,7)
R_{12}	sum of matter radii for nucleus–nucleus system (6,7)
R_1, R_2	initial and final bound-state radius in transfer reaction (4,5)
R_i, R_f	initial and final channel radius in transfer reaction (4)
R_C	effective interaction distance, defined for pure Coulomb interaction at $r > R_C$ (1,3)
	Coulomb radius of optical model potential (2)
R_{C1}, R_{C2}	charge radius of individual nucleus, defined by equivalent homogeneous charge distribution (3)
R_{int}	interaction distance of nucleus–nucleus system (1,3)
R_S	safe distance for pure Coulomb excitation (3)
R_U	potential radius of nucleus–nucleus system (3)
R_V	radius of real potential (2,3,6)
R_W	radius of imaginary potential (2,3)
R_λ	interaction distance in channel λ (8)
$\langle R\beta \rangle$	deformation length in inelastic scattering (3)

Symbol	Significance
s	surface-separation coordinate of nucleus–nucleus system (6,7)
S	channel spin (1,2,7,8)
	surface area of nucleus (7)
	nuclear cross-section factor (7)
$\tilde{S}$	nuclear cross section factor with size correction (7)
$S_A(l_1,j_1)$	spectroscopic factor for initial bound state in transfer reaction (5)
$S_B(l_2,j_2)$	spectroscopic factor for final bound state in transfer reaction (5)
S_l	nuclear scattering coefficient for partial wave l (1,3)
$\bar{S}_l$	total scattering coefficient for partial wave l (1,3)
S_λ	separation energy in channel λ (8)
t	surface thickness of interaction region (1)
	time (6)
	nuclear temperature (8)
T	kinetic energy (6)
	nuclear temperature (6)
	transmission coefficient (7,8)
	transition amplitude in inelastic scattering (3)
	or transfer reaction (5)
$U(r)$	complex optical model potential (2,3,5)
U_{34}	total excitation energy of intermediate complex (6)
v	relative velocity of nucleus–nucleus system (1,3)
v_1–v_4	velocity of particles 1–4 in the centre-of-mass system (1)
v_{CM}	velocity of centre of mass in the laboratory system (1)
v_{Lab}	particle velocity in the laboratory system (1)
$v(x,t)$	drift coefficient for nucleon diffusion in deep-inelastic collisions (6)
V	depth parameter of real part of Woods–Saxon potential (2,3,5)
$V(r)$	real part of optical model potential (2,3,5)
	classical nucleus–nucleus potential (6,7)
V_{12},V_{34}	potential energy of intermediate complex (6)
V_i,V_f	potential energy before and after transfer (4)
V_{if}	effective interaction in DWBA transition amplitude (3,5)
$w(x,x',t)$	microscopic transition probability (6)
W	depth parameter of imaginary part of Woods–Saxon potential (2,3)
$W(r)$	imaginary part of optical model potential (2,3,4)
$W_J(\varepsilon,\Theta)$	function describing angular distribution of evaporated particles (8)
$\bar{W}_J(\Theta)$	function describing angular distribution of fission fragments (8)
x	Cartesian coordinate (6)

Symbol	Significance
x	fissility parameter of liquid-drop model (7)
	index denoting number of evaporated neutrons (8)
x	index denoting transferred particle (2,4,5)
x_{12}	generalized fissility parameter for nucleus–nucleus system (7)
y	rotational parameter of liquid-drop model (7)
y_{12}	generalized rotational parameter for nucleus–nucleus system (7)
Y_{lm}	spherical harmonic (3,5,7)
z	charge number of transferred particle (2,4,5)
Z	nuclear charge number, atomic number (1–4, 6, 7)

α	index denoting initial channel (3,7,8)
$\alpha_2, \alpha_3, \alpha_4$	shape coordinates of nucleus–nucleus system (7)
α_l, α_L	Coulomb plus nuclear scattering phase shift (3,4)
β	index denoting final channel (3,7,8)
	coordinate describing axially symmetric quadrupole deformation (3)
	index denoting quadrupole-vibrational degree of freedom (3)
β_C	deformation of charge distribution (3)
β_V, β_W	deformation of optical model potential (3)
$\beta_{\lambda\mu}$	deformation of multipole order λ, μ (3)
γ	surface energy per unit area in the liquid-drop model (7)
	index denoting emission of γ radiation (8)
γ_3	coefficient in the transformation of kinematic quantities from the centre-of-mass to the laboratory system (1)
γ_V	coefficient related to change of potential energy in transfer reaction (4)
γ_L	coefficient related to change of rotational energy in transfer reaction (4)
Γ	total decay width of compound nucleus (6)
Γ_J	total decay width of compound nucleus with angular momentum J (7,8)
$\Gamma_J(\lambda)$	partial decay width into channel λ (7,8)
δ_l, δ_L	nuclear scattering phase shift (1–3)
$\delta(n), \delta(p)$	pairing correction related to the number of transferred neutrons or protons (6)
δ_λ	pairing correction related to the final nucleus in channel λ (8)
$\Delta E, \Delta E_{diss}$	loss of kinetic energy in deep-inelastic collision (6)
ΔE_{if}	energy transfer in inelastic excitation of discrete level (3)
ΔR	change of radial coordinate (recoil) in transfer reaction (4,5)

Symbol	Significance
$\Delta V, \Delta V_C$	change of potential energy in transfer reaction (4,6)
$\Delta \alpha$	phase difference between initial and final wave functions in transfer reaction (5)
ε	fragment kinetic energy per nucleon in the laboratory system (1,3)
ε_{if}	symmetrized kinetic energy per nucleon (3)
ε_λ	relative kinetic energy in channel λ in the centre-of-mass system (8)
ϑ	scattering angle, emission angle in the laboratory system (1)
$\vartheta_{1,2}$	polar angle of radius vector describing nuclear surface (3)
ϑ_{CM}	scattering angle in the centre-of-mass system (3)
Θ	scattering angle, emission angle in the centre-of-mass system (1–6)
$\Theta_{1/4}$	quarter-point angle (3)
Θ_{gr}	grazing angle (1,6)
$\Theta(L)$	classical deflection function (3,6)
Θ_λ	angle of particle emission in channel λ (8)
κ	decay length of bound-state wave function (2,5)
λ	multipole order (angular momentum) of collective mode of excitation (3)
	index denoting arbitrary two-body channel (8)
$\lambdabar$	de Broglie wave length of relative motion (1,3,7,8)
μ	reduced mass of nucleus–nucleus system (1, 3–8)
	angular momentum projection (3)
$\mu_2(x,t)$	second moment of microscopic transition probability $w(x,x',t)$ with respect to $(x-x')$ (6)
μ_A, μ_B	reduced mass of initial or final bound state in transfer reaction (5)
μ_i, μ_f	reduced mass of initial or final scattering state in transfer reaction (4,5)
μ_λ	reduced mass in channel λ (8)
ξ	adiabaticity parameter in Coulomb excitation (3)
	kinematic variable related to angular momentum (4)
ξ_{if}	symmetrized adiabaticity parameter (3)
ξ_j	strength parameter for gamma emission with multipole order j (8)
Π	parity (5)
$\rho(r)$	total nucleon density at position r (2)
$\rho_1(r), \rho_2(r)$	partial nucleon density arising from fragment 1 or 2 at position r (6)
$\rho(E,I), \rho(E,J)$	level density as a function of energy and angular momentum (6,8)

Symbol	Significance
$\bar{\rho}(E)$	level density as a function of energy only (6)
$\rho(x)$	statistical weight (density of states) as a function of variable x (6)
$\rho_\lambda(E,I),\bar{\rho}_\lambda(E)$	level denstiy of final nucleus in channel λ (8)
σ	angle-integrated cross section (1, 3–8)
	spin cut-off parameter in level density formula (2)
σ_A	variance of mass distribution (6)
σ_{cl}	classical cross section (3,4)
σ_f	fission cross section (8)
σ_{fu}	fusion cross section (7)
$\sigma_{if},\sigma_{\alpha\beta}$	integrated cross section connecting specified entrance and exit channels (3,7,8)
$\sigma_J,\sigma_J(\alpha)$	partial fusion cross section for total angular momentum J (7,8)
σ_R	total reaction cross section (1,3,7)
σ_Z	variance of charge distribution (6)
σ_λ	integrated cross section for Coulomb excitation with multipole order λ (3)
	spin cut-off parameter for residual nucleus in channel λ (8)
$d\sigma/d\Theta$	differential cross section referred to angle (6)
$d\sigma/d\Omega$	differential cross section referred to solid angle (1–6)
$(d\sigma/d\Omega)_c$	Coulomb (Rutherford) scattering cross section (1,3,5)
τ	interaction time, lifetime of intermediate complex (6)
τ_0,τ_1	characteristic nuclear time constants (6)
φ_1,φ_2	azimuthal angle of radius vector describing nuclear surface (3)
ϕ_A,ϕ_B	radial part of initial or final bound-state wave function in transfer reaction (5)
χ	kinematic variable related to angular momentum (4)
$\chi_i^{(+)},\chi_f^{(-)}$	initial or final scattering wave function, generated by optical model (3,5)
χ_λ	strength parameter in Coulomb excitation (3)
$\psi(r)$	wave function of relative motion (1)
ψ_A,ψ_B	initial or final bound state wave function in transfer reaction (5)
ψ_i,ψ_f	initial or final internal wave function in inelastic scattering (3)
ω	oscillator frequency of single-particle shell model (2)
$\omega,\bar{\omega}$	relative angular velocity of nucleus–nucleus system (6)
ω_1,ω_2	intrinsic angular velocity of fragment 1 or 2 (6)
ω_f	oscillator frequency characterizing fission barrier (8)
ω_l	oscillator frequency characterizing fusion barrier for partial wave l (7)
$d\Omega$	element of solid angle in the centre-of-mass system (1–6, 8)

B: Nonrelativistic Two-Body Kinematics

$A_1 + A_2 \rightarrow A_3 + A_4$ (A_3 observed)
Laboratory energies and angles: E_1, E_3, ϑ_3, $d\omega_3$
Centre-of-mass energies and angles: E_i, $E_f = E_i + Q$, Θ, $d\Omega$

1. Transformation from the centre-of-mass system to the laboratory system.

$$\text{Define:} \quad \gamma_3 = \left(\frac{A_1 A_3}{A_2 A_4} \frac{E_i}{E_i + Q} \right)^{1/2}, \tag{B.1}$$

$$\tan \vartheta_3 = \frac{\sin \Theta}{\cos \Theta + \gamma_3}, \tag{B.2}$$

$$E_3 = \frac{A_1 A_3 E_i}{A_2 (A_1 + A_2)} \left(\frac{1 + \gamma_3^2 + 2\gamma_3 \cos \Theta}{\gamma_3^2} \right), \tag{B.3}$$

$$\frac{d\sigma}{d\omega_3} = \frac{d\sigma}{d\Omega} \frac{(1 + \gamma_3^2 + 2\gamma_3 \cos \Theta)^{3/2}}{|1 + \gamma_3 \cos \Theta|}. \tag{B.4}$$

2. Calculation of centre-of-mass angle, laboratory energy, and cross section for a given laboratory angle and Q value (γ_3).

$$\sin \Theta = \sin \vartheta_3 \, [\gamma_3 \cos \vartheta_3 \pm (1 - \gamma_3^2 \sin^2 \vartheta_3)^{1/2}], \tag{B.5}$$

$$E_3 = \frac{A_1 A_3 E_i}{A_2 (A_1 + A_2)} \frac{[\gamma_3 \cos \vartheta_3 \pm (1 - \gamma_3^2 \sin^2 \vartheta_3)^{1/2}]^2}{\gamma_3^2}, \tag{B.6}$$

$$\frac{d\sigma}{d\omega_3} = \frac{d\sigma}{d\Omega} \frac{[\gamma_3 \cos \vartheta_3 \pm (1 - \gamma_3^2 \sin^2 \vartheta_3)^{1/2}]^2}{(1 - \gamma_3^2 \sin^2 \vartheta_3)^{1/2}}. \tag{B.7}$$

3. Transformation from the laboratory system to the centre-of-mass system.

$$\text{Define:} \quad \delta_3 = \left(\frac{A_1 A_3}{(A_1 + A_2)^2} \frac{E_1}{E_3} \right)^{1/2}, \tag{B.8}$$

$$\sin \Theta = \frac{\sin \vartheta_3 [\delta_3 \cos \vartheta_3 \pm (1 - \delta_3 \cos \vartheta_3)]}{(1 + \delta_3^2 - 2\delta_3 \cos \vartheta_3)^{1/2}}, \tag{B.9}$$

$$\gamma_3 = \frac{\delta_3}{(1 + \delta_3^2 - 2\delta_3 \cos \vartheta_3)^{1/2}}, \tag{B.10}$$

$$Q = \frac{A_1 A_3 E_1}{(A_1 + A_2) A_4} \left(\frac{(1 + \delta_3^2 - 2\delta_3 \cos \vartheta_3)}{\delta_3^2} - \frac{A_2 A_4}{A_1 A_3} \right), \tag{B.11}$$

$$\frac{d\sigma}{d\Omega} = \frac{d\sigma}{d\omega_3} |1 - \delta_3 \cos \vartheta_3| (1 + \delta_3^2 - 2\delta_3 \cos \vartheta_3)^{1/2}. \tag{B.12}$$

4. Special cases and restrictions

If $\gamma_3 \geqslant 1$, $\vartheta_3 \leqslant \arcsin\left(\frac{1}{\gamma_3}\right)$.

If $\gamma_3 = 1$, $\vartheta_3 = \frac{\Theta}{2}$.

Elastic scattering, observation of scattered particle

$$\left(A_3 = A_1,\ A_4 = A_2;\ \gamma_3 = \frac{A_1}{A_2}\right):$$

If $A_1 > A_2$, $\vartheta_3 \leqslant \arcsin\left(\frac{A_2}{A_1}\right)$.

Elastic scattering, observation of recoiling target nucleus

$(A_4 = A_1,\ A_3 = A_2;\ \gamma_3 = 1):$

$$E_3 = \frac{4A_1E_t}{(A_1 + A_2)}\cos^2\left(\frac{\Theta}{2}\right) = \frac{4A_1A_2E_1}{(A_1 + A_2)^2}\cos^2\vartheta_3 .$$

C: Coulomb Scattering

1. Useful numbers and formulae
   (The numerical coefficients in (C.1–3) are for ε in MeV)

$$e^2 = 1.44 \text{ MeV fm}; \quad \frac{\hbar^2}{2m} = 20.9 \text{ MeV fm}^2,$$

$$a = \frac{1}{2} D(\pi) = \frac{e^2}{2} \frac{Z_1 Z_2}{A_{12}\varepsilon} \qquad = 0.72 \frac{Z_1 Z_2}{A_{12}\varepsilon} \text{ fm}. \tag{C.1}$$

$$k = (\lambdabar)^{-1} \quad = \left(\frac{\hbar^2}{2m}\right)^{-1/2} A_{12}\varepsilon^{1/2} = 0.219 \, A_{12}\varepsilon^{1/2} \text{ fm}^{-1}. \tag{C.2}$$

$$n = ka \quad = \frac{e^2}{2}\left(\frac{\hbar^2}{2m}\right)^{-1/2} \frac{Z_1 Z_2}{\varepsilon^{1/2}} = 0.157 \frac{Z_1 Z_2}{\varepsilon^{1/2}}. \tag{C.3}$$

$$\left(\frac{d\sigma}{d\Omega}\right)_c = \frac{a^2}{4} \sin^{-4}\left(\frac{\Theta}{2}\right). \tag{C.4}$$

$$\varepsilon_C = \frac{E_C}{A_{12}} = \frac{Z_1 Z_2 e^2}{A_{12} R_C}. \tag{C.5}$$

$$\Theta_{gr} = 2 \arcsin\left(\frac{\varepsilon_C}{2\varepsilon - \varepsilon_C}\right) = 2 \arcsin\left(\frac{a}{R_C - a}\right). \tag{C.6}$$

$$L_{gr} = kR_C\left(\frac{\varepsilon - \varepsilon_C}{\varepsilon}\right)^{1/2} = \left(\frac{\hbar^2}{2m}\right)^{-1/2} A_{12} R_C (\varepsilon - \varepsilon_C)^{1/2}. \tag{C.7}$$

$$\sigma_R = \pi\lambdabar^2 L_{gr}^2 = \pi R_C^2\left(\frac{\varepsilon - \varepsilon_C}{\varepsilon}\right). \tag{C.8}$$

$$R_C \approx [1.12\,(A_1^{1/3} + A_2^{1/3}) - 0.94\,(A_1^{-1/3} + A_2^{-1/3}) + 3] \text{ fm}. \tag{C.9}$$

2. Relationship between kinematic variables

Required quantity	Formula in terms of		
	D	L	Θ
distance of closest approach D		$a\left[1 + \left(1 + \frac{L^2}{n^2}\right)^{\frac{1}{2}}\right]$	$a\left[1 + \sin^{-1}\left(\frac{\Theta}{2}\right)\right]$
angular momentum L	$n\left[\frac{D}{a}\left(\frac{D}{a} - 2\right)\right]^{\frac{1}{2}}$		$n \cot\left(\frac{\Theta}{2}\right)$
scattering angle Θ	$2\sin^{-1}\left(\frac{a}{D - a}\right)$	$2\tan^{-1}\left(\frac{n}{L}\right)$	

D: List of Conference Proceedings

AR 71 *Proceedings of the Symposium on Heavy-Ion Scattering,* Argonne National Laboratory, March 1971; Report ANL-7837 (1971)

AR 73 *Proceedings of the Symposium on Heavy-Ion Transfer Reactions,* Argonne National Laboratory, March 1973; Informal Report PHY-1973B

AR 76 *Proceedings of the Symposium on Macroscopic Features of Heavy-Ion Collisions* Argonne National Laboratory, April 1976; Report ANL-PHY-76-2 (1976)

AS 63 *Proceedings of the Third Conference on Reactions between Complex Nuclei,* Asilomar, April 1963, ed. by A. Ghiorso, R.M. Diamond, H.E. Conzett (University of California Press, Berkeley, Los Angeles 1963)

CA 76 *Proceedings of the European Conference on Nuclear Physics with Heavy-Ions,* Caen, September 1976, ed. by B. Fernandez et al., Vol. 1: Contributions

CO 75 *Proc. of the Second International Conference on Clustering Phenomena in Nuclei,* College Park, Maryland, April 1975, ed. by D.A. Goldberg, J.B. Marion, S.J. Wallace; ERDA Report ORO-4856-26 (US National Technical Information Service, Springfield, Virginia 1975)

HE 66 *Proceedings of the Symposium on Recent Progress in Nuclear Physics with Tandems,* Heidelberg, July 1966; Invited Papers, ed. by W. Hering

HE 69 *Proceedings of the International Conference on Nuclear Reactions Induced by Heavy Ions,* Heidelberg, July 1969, ed. by R. Bock, W.R. Hering (North-Holland-American Elsevier 1970)

HE 74 *Proceedings of the Symposium on Classical and Quantum Mechanical Aspects of Heavy Ion Collisions,* Heidelberg, October 1974, ed. by H.L. Harney, P. Braun-Munzinger, C.K. Gelbke (Berlin, Heidelberg, New York: Springer 1975)

MU 73 *Proceedings of the International Conference on Nuclear Physics,* Munich, August/September 1975, ed. by J. de Boer, H.J. Mang (North-Holland—American Elsevier 1975)

NA 74 *Proceedings of the International Conference on Reactions between Complex Nuclei,* Nashville, June 1974, ed. by R.C. Robinson, F.K. McGowan, J.B. Ball, J.H. Hamilton (North-Holland-American Elsevier 1974)

RO 73 *Proceedings of the Third IAEA Symposium on the Physics and Chemistry of Fission;* Rochester, New York, August 1973 (IAEA, Vienna 1974)

RO 74 *Super-Heavy Elements—Theoretical Prediction and Experimental Generation;* Proc. 27th Nobel Symposium, Ronneby, Sweden, June 1974, ed by S.G. Nilsson, N.R. Nilsson (Almqvist and Wiksell, Stockholm 1974) Phys. Scr. *10A,* 1–187 (1974)

Subject Index